Rotor Balancing

Hatto Schneider

Rotor Balancing

Fundamentals for Systematic Processes

Springer Vieweg

Hatto Schneider
Rotor Balancing Consulting
Heppenheim, Germany
hatto.schneider@t-online.de

ISBN 978-3-662-66048-5 ISBN 978-3-662-66049-2 (eBook)
https://doi.org/10.1007/978-3-662-66049-2

Translation of the 9. German original edition published by Springer Fachmedien Wiesbaden, Wiesbaden, Germany, 2020

Responsible Editor: Eric Blaschke
This Springer Vieweg imprint is published by the registered company Springer-Verlag GmbH, DE, part of Springer Nature.
The registered company address is: Heidelberger Platz 3, 14197 Berlin, Germany

Preface

This book **Rotor balancing** is the translation of the German book **Auswuchttechnik** from its 9th edition, published 2020 by Springer Berlin Heidelberg, Germany.

Rotor balancing is only a little step in the production of a wide range of rotors but necessary to guarantee smooth running and long-lasting operation. Nevertheless, often the processes applied and criteria used for the final check are outdated, following certain established in-house rules.

This translation was initiated by the author, because rotor balancing today has reached a level in development and standardization, which should be known, discussed and used worldwide. Additional recommendations came from the ISO working group concerned with rotor balancing.

More details on the actual situation in the very special field of rotor balancing, as well as on the task and intentions of the author will be found in the preface of the 9th edition (translated into English).

For the first edition of the book Auswuchttechnik, Helmut Schleeger had helped me a lot to compare the print proofs with the manuscript.

Now, with the English edition, important helpers were (in alphabetical order): Andreas Buschbeck, Carolina Montag and Ulrich Pabst.

We hope to have created a passably comprehensible book for the English-speaking world.

Preface to the 9th (German) Edition

The book **AUSWUCHTTECHNIK** (English: Rotor Balancing) was first published in 1972 by VDI Verlag as paperback T 29 with 168 pages.

Today it is published in its 9th edition by Springer Berlin Heidelberg in cooperation with VDI as a more than twice as extensive hardcopy, but also available as an eBook.

It can therefore probably be called the reference on balancing technology in the German-speaking world. The international interest is shown by translations of some editions into several languages: English, French, Chinese and Japanese.

The 9th edition has been changed significantly compared to the 8th edition in order to describe the progress in standardisation since then and the current view on balancing technology.

Major Innovations

All balancing standards have been combined in a single series of standards—ISO 21940[1]; the various parts are arranged according to subject areas (see Table 1.2).

Of particular importance are:

- Part 1: Introduction to balancing technology,
- Part 2: Vocabulary,
- Part 11: Balancing procedures and tolerances for rotors with rigid behaviour,
- Part 12: Balancing procedures and tolerances for rotors with flexible behaviour,
- Part 14: Procedures for assessing balance errors,
- Part 21: Description and evaluation of balancing machines.

[1]ISO 21940: Mechanical vibration—Rotor balancing.

The German Standards Sub-Committee NALS NA 001-03-06[2] had further systematised the still somewhat scattered information on shaft-elastic rotor behaviour and summarised it in draft E-VDI 3835 in 2009. This draft of a guideline was subsequently fundamentally revised and now published as Supplement 1 to DIN ISO 21940-12[3].

According to this, the rotor behaviour is determined by a whole bundle of jointly acting criteria:

- The development of the unbalances over the speed.
- The types and numbers of unbalances to be corrected or controlled.
- The ability of the rotor to maintain the position of its mass elements and their mass centres in relation to each other.
- Rotors with *flexible behaviour* differ in the way the position of the mass elements changes over the speed range.

The combination of these criteria results in the respective rotor behaviour, which is then assigned to the different balancing methods described in ISO 21940-11[4] (formerly ISO 1940-1) and ISO 21940-12[5] (formerly ISO 11342).

Through the work on these two standards, however, it also became clear that balancing of rotors with shaft-elastic behaviour often *falls short* of a final step: Even with rotors that run through bending resonances up to the service speed, one or two flexural resonances *above* the service speed must be considered also. As a consequence of this fact, *systematic* balancing of rotors with shaft-elastic behaviour *definitely* requires correction of the low-speed unbalances.

In addition to *unbalances* of rotors with shaft-elastic behaviour, ISO 21940-12 also recommends *vibrations* as a quality criterion in balancing machines—they are still widely used in practice today. However, it has been shown that they cannot be a reliable criterion.

---

[2]NALS: Normenausschuss Akustik, Lärmminderung und Schwingungstechnik im DIN und VDI. C6: Auswuchten und Auswuchtmaschinen (English: Standards Committee for Acoustics, Noise Reduction and Vibration Control in DIN and VDI. C6: Rotor balancing and balancing machines).

[3]DIN ISO 21940: Mechanische Schwingungen—Auswuchten von Rotoren—Teil 12: Verfahren und Toleranzen für Rotoren mit nachgiebigem Verhalten, Beiblatt 1 (2015): Verfahren zum Auswuchten bei mehreren Drehzahlen. (English: DIN ISO 21940: Procedures and tolerances for rotors with flexible behaviour, Supplement 1 (2015): Methods for balancing at multiple speeds).

[4]ISO 21940 Mechanical Vibration—Rotor balancing, Part 11 (2016): Procedures and tolerances for rotors with rigid behaviour.

[5]ISO 21940 Mechanical Vibration—Rotor balancing, Part 12 (2016): Procedures and tolerances for rotors with flexible behaviour.

Gaps

During the debate in the ISO committees about *unbalance tolerances* in ISO Standards and some API Standards[6], it became clear that certain ISO principles have not yet been sufficiently adopted in practice, e.g.:

ISO 21940-14[7] distinguishes between the

- permissible residual unbalance for *delivery*, the value which is agreed between manufacturer and customer, and the
- permissible residual unbalance indication for the *balancing process.*

The second value is smaller by the errors of the balancing process—intended and unintended errors—and is thus a characteristic of the balancing process itself and the means used.

In the current version of ISO 21940-11 (2016), this difference is worked out more clearly; a supplement to ISO 21940-14 is in progress.

Rotors with shaft-elastic behaviour can be balanced at low speed in certain cases only, but always at high speed. For high-speed balancing, there are sufficient proposals on how to determine unbalance tolerances.

For low-speed balancing of rotors with shaft elastic behaviour, however, there are no such suggestions. The author presents his ideas on this subject in Sect. 8.3.

Conclusion

Meanwhile the new system of balancing technology has been fixed in standards, so that it can be explained and can be applied as state of the art in this book.

Despite the progress that has been made, I would like to emphasise that there are still various topics that need attendance: Further work on the balancing theory, on research on the possibilities of implementation, but above all on the application of the state of knowledge that has already been fixed in standards by all the people engaged in rotor balancing.

This book intends to be a reliable support in this respect.

Heppenheim, Germany
Spring 2020

Hatto Schneider

---

[6]API: American Petroleum Institute, USA.

[7]ISO 21940 Mechanical Vibration—Rotor balancing, Part 14 (2012): Procedures for assessing balance errors.

Contents

Introduction

1

Contents

1.1 Preliminary Note

Balancing is a process in which the mass distribution of a rotor is checked and—if necessary—improved to such an extent that the unbalances are within permissible limits. As rotors in this sense are not only all those parts that rotate in the operating state, but also those that are rotatingly mounted for functional reasons and are subjected to accelerations.

Rotors can have extremely different properties and thus can perform extremely different tasks Table 1.1.

H. Schneider, *Rotor Balancing*, https://doi.org/10.1007/978-3-662-66049-2_1

Table 1.1 Rotor bandwidth

Criterion	Lower limit	Upper limit
	Example	Example
Mass	<1 g	>300 t
	Balance wheel	LP steam turbine
Diameter	<3 mm	>6 m
	Textile spindle	Water turbine
Length	<10 mm	>20 m
	Rotor model trains	Turbogenerator
Service speed	0 min^{-1}	>1,000,000 min^{-1}
	Green grinding wheel	Turbine dental drill
Unbalance tolerance as centre of gravity eccentricity	<0.01 μm	>0.5 mm
	Aviation gyro	Railway wheel
Value of a rotor	<1 €	>500 million €
	Toy rotor	Communication satellite
Balancing of rotors on a machine	<10 pieces per year	>3 million pieces per year
	Satellite	Small motor armature for automobile

1.2 Importance and Quality of Balancing

Today, balancing is considered absolutely necessary for almost all rotors, whether it is to improve the function of the machine, to extend its service life, or to obtain special application possibilities or an additional sales argument through low-vibration or low-noise operation.

Although most of those responsible are aware of its existence, the balancing process still frequently takes place in a niche and is only subjected in some companies to one of the widespread quality management systems, for example—e.g. ISO EN 9000.

According to the specifications of the current quality management systems would have to be defined:

- The quality objective,
- the procedure(s) to achieve this objective,
- and the verification of the result.

For balancing, these are—and for all rotors—whether with rigid behaviour, or with flexible behaviour:

- The balancing quality,
- the balancing procedure with all details,
- and checking the balance quality achieved.

Whereas for other operations, such as turning, all the important data are specified—the machine tool, the fixture for the workpiece, the turning steel, cutting speed, feed rate, cutting depth, set-up time and piece time—balancing often escapes planning and control. Many details are therefore left to the "balancing area" and the people working there. They then decide on the basis of experience or their own ideas what should be done and how it should be done.

This is mainly due to the fact that the basic knowledge of balancing technology has not yet become sufficiently common knowledge. Sometimes the extent to which insights and methods have been further developed in the meantime is also underestimated. People work with traditional procedures and standards, so that the possibilities available today are not fully exploited.

It is also sometimes misunderstood that the essential prerequisites for a feasible and cost-effective balancing process must already be created when designing a rotor.

Similarly, there is often a lack of clarity as to how the various balancing problems can be solved most appropriately, what options the balancing machine market offers today and how the balancing machines—which act as measuring equipment—can be regularly checked.

This book is intended to contribute to the understanding of balancing technology, to serve as an introduction to this subject for beginners, but above all to enable employees in industry and research to independently assess the balancing problems at hand.

1.3 Development of Balancing Technology and Balancing Machines

1.3.1 Unbalance Types

It can be assumed that the problem of "balancing" arose many thousands of years ago with the first water and wind wheels. If these impellers were not built sufficiently symmetrically, or if care was not taken when selecting the material to ensure equal density and identical dimensions, difficulties arose: the wheel liked to turn into a certain position (heavy spot downwards) and did not start at all in weak currents.

This **static unbalance** could be empirically corrected by additional masses m on the radius r (at rest above the axis), so that the impeller subsequently ran "round" (Fig. 1.1).

In the course of time, the aids were improved and by the beginning of the nineteenth century, static unbalance was sufficiently under control: the rotors were balanced on blades or rollers with a great deal of skill and sensitivity. Later, they were even balanced on special balancing scales.

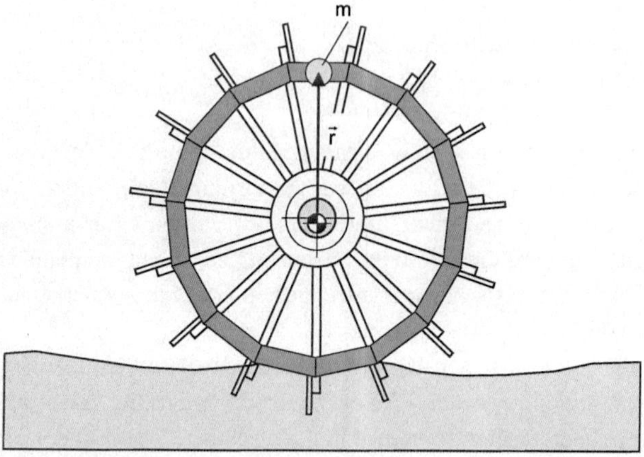

Fig. 1.1 A problem for many millennia, a **static unbalance** on a waterwheel: The centre of gravity is below the axis at rest. The static unbalance can be corrected by a correction mass *m* at the radius *r*

Sometimes, however, they had to be further corrected in the operating condition in order to achieve smooth and trouble-free running. For this purpose, balancing masses were placed in different positions and conclusions were drawn about optimum balancing from these results.

With the first high-speed machines in the second half of the nineteenth century and the triumphant advance of electric machines, another, previously unknown unbalance problem arose; the tried and tested balancing methods were suddenly no longer sufficient. Another type of unbalance was discovered the **moment unbalance** (Fig. 1.2) and the experts learned that it can only be detected under rotation, since the effects cancel each other out at rest.

Due to the growing number of steam turbines, generators, electric motors, centrifugal pumps and compressors, this problem became more and more pronounced. In operating condition or in simple racks and with simple marking means—chalk, pencil—balancing was carried out in two balancing planes. It was an iterative procedure, i.e. one came closer to the goal only in small steps. In most cases, each manufacturer of rotating machines had his own balancing devices, his own "secret recipe" and special balancing experts for this "secret science".

In the first decades of the twentieth century, new problems in balancing again emerged. Rotors that were balanced with the experience of the past showed serious vibration problems. These were always rotors whose operating speed was just below or even above a bending resonance, i.e. which exhibited typical *resonance phenomena.* For these rotors, additional or very special balancing procedures were required, whereby one usually run close to these resonances in order to reduce the deflections by means of corrections in several planes.

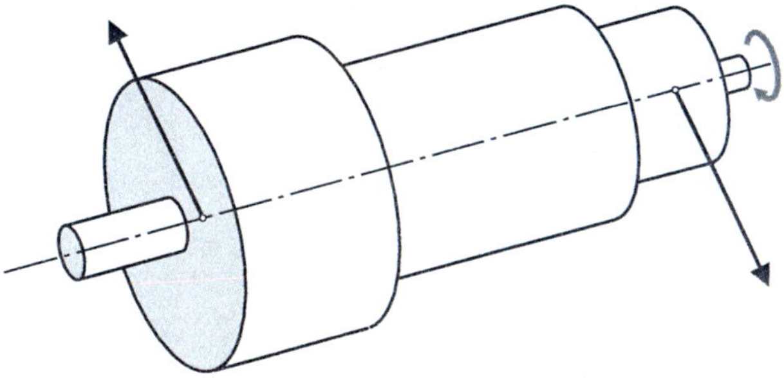

Fig. 1.2 A hitherto unknown problem, the **moment unbalance:** Shown here as an **couple unbalance**: two equal but opposite unbalances in two different radial planes. A moment unbalance can only be detected during rotation

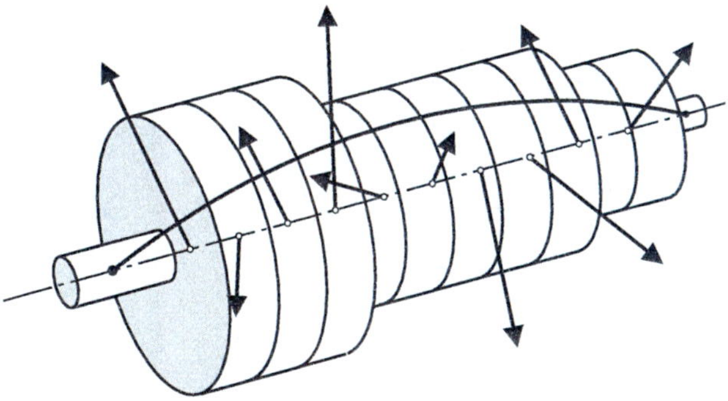

Fig. 1.3 Modal unbalance: The individual unbalances along the rotor are weighted by a flexural mode (here the 1st flexural mode). More than 2 correction planes are generally required to correct modal unbalance

Later, these particular unbalances were called **modal unbalances** (Fig. 1.3).

1.3.2 Balancing Machines

An early patent dealing with balancing was registered in Canada in 1870—i.e. 4 years after the invention of the dynamo machine by W. VON SIEMENS—by H. MARTINSON (Fig. 1.4). It dealt with the drive by a cardan shaft and shows the model of a balancing machine, which, however, was not yet adapted to the needs of industry.

Around the turn of the century (to the twentieth century), balancing technology received new impulses from N. W. AKIMOFF in the USA and A. STODOLA in Switzerland.

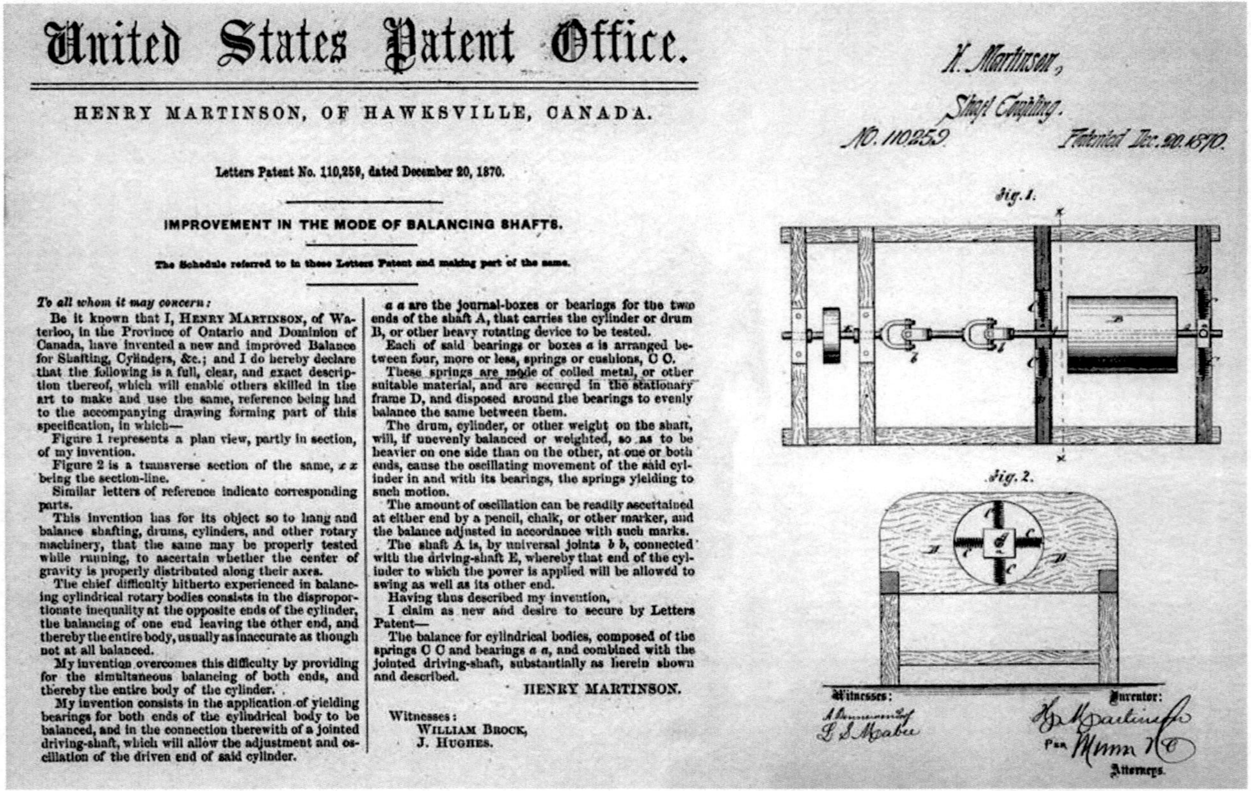

United States Patent Office.

HENRY MARTINSON, OF HAWKSVILLE, CANADA.

Letters Patent No. 110,259, dated December 20, 1870.

IMPROVEMENT IN THE MODE OF BALANCING SHAFTS.

The Schedule referred to in these Letters Patent and making part of the same.

To all whom it may concern:

Be it known that I, HENRY MARTINSON, of Waterloo, in the Province of Ontario and Dominion of Canada, have invented a new and improved Balance for Shafting, Cylinders, &c.; and I do hereby declare that the following is a full, clear, and exact description thereof, which will enable others skilled in the art to make and use the same, reference being had to the accompanying drawing forming part of this specification, in which—

Figure 1 represents a plan view, partly in section, of my invention.

Figure 2 is a transverse section of the same, *x x* being the section-line.

Similar letters of reference indicate corresponding parts.

This invention has for its object so to hang and balance shafting, drums, cylinders, and other rotary machinery, that the same may be properly tested while running, to ascertain whether the center of gravity is properly distributed along their axes.

The chief difficulty hitherto experienced in balancing cylindrical rotary bodies consists in the disproportionate inequality at the opposite ends of the cylinder, the balancing of one end leaving the other end, and thereby the entire body, usually as inaccurate as though not at all balanced.

My invention overcomes this difficulty by providing for the simultaneous balancing of both ends, and thereby the entire body of the cylinder.

My invention consists in the application of yielding bearings for both ends of the cylindrical body to be balanced, and in the connection therewith of a jointed driving-shaft, which will allow the adjustment and oscillation of the driven end of said cylinder.

a a are the journal-boxes or bearings for the two ends of the shaft A, that carries the cylinder or drum B, or other heavy rotating device to be tested.

Each of said bearings or boxes *a* is arranged between four, more or less, springs or cushions, C C.

These springs are made of coiled metal, or other suitable material, and are secured in the stationary frame D, and disposed around the bearings to evenly balance the same between them.

The drum, cylinder, or other weight on the shaft, will, if unevenly balanced or weighted, so as to be heavier on one side than on the other, at one or both ends, cause the oscillating movement of the said cylinder in and with its bearings, the springs yielding to such motion.

The amount of oscillation can be readily ascertained at either end by a pencil, chalk, or other marker, and the balance adjusted in accordance with such marks.

The shaft A is, by universal joints *b b*, connected with the driving-shaft E, whereby that end of the cylinder to which the power is applied will be allowed to swing as well as its other end.

Having thus described my invention,

I claim as new and desire to secure by Letters Patent—

The balance for cylindrical bodies, composed of the springs C C and bearings *a a*, and combined with the jointed driving-shaft, substantially as herein shown and described.

HENRY MARTINSON.

Witnesses:
WILLIAM BROCK,
J. HUGHES.

Fig. 1.4 An extract from H. MARTINSON's patent on a balancing machine, 1870. This is a physical model rather than a solution for industrial rotor balancing

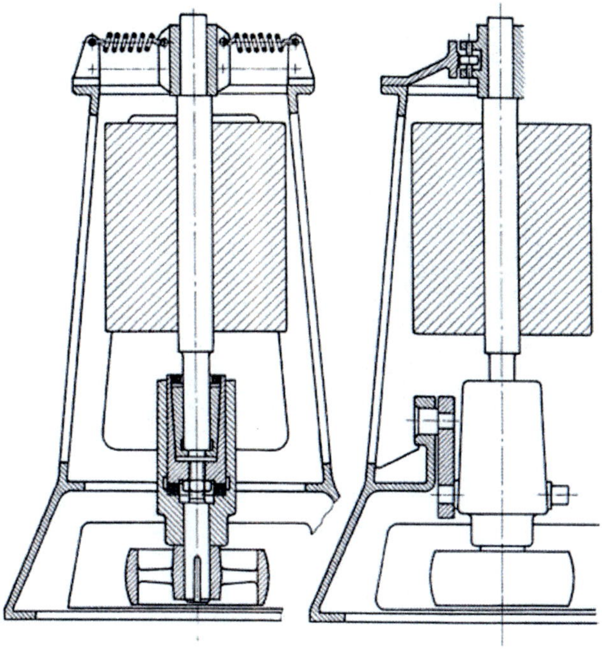

Fig. 1.5 Extract from patent LAWACZEK (1907): Balancing machine with vertical arrangement of the rotor

In Germany, a machine for balancing in two planes was patented in 1907 by F. LAWACZEK (Fig. 1.5) and built at Carl Schenck, Darmstadt. The first version still caused some problems, but the idea was further developed (patent on horizontal balancing machine 1912) and successfully modified by the work of H. HEYMANN.

These machines were delivered to companies all over the world and represented the beginning of an industrial production of balancing machines.

Machines from the early years of the twentieth century have very little in common with the modern balancing machines of the twenty-first century. Although the rotor also had to be supported and driven—basically with similar elements as today—the measuring technology was still in its infancy.

For industrial use, one was dependent on robust and easy-to-use solutions and thus on purely mechanical measuring equipment.

In order to increase the measurement sensitivity, measurements were made during the run-out in the support resonance, whereby a quite good frequency selectivity (suppression of interfering signals) was obtained as a by-product.

At the beginning, however, only assumptions could be made about the angular position, and an exact assignment of the measured values to the desired correction planes (plane separation) was also not yet possible.

With a wealth of new ideas and patents, the machines were completed and improved in the following decades, and variants or new systems were developed (Fig. 1.6). The main

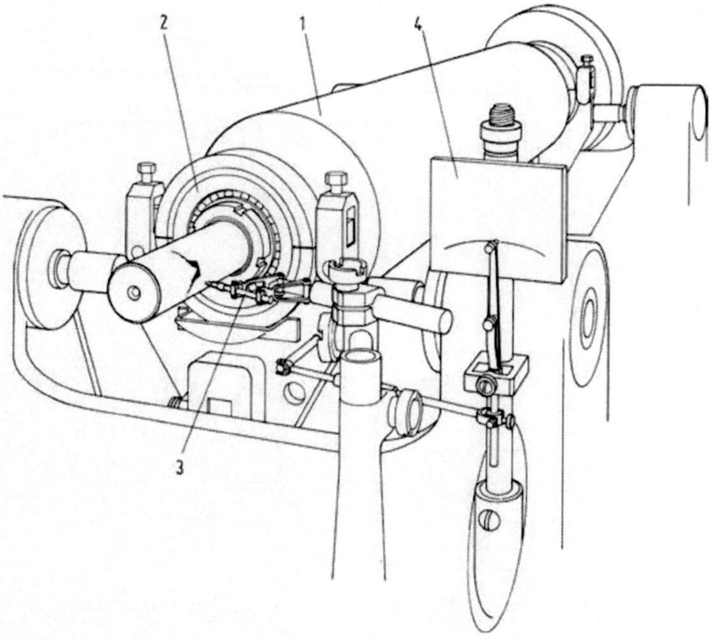

Fig. 1.6 A LAWACZEK-HEYMANN BALANCING MACHINE: With horizontally mounted rotor **(1)**, self-aligning ball bearing **(2)**, marker for the unbalance angle **(3)** and for the unbalance amount **(4)**

aims were always to improve accuracy in order to meet the increasing requirements and to increase economic efficiency, which could be achieved above all by shortening the piece times. At that time, all these advances took place only on the mechanical engineering side.

This changed somewhat as early as 1930 with the introduction of mechanical–electrical transducers, but the fundamental change came only after the Second World War with the rapid development of electronic measuring technology, semiconductor technology and the introduction of computers in all areas of industry.

With the greater emphasis on the measuring side, the mechanics of the balancing machine could be simplified again and—with the exception of special machines—has returned to the clear design of the early years (Fig. 1.7).

All important tasks like: sensitivity, frequency selection, plane separation, correction instruction, etc. are now performed by the measuring device.

Nevertheless, mechanics is still of great importance today (as can be seen from details in Chap. 12), because ultimately it is always about the harmonious interaction of all components: the mechanics, the drive technology and the measurement technology.

Even though older balancing machines are still occasionally in operation today, only modern concepts are used as the basis and described in the following chapters.

Fig. 1.7 Current balancing machine for universal application: With cardan shaft drive and protection against flying off parts by telescopic casing according to ISO 21940, part 23 (2012), class C

1.4 Standards and Guidelines

1.4.1 Historical Course

The first efforts to obtain uniform standards concerned *machine vibrations.* In Germany, in the mid-fifties of the twentieth century, a working committee of the VDI specialist group[1] "Schwingungstechnik" (vibration technology) started work on guideline VDI 2056 (1964) "Beurteilungsmaßstäbe für mechanische Schwingungen von Maschinen", (Assessment standards for mechanical vibrations of machines).

Considerations on *rotor balancing* led to the guideline VDI 2060 (1966) "Beurteilungsmaßstäbe für den Auswuchtzustand rotierender starrer Körper".

The VDI Guideline 2060 was submitted to the responsible ISO secretariat[2] as a proposal. It was an essential basis for ISO 1940 "Balance quality of rotating rigid bodies" (1973).

[1] VDI: Verein Deutscher Ingenieure (English: Association of German Engineers).

[2] ISO: International Organization for Standardization.

The ISO 1925 "Balancing—Vocabulary" (1974) became an essential aid to understanding in the field of balancing technology. In it, the most important terms of balancing technology were determined and defined.

Guidance for the complete description and correct evaluation of *balancing machines for universal use* was provided by ISO 2953 "Balancing machines—Description and evaluation" (1975).

ISO 5406 "The mechanical balancing of flexible rotors" (1980) described different types of flexible rotors and assigned low-speed and high-speed balancing methods to their behaviour. Balancing tolerances for this area were described in ISO 5343 "Criteria for evaluating flexible rotor balance" (1983). Both were later combined and updated to ISO 11342.

ISO 7475 "Balancing machines—Enclosures and other safety measures" (1984) was the first to comment on the hazards associated with rotor balancing and recommended stepped measures.

An important detail in the balancing of individual parts—*the treatment of parallel keys*—was defined in ISO 8821 (1989).

Guidance on the proper design of rotors with shaft-elastic behaviour, depending on the characteristics of the rotors, their service conditions and the modal damping can be found in ISO 10814 (1996).

After the early years, when each country set up its own standards and classifications, by around the turn of the millennium the work leading the way was carried out at ISO—supported by the main industrialized countries—so that understanding at the international level also became easier in this area.

In 2004, ISO 20806 was published: "Guidance for the in-situ balancing of medium and large rotors".

In addition, ISO 19499 was published in 2007, the result of a lengthy discussion. This standard provided an up-to-date introduction to balancing technology and an overview of balancing standards.

1.4.2 Current Situation

In 2010, the relevant ISO committees decided to group all balancing standards under one number—21940—and to assign part numbers according to an agreed systematic.

ISO 2041 is an exception to this rule, as it applies not only to balancing technology, but also to vibration and shock in general.

In Table 1.2 these new numbers are shown in bold, the old numbers in italics for comparison.

Table 1.2 Overview of all ISO balancing standards and the classification system: New designations in **bold,** old designations in *italics* for comparison

Subject	ISO Standard		
Introduction	Balancing technology **ISO 21940—Part 1** To be published presumably in 2020 *ISO 19499*		
Terms	Balancing **ISO 21940—Part 2** *ISO 1925*	Vibrations, shocks **ISO 2041** *Unchanged*	
Balancing procedures and tolerances	Rotors with rigid behaviour	Rotors with flexible behaviour	In-situ balancing
	Procedures and tolerances **ISO 21940—Part 11** *ISO 1940-1*	Procedures and tolerances **ISO 21940—Part 12** *ISO 11342*	In-situ balancing of medium and large rotors **ISO 21940—Part 13** *ISO 20806*
	Procedures for assessing balance errors **ISO 21940—Part 14** *ISO 1940-2*	Verfahren zum Auswuchten bei mehreren Drehzahlen **DIN ISO 21940—Part 12** Beiblatt 1[3] *E-VDI 3835*	
Balancing machines	Description and evaluation of balancing machines **ISO 21940—Part 21** *ISO 2953*	Symbols for balancing machines and associated instrumentation ISO 3719 **Retracted**	Enclosures and other protective measures for the measuring station of balancing machines **ISO 21940—Part 23** *ISO 7475*
Machine design for balancing	Susceptibility and sensitivity of machines to unbalances **ISO 21940—Part 31** *ISO 10814*	Shaft and fitment key convention **ISO 21940—Part 32** *ISO 8821*	

[3]DIN ISO 21940: Mechanische Schwingungen—Auswuchten von Rotoren—Teil 12: Verfahren und Toleranzen für Rotoren mit nachgiebigem Verhalten, Beiblatt 1 (2015): Verfahren zum Auswuchten bei mehreren Drehzahlen. (English: DIN ISO 21940—Part 12: Procedures and tolerances for rotors with flexible behaviour, Supplement 1 (2015): Methods for balancing at multiple speeds).

The necessary rework ranged from purely editorial measures to complete revisions of the technical content.

In the process, the partly scattered terms and definitions were combined in ISO 21940—Part 2, which is actually responsible for this area, and deleted in the other parts.

However, some details were no longer seriously worked on—around 2000—so that the German Committee C 6 in NALS[4] set up two working groups to further develop important topics:

- AK 1: Statistical methods for the balancing technology
  - Quality capability parameters for the evaluation of the unbalance measurement process.
- AK 2: Balancing of rotors with shaft-elastic behaviour
  - Methods for balancing at several speeds.

AK 1: The fundamentals for the evaluation of measurement processes with ***multivariate*** normally distributed measurement results are described in various standards, but this is not sufficient in detail for a targeted application. Therefore, Working Group 1 attempts to explain the possible and useful procedures for rotor balancing. A publication as supplement 1 to DIN ISO 21940—Part 21 is in preparation, title (English: Statistical quality capability parameters for the evaluation of the unbalance measurement process).

AK 2: ISO 11342 (1998) describes 9 different procedures for balancing rotors with shaft-elastic behaviour, 6 low-speed and 3 high-speed procedures. For the most important and universal method G—high-speed balancing at multiple speeds—the standard specifies 2 different criteria for the balancing quality and two different procedures for the balancing process.

Based on recent findings, Working Group 2 of NALS C 6 came to the conclusion that only *one* specific combination is viable and has based and described this in a VDI guideline. The draft E-VDI 3835 was published in Sept. 2009. After a thorough revision, it was published again in 2015 as Beiblatt 1 to DIN ISO 21940-12.

[4]NALS: Normenausschuss Akustik, Lärmminderung und Schwingungstechnik im DIN und VDI. (English: Standards Committee for Acoustics, Noise Reduction and Vibration Control in DIN and VDI).

1.5 List of Current Standards

ISO 19499 (2007): Mechanical vibration—Balancing—Guidance on the use and application of balancing standards. It will be replaced by:

- ISO 21940: Mechanical vibration—Rotor balancing—Part 1: **Introduction.**
- ISO 21940: Mechanical vibration—Rotor balancing—Part 2 (2017): **Vocabulary.**
- (replacement for ISO 1925).
- ISO 21940: Mechanical vibration—Rotor balancing—Part 11 (2016): **Procedures and tolerances for rotors with rigid behaviour.**
- (replacement for ISO 1940-1).
- ISO 21940: Mechanical vibration—Rotor balancing—Part 12 (2016): **Procedures and tolerances for rotors with flexible behaviour.**
- (replacement for ISO 11342).
- DIN ISO 21940: Mechanische Schwingungen—Auswuchten von Rotoren—Teil 12: **Verfahren und Toleranzen für Rotoren mit nachgiebigem Verhalten**, Beiblatt 1 (2015): Verfahren zum Auswuchten bei mehreren Drehzahlen. (English: DIN ISO 21940—Part 12: Procedures and tolerances for rotors with flexible behaviour, Supplement 1 (2015): Methods for balancing at multiple speeds).
- ISO 21940: Mechanical vibration—Rotor balancing—Part 13 (2012): **Criteria and safeguards for in-situ balancing of medium and large rotors.**
- (replacement for ISO 20806).
- ISO 21940: Mechanical vibration—Rotor balancing—Part 14 (2012): **Procedures for assessing balance errors.**
- (replacement for ISO 1940-2).
- ISO 21940: Mechanical vibration—Rotor balancing—Part 21 (2019): **Description and evaluation of balancing machines.**
- (replacement for ISO 2953).
- ISO 21940: Mechanical vibration—Rotor balancing—Part 23 (2012): **Enclosures and other protective measures for the measuring station of balancing machines.**
- (replacement for ISO 7475).
- ISO 21940: Mechanical vibration—Rotor balancing—Part 31 (2013): **Susceptibility and sensibility of machines to unbalances.**
- (replacement for ISO 10814).
- ISO 21940: Mechanical vibration—Rotor balancing—Part 32 (2012): **Shaft and fitment key convention.**
- (replacement for ISO 8821).

Physical Basics

2

Contents

H. Schneider, *Rotor Balancing*, https://doi.org/10.1007/978-3-662-66049-2_2

2.1 Preliminary Note

The theory of balancing technology is based on general physical principles. In order to avoid having to painstakingly gather the individual derivations and explanations from other books, the most important points for balancing have been compiled in the following sections.

2.2 Physical Quantities

Physical facts are described by equations between physical quantities. The essential characteristic of a quantity is its measurability. A distinction is made between basic quantities, which are not linked back to others, already defined quantities by equations, and derived quantities, which arise from the connection between the basic quantities.

Every physical ***quantity*** is composed of a ***numerical value*** and a ***unit***, e.g. a specification for the distance s:

$$s = 12\,\mathrm{m}$$

The unit is an arbitrarily chosen and agreed reference value for the physical quantity. To avoid too large or too small numerical values, decimal multiples and sub-multiples of the units are used (see Sect. 20.2), for distance also e.g. km, mm and µm.

Note

Only rarely, e.g. for multiples of the time unit second, non-decadal multiples (minute, hour, day, year) are allowed.

2.3 Scalar and Vector

There are undirected quantities, the ***scalars,*** and directed quantities, the ***vectors.***

A typical *scalar* is the mass: The value 7.5 kg is sufficient to describe the situation.

The property of a *vector* can be illustrated, for example, by the length: The indication 12 m obviously is not sufficient. In colloquial speech, one usually adds: high, long, far or similar; for a given object or process, this means an indication of direction.

To illustrate physical facts or processes, one uses coordinate systems (reference systems) and indicates the position of the vectors there. Vectors are best represented by arrows pointing into the desired direction, with the length corresponding to the magnitude. In equations, vectors are marked with a transverse arrow above the abbreviation, e.g. the travel is marked $\vec{s}$.

For calculations, scalars and vectors show significant differences.

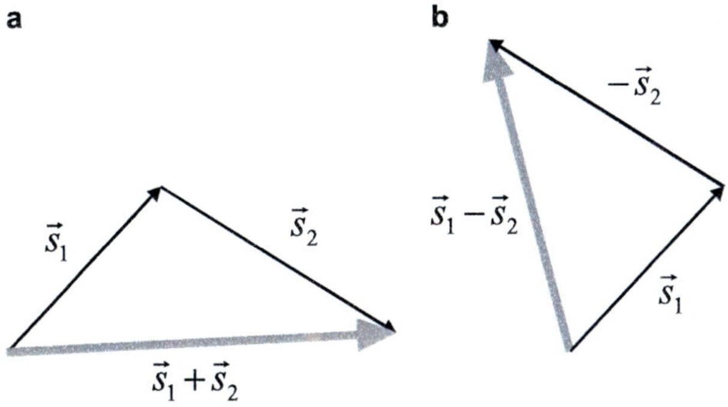

Fig. 2.1 Addition and subtraction of vectors: **(a)** Addition, **(b)** Subtraction

2.3.1 Addition

Basically, only quantities with the same unit may be added or subtracted. However, while with scalars the numerical values are just added together taking into account their signs (3 kg+9 kg=12 kg), vectors must be added "vectorially": The vector $\vec{s}_2$ is added to the end point of the vector $\vec{s}_1$, the vectorial sum is the vector from the beginning of the vector $\vec{s}_1$ to the end of the vector $\vec{s}_2$ (Fig. 2.1).

The difference $\vec{s}_1 - \vec{s}_2$ is formed by plotting $\vec{s}_2$ in the opposite direction and then following the same scheme (Fig. 2.1): $\vec{s}_1 + (-\vec{s}_2)$.

2.3.2 Multiplication

The multiplication of a *scalar with another scalar* results in another scalar, e.g.

$$Pt = W \qquad \text{power} \cdot \text{time} = \text{work} \tag{2.1}$$

If a *scalar* is multiplied *by a vector* a new vector is created, which generally has a different magnitude, but always the same line of action as the original vector, e.g.

$$\vec{v}t = \vec{s} \qquad \text{velocity} \cdot \text{time} = \text{travel} \tag{2.2}$$

For the multiplication of *two vectors* on the other hand, there are two fundamentally different types:

The **scalar product** has (as the name implies) a scalar as result, the equation is e.g.

$$\vec{F}\vec{s} = W; \qquad \text{force} \cdot \text{travel} = \text{work} \tag{2.3}$$

As long as the force is in line with the travel, one can also write: $Fs = W$ (without taking into account the vector character of the force and the travel, but the signs must be observed).

If the force is perpendicular to the path, the work is zero. Therefore, only the component into the direction of the travel is to be considered (Fig. 2.2).

With this component, the quantities can be treated as scalars.

$$Fs\cos\varphi = W \qquad \mathrm{N\cdot m} \tag{2.4}$$

For a **vector product** the result is again a vector that has a certain position in relation to the original vectors, e.g.

$$\vec{r}\times\vec{F} = \vec{M}; \qquad \text{radius “cross” force} = \text{torque} \tag{2.5}$$

Contrary to the scalar product, the result is particularly large if there is a right angle between the radius vector and the force vector, and the result is zero if both vectors point into the same direction. Numerically this means:

$$r\,F\sin\varphi = M \qquad \mathrm{N\cdot m} \tag{2.6}$$

The direction, one has to rotate the radius vector, to bring it into the same direction as the force vector by the shortest path indicates the direction of rotation of the moment (Fig. 2.3).

The vector of the torque is perpendicular to the plane in which $\vec{r}$ and $\vec{F}$ lie (so here perpendicular to the plane of the picture), the tip points downwards. So you look at its back end (it is also said that it indicates the direction of movement of a right-hand screw under the rotation of the moment).

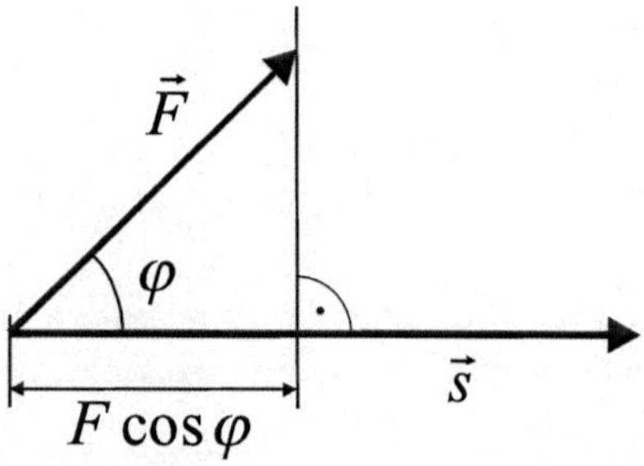

Fig. 2.2 Example of a **scalar product:** work equals force · travel

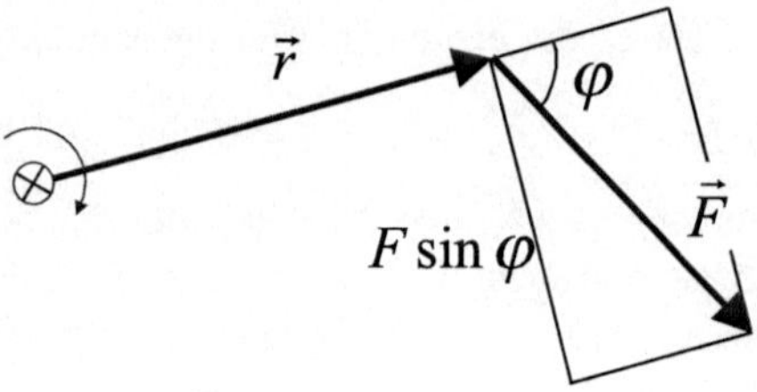

Fig. 2.3 Example for a **vector product:** Torque equals lever arm × force

From this it follows that $\vec{r}$ and $\vec{F}$ may not simply be interchanged (as this would result in a different sense of rotation); written as an equation this is

$$\vec{r} \times \vec{F} = -\vec{F} \times \vec{r} \qquad \mathrm{N} \cdot \mathrm{m} \tag{2.7}$$

The specification of the torque vector contains three statements: the axis of rotation, the magnitude and the direction of rotation of the torque.

2.4 System of Units

2.4.1 Basic Quantities

The International System of Units SI (ISO 80000) defines seven basic quantities. Of these basic quantities, only the following are of interest for the balancing technology:

- length s with unit meter m
- time t with unit second s
- mass m with unit kilogram kg

While length and time are easily to be understood, the third basic quantity needs a little more explanation.

The mass is a body property. It is independent of location and can be assumed to be constant in this context (when balancing rotors). A mass is usually measured on a balance (scale)—comparing it to calibrated masses or by direct indication.

2.4.2 Derived Quantities

Important derived quantities for rotor balancing are:

Speed $\vec{v}$ is the quotient of the distance covered and the time required.

$$\vec{v} = \vec{s}/t \qquad \text{in m/s} \tag{2.8}$$

If the speed $\vec{v}$ is not constant, this value corresponds to the average speed. If the instantaneous speed is to be specified, it must be written:

$$\vec{v} = d\vec{s}/dt \qquad \text{in m/s} \tag{2.9}$$

By $d\vec{s}$ and dt infinitely short path and time intervals (differentials) are to be understood. Velocity and displacement vectors always have the same direction.

The **acceleration** $\vec{a}$ indicates how fast the speed changes.

$$\vec{a} = d\vec{v}/dt \qquad \text{in (m/s)/s} \quad \text{resp.} \quad \mathrm{m/s^2} \tag{2.10}$$

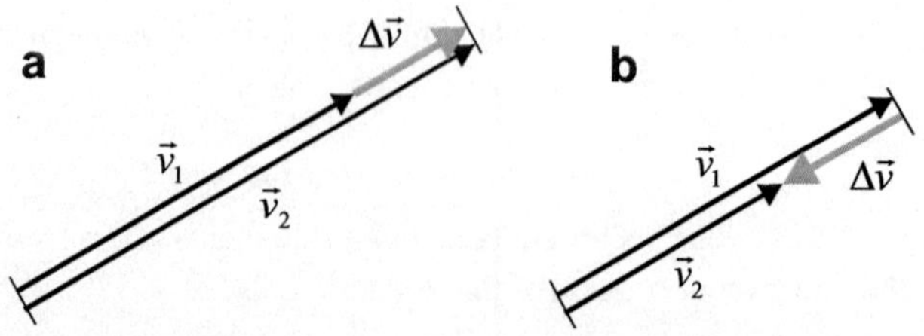

Fig. 2.4 Location of velocity vectors: **(a)** During acceleration, where $\Delta\vec{v}$ has the same direction as $\vec{v}_1$: $\vec{a}$ becomes positive. **(b)** During deceleration (braking), where $\Delta\vec{v}$ has the opposite direction to $\vec{v}_1$: $\vec{a}$ becomes negative

If the velocity increases, $\vec{a}$ becomes positive; if it decreases, $\vec{a}$ becomes negative (Fig. 2.4).

In colloquial language, these two alternatives are named differently: in one case with acceleration, in the other with deceleration resp. braking.

2.5 Physical Laws

In order to understand the theory of balancing technology and its practical handling, two physical laws are essential which are to be explained briefly.

2.5.1 Newton's 2nd Law

Newton's 2nd law (the dynamic law of motion) is:

$$\vec{F} = m\frac{d\vec{v}}{dt} = m\vec{a} \qquad \text{force} = \text{mass} \cdot \text{acceleration} \tag{2.11}$$

For a body with mass *m,* the velocity vector changes due to an applied force $\vec{F}$. The force is a vector and has the same direction as $d\vec{v}$ resp. $\vec{a}$. The unit of force is obtained when a mass of 1 kg is accelerated at 1 m/s^2, it is called newton:

$$1\,\text{kg} \cdot 1\,\frac{\text{m}}{\text{s}^2} = 1\,\text{N} \qquad 1\text{ newton} \tag{2.12}$$

The force acting on a body under the acceleration $\vec{g} = 9.81$ m/s^2 due to gravity is called its weight $\vec{G}$:

$$\vec{G} = m\,\vec{g} \qquad \text{N} \tag{2.13}$$

The weight of a mass of 1 kg is

$$G = 1\,\text{kg} \cdot 9.81\,\text{m/s}^2 = 9.81\,\text{N} \tag{2.14}$$

For approximate calculations, the acceleration can be set to $g \approx 10\ \mathrm{m/s^2}$, so that the weight G of a mass of 1 kg ≈ 10 N.

2.5.2 Mass Attraction

The gravity of the earth and thus the weight is a special case of the law of mass attraction according to which two masses attract each other; the result is:

$$\vec{F} = a\frac{m_1\, m_2}{r^2} \qquad a \text{ is a constant in } \mathrm{N \cdot m^2/kg^2} \tag{2.15}$$

Here m_1 and m_2 are the two masses and r is the distance between their mass centres. With the mass of the body m_1 and the earth m_2 it becomes obvious that the weight on the earth is different than e.g. on the moon, so it cannot be a constant unit of the body.

2.6 Circular Motion

All bodies, for which rotor balancing is important, rotate or are at least rotatably mounted. The rotary motion and all related terms and formulas are therefore very important for the balancing technology. In the next sections, the important basics will be explained and the physical quantities required in balancing will be derived. This is more about basic understanding than exact general mathematics.

2.6.1 Plane Angle

If a point moves on a circular path with radius r from 1 to 2, it has travelled the distance b (Fig. 2.5).

The quotient

$$\frac{b}{r} = \varphi \qquad \text{with radiant (rad) as unit} \tag{2.16}$$

is called the ***plane angle***.

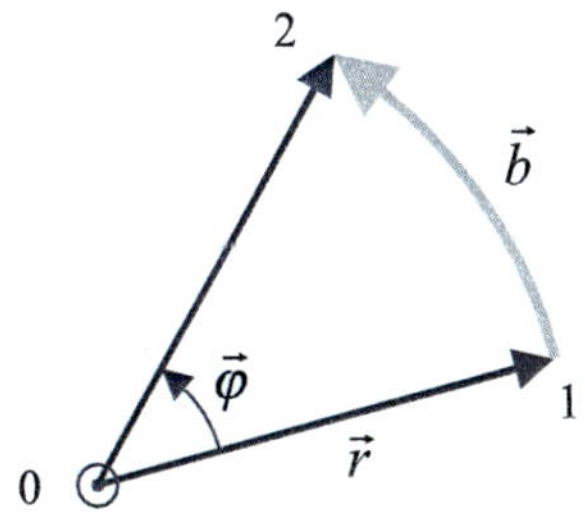

Fig. 2.5 Relationship between arc $\vec{b}$, radius $\vec{r}$ and angle $\vec{\varphi}$

The plane angle is a vector that simultaneously specifies the axis of rotation, the sense of rotation, and the angle of rotation.[1] In Fig. 2.5 it points forward from the image plane at centre 0 (right-hand screw). For $b = r$, becomes $\varphi = 1$ rad; for a full revolution, $b = 2\pi$ r and thus $\varphi = 2\pi$ rad.

It can be seen that the plane angle is an analogous indication to the angle in degrees: both terms express the rotation—the radius vector has made during its migration from 1 to 2—in a different way. A complete revolution, expressed in angular degrees, is 360°, as a plane angle it is 2π rad; it follows:

$$1\ \text{rad} = \frac{360^\circ}{2\pi} \cong 57.3^\circ \tag{2.17}$$

Now back to the plane angle. If Equation 2.16 is solved for b and written as a vector product, one obtains:

$$\vec{b} = \vec{\varphi} \times \vec{r} \qquad \text{with unit} \quad \text{m} \tag{2.18}$$

a very simple way of calculating the distance travelled on the circular path.

2.6.2 Angular Frequency

With a *uniform* circular motion the vector $\vec{\varphi}$ increases continuously. If the plane angle by $\vec{\varphi}$ is divided by the time t which elapses during the rotation, the angular frequency $\vec{\omega}$ is obtained:

$$\vec{\omega} = \frac{\vec{\varphi}}{t} \qquad \text{with} \quad \text{rad/s} \quad \text{as unit} \tag{2.19}$$

In a *non-uniform* circular motion the angular frequency changes. In order to capture the instantaneous value of the angular frequency, infinitely small movements and small times (differentials) must be taken as a basis, i.e.

$$\vec{\omega} = \frac{d\vec{\varphi}}{dt} \qquad \text{rad/s} \tag{2.20}$$

The angular frequency $\vec{\omega}$ indicates how many radians are covered per second and thus corresponds in the statement to the rotational speed n, which describes the revolutions per minute (min^{-1}) and the frequency f, which indicates how many revolutions are made per second (s^{-1} or Hz).

[1] In the standard ISO 80000-1, the plane angle is correctly given a dimension. However, the system of measurement unfortunately is not completely uniform; because the dimension *rad* is dropped again in subsequent calculations (see e.g. angular velocity, angular acceleration).

The equational relationship can be easily established: Eq. 2.19 is valid in general, i.e. also for a complete revolution, where $\varphi = 2\pi$ (rad) and $t = T$ (with T as the period in seconds). This means: $\omega = 2\pi/T$ (rad/s). The period T is inversely proportional to the frequency f, i.e. $T = 1/f$, and thus

$$\omega = 2\pi f \qquad \text{rad/s} \tag{2.21}$$

The usual term in mechanical engineering is the speed n. With $n = 60\,f$ or $f = n/60$ it results in:

$$\omega = \frac{2\pi n}{60} = \frac{\pi n}{30} \approx \frac{n}{10} \qquad \text{rad/s} \tag{2.22}$$

The approximate value $n/10$ is sufficiently accurate for all rough calculations.

Note

In addition to the size, the vector $\vec{\omega}$ contains the position of the axis of rotation and the direction of rotation also.

2.6.3 Circular Speed

If Equation 2.16 is solved for b and divided by the time, one obtains:

$$\frac{\vec{b}}{t} = \frac{\vec{\varphi}}{t} \times \vec{r} \qquad \text{m/s} \tag{2.23}$$

Here, the covered arc $\vec{b}$ as well as the plane angle $\vec{\varphi}$ are divided by the required time t; the result is the circular speed $\vec{v}$ of the point:

$$\vec{v} = \vec{\omega} \times \vec{r} \qquad \text{m/s} \tag{2.24}$$

Since $\vec{\omega}$ always is perpendicular to $\vec{v}$, one can also write (simplified, without vector arrows and vector product):

$$v = \omega\, r \qquad \text{m/s} \tag{2.25}$$

Example

We are looking for the circular velocity v of a point on the radius $r = 1.5$ m at the speed $n = 1000\ \text{min}^{-1}$. ◀

Solution

With $\omega \approx 100$ rad/s according to Eq. 2.22: $v \approx 100 \cdot 1.5\ \text{m/s} = 150\ \text{m/s}$.

2.6.4 Angular Acceleration

If the angular frequency changes with time, e.g. when a machine is started up or braked down, the *angular acceleration* $\vec{\alpha}$ can be determined (by the differentials) to:

$$\vec{\alpha} = \frac{d\vec{\omega}}{dt} \qquad \text{unit rad/s}^2 \tag{2.26}$$

2.6.5 Circular Acceleration

Analogous to Eq. 2.24 the *circular acceleration* of a point is (see footnote to Sect. 2.6.1):

$$\vec{a} = \vec{\alpha} \times \vec{r} \qquad \text{in} \quad \text{m/s}^2 \tag{2.27}$$

The *circular acceleration* $\vec{a}$ is also called *tangential acceleration* $\vec{a}_t$ (in contrast to the radial acceleration, see Sect. 2.6.8).

Example

How large is the circular acceleration, if a point on the radius 1.5 m is accelerated uniformly so that the speed 1000 min^{-1} is reached in a time of 5 s? ◀

Solution

$\omega \approx 100\,\text{rad/s}$ (Eq. 2.22); $\alpha \approx \frac{100}{5}\,\frac{\text{rad/s}}{\text{s}} = 20\,\text{rad/s}^2$; $a_t = 20 \cdot 1{,}5\,\frac{\text{m}}{\text{s}^2} = 30\,\frac{\text{m}}{\text{s}^2}$.

2.6.6 Torque

If the point has a mass *m*, then a circumferential force $\vec{F}$ must act on it from the drive in order to accelerate it. With $\vec{F} = m\vec{a}$ and Eq. 2.27 it becomes $\vec{F} = m(\vec{\alpha} \times \vec{r})$. This force acts on the radius $\vec{r}$, so that the torque *M* becomes (here without vector sign and vector product):

$$M = r\,F = m\,\alpha\,r^2 \qquad \text{N} \cdot \text{m} \tag{2.28}$$

Here $m\,r^2$ is the *moment of inertia of* the point, referred to the axis of rotation. For a general rotor with moment of inertia *J* (about the axis of rotation) this leads to:

$$\vec{M} = J\vec{\alpha} \qquad \text{N} \cdot \text{m} \tag{2.29}$$

Example

How large is the torque if, in the example given in Sect. 2.6.5, the mass of the point is 1 kg? ◀

Solution

According to Eq. 2.28: $M \approx 1 \cdot 20 \cdot 1.5^2\,\mathrm{kg}\frac{1}{\mathrm{s}^2}\mathrm{m}^2 = 45\,\mathrm{N} \cdot \mathrm{m}$ (see footnote to Sect. 2.6.1)

2.6.7 Moment of Inertia

The *axial moment of inertia* indicates the resistance of a rotor to a change in speed (an angular acceleration) due to its mass distribution.

It corresponds to the effect of the mass of a body in linear motion. The equations (Eq. 2.11) and (Eq. 2.29) are structured therefore quite similar.

For a general body, the moment of inertia results from the sum of the products of all mass parts with the square of their distance to the axis of rotation:

$$J = \int r^2 dm \qquad \mathrm{kg} \cdot \mathrm{m}^2 \tag{2.30}$$

For easier understanding, one can imagine that the whole mass of the body is contracted into a narrow ring of radius r_i (radius of inertia) without changing the value of the moment of inertia. Then, from Eq. 2.30 follows:

$$J = m r_i^2 \qquad \mathrm{kg} \cdot \mathrm{m}^2 \tag{2.31}$$

2.6.8 Radial Acceleration

The statement in Eq. 2.11 can be described like this: Every body remains in a state of rest or uniform motion on a straightforward path as long as no external force acts on it.

For a circular orbit with constant angular frequency, the magnitude of the circular velocity of the point does not change (Eq. 2.24), but its direction does. The velocity vector, which is always perpendicular to the radius, i.e., tangential to the circular movement, changes continuously; **radial acceleration** occurs (Fig. 2.6).

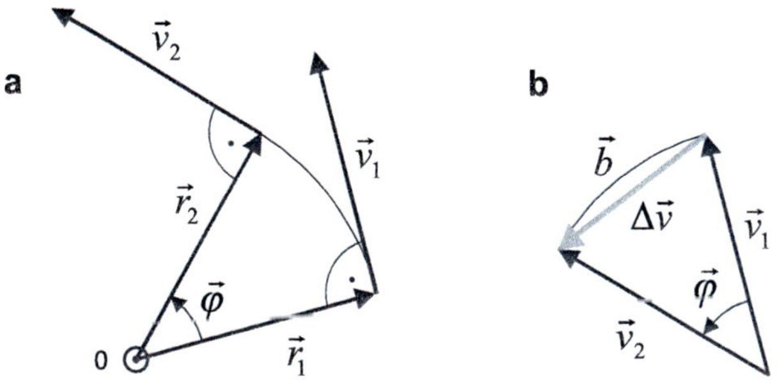

Fig. 2.6 Derivation of radial acceleration: **(a)** motion on the circle, **(b)** vector difference

Herein are $\vec{r}_1$ and $\vec{v}_1$ radius and velocity at point of time 1, $\vec{r}_2$ and $\vec{v}_2$ at point of time 2 (Fig. 2.6a). The difference between $\vec{v}_1$ and $\vec{v}_2$ is $\Delta\vec{v}$ (Fig. 2.6b), the arc $\vec{b}$ is: $\vec{b} = \vec{\varphi} \times \vec{v}_1$. For very small angles $\Delta\vec{\varphi}$ the arc $\vec{b}$ is sufficiently equal to the chord $\Delta\vec{v}$, so that we can write $d\vec{v} = d\vec{\varphi} \times \vec{v}_1$. With *dt,* the time required, it becomes:

$$\vec{a} = \frac{d\vec{v}}{dt} = \frac{d\vec{\varphi} \times \vec{v}_1}{dt} = \vec{\omega} \times \vec{v}_1 \qquad \text{m/s}^2 \tag{2.32}$$

With Eq. 2.24 the radial acceleration is given by:

$$\vec{a}_r = \vec{\omega} \times (\vec{\omega} \times \vec{r}_1)\ \text{m/s}^2 \tag{2.33}$$

or, simplified (without vector signs and vector product):

$$a_r = \omega^2 r = \frac{v^2}{r} \qquad \text{m/s}^2 \tag{2.34}$$

The force that causes this acceleration is directed towards the axis; it is called ***centripetal force.***

2.6.9 Centrifugal Force

The opposite force of same magnitude, the inertia force of the mass, is called ***centrifugal force***:

$$\vec{F} = -m\vec{a}_r = -m\vec{r}\omega^2 \tag{2.35}$$

Or, with the circumferential velocity v (Eq. 2.24), only as an amount:

$$F = m\frac{v^2}{r} \tag{2.36}$$

Example

How large is the centrifugal force of a mass of 1 kg on a radius of 1.5 m at 1000 min^{-1}? ◀

Solution

$\omega \approx 100$ rad/s; $F = 1.5 \cdot 100^2\ \text{kg} \cdot \text{m}\ (1/\text{s})^2 = 15{,}000$ N.

2.7 Vibration

The change of a physical quantity with time is called vibration.

Of the various types of vibrations, ***periodic vibrations*** are of particular interest in connection with rotor balancing technology. Here, a physical quantity changes with time in such a way that the same course of change begins again after the period duration T.

The simplest case of a periodic oscillation is the ***harmonic oscillation,*** where the change in the physical quantity over time can be described by a sinewave equation:

$$x = \hat{x}\ \sin\left(\omega\, t + \varphi_0\right) \tag{2.37}$$

The harmonic oscillation can be thought of as being created by projecting a uniform circular motion onto an axis (lying in the plane of the circular motion).

All other periodic oscillations can be described sufficiently accurately by a finite number of superimposed sinewaves of different frequency and amplitude.

2.7.1 Single Mass Oscillator with Centrifugal Excitation

A mass m is guided in such a way that it can only move in the direction x. In addition, it is supported by a spring with stiffness c and a damper b (proportional to velocity) with damping ratio D.

At the mass an unbalance $u \cdot r$ (s. Chap. 4) rotates with the variable angular frequency Ω (Fig. 2.7).

Note

Ω is always used in the following for the angular frequency of the service speed.

The amplitude of the mass has the following characteristic course (Fig. 2.8): For damping ratio $D = 0$, the *specific amplitude* $A = \hat{x}/(u \cdot r/m)$ first increases quadratically with the speed (from the value 0), then increases faster, becomes infinite, then decreases again

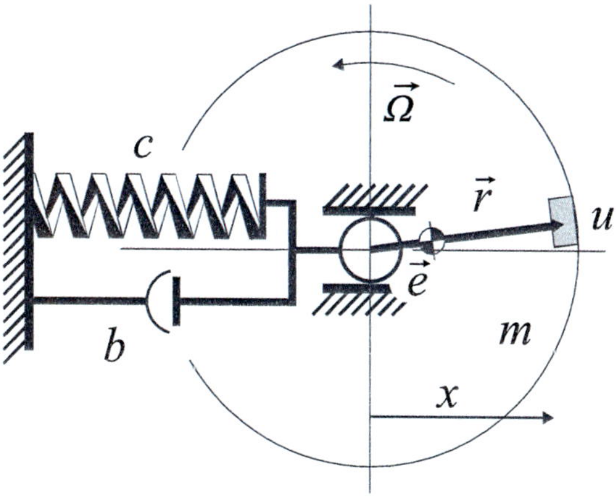

Fig. 2.7 Unbalance-excited single-mass oscillator with damping, guided in the direction x

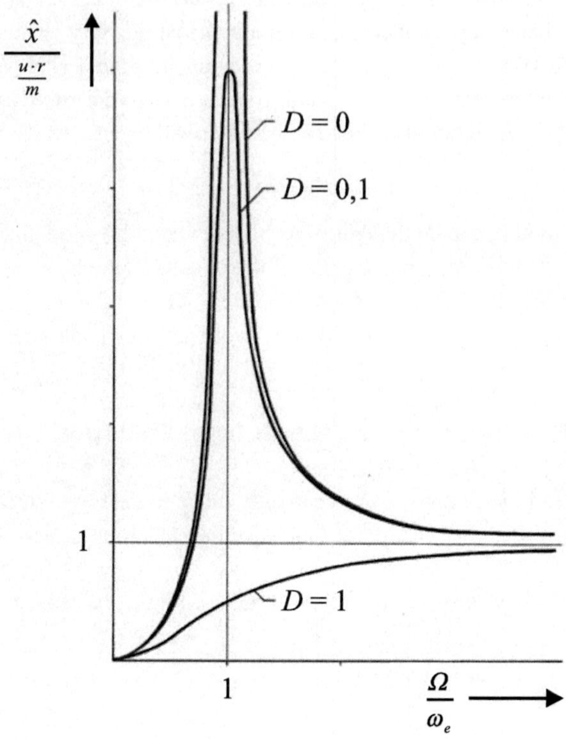

Fig. 2.8 Amplitude response of the unbalanced single-mass oscillator for different damping ratios D

and slowly approaches the value 1. The three regions are called ***subcritical, critical*** (resonance area) and ***supercritical***.

The natural angular frequency of the oscillating, undamped system is:

$$\omega_e = \sqrt{\frac{c}{m}} \quad \text{rad/s} \tag{2.38}$$

As the damping ratio increases, the maximum amplitude shifts from $\Omega/\omega_e = 1$ to higher values.

With increasing angular frequency, the angular position between the exciting centrifugal force and the oscillation also changes (Fig. 2.9): the oscillation lags behind the excitation by the phase shift angle φ.

From the plots Fig. 2.8 and 2.9 we can extract two features that are general:

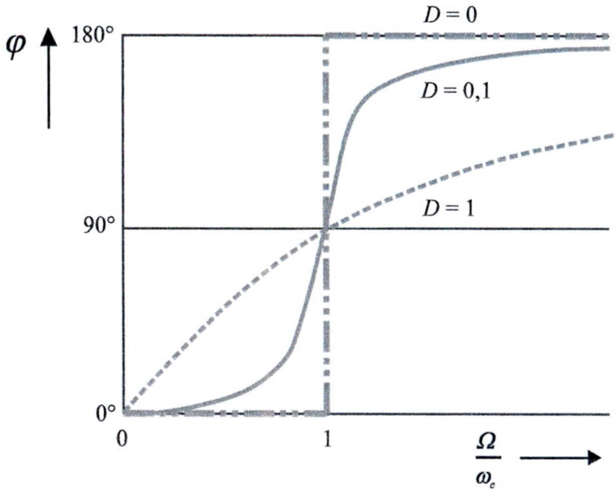

Fig. 2.9 Phase shift angle φ of the oscillation in relation to the excitation of the single-mass oscillator

- In resonance ($\Omega/\omega_e = 1$) the phase shift is always 90°, independent of the damping.
- Phase change from 0° to 180° occurs over a wider frequency range as the damping increases.

Figure 2.10 is a summary of Fig. 2.8 and 2.9: It is the representation in polar coordinates for two different damping ratios D.

The numbers on the curves are the values for Ω/ω_e. The distance from the zero point to a curve point indicates the specific amplitude and the direction of this connecting line indicates the phase position (the exciting centrifugal force is in the direction 0°).

The general equation for the **specific vibration amplitude** for centrifugal force excitation ($F = ur\omega^2$) is:

$$\frac{\hat{x}}{\frac{ur}{m}} = \frac{\left(\frac{\Omega}{\omega_e}\right)^2}{\sqrt{\left[1 - \left(\frac{\Omega}{\omega_e}\right)^2\right]^2 + 4D^2\left(\frac{\Omega}{\omega_e}\right)^2}} \qquad \text{dimensionless} \tag{2.39}$$

The general equation for the **phase shift angle** for centrifugal force excitation ($F = ur\omega^2$) is:

$$\varphi = \arctan \frac{2D\frac{\Omega}{\omega_e}}{1 - \left(\frac{\Omega}{\omega_e}\right)^2} \qquad \text{rad} \tag{2.40}$$

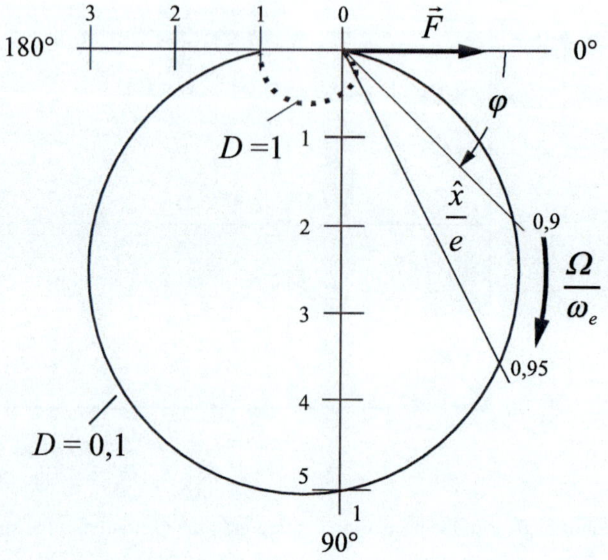

Fig. 2.10 Polar diagram (polar representation of the amplitude and phase) of a single-mass oscillator for two damping ratios D

In order to recognize principle relationships, one sets the degree of damping $D=0$; Eq. 2.39 then becomes:

$$\frac{\hat{x}}{\frac{ur}{m}} = \frac{\left(\frac{\Omega}{\omega_e}\right)^2}{1 - \left(\frac{\Omega}{\omega_e}\right)^2} \quad \text{dimensionless} \tag{2.41}$$

2.7.1.1 Subcritical Area

For very small excitation frequencies with respect to the natural frequency (i.e. $\Omega \ll \omega_e$), the term $(\Omega/\omega_e)^2$ in the denominator can be neglected towards 1, thus:

$$\frac{\hat{x}}{\frac{ur}{m}} = \frac{\Omega^2}{\omega_e^2}; \quad \text{with } \omega_e^2 = \frac{c}{m}; \quad \frac{\hat{x}}{\frac{ur}{m}} = \frac{\Omega^2 m}{c}; \quad \hat{x} = \frac{ur\Omega^2}{c} = \frac{F}{c} \tag{2.42}$$

The max. deflection $\hat{x}$ thus corresponds to the centrifugal force divided by the stiffness: as if not an oscillating force but a static (constant) force were acting.

In Eq. 2.40, for very small D and $\Omega<\omega_e$, the arctan (arc of the angle given by the tangent) of a very small positive value is sought, i.e. $\varphi \cong 0°$. That is: excitation and oscillation have the same phase in the (far) **subcritical region.**

2.7.1.2 Resonance Area

If the excitation frequency is equal to the natural frequency, i.e. $\Omega = \omega_e$, then the denominator in Eq. 2.39 becomes zero, the quotient—and thus the amplitude of the oscillator—is infinitely large.

The argument of the arctan is indefinite and so is the phase shift angle: it jumps from 0° to 180° at this point.

If damping is present, the denominator in Eq. 2.39 can never become zero, the amplitude is limited to finite values.

The phase shift angle obtained from Eq. 2.40 is $\varphi = \arctan \infty = 90°$, independent of the magnitude of the damping factor D.

Note

The phase difference of 90° between excitation and oscillation can therefore be regarded as a typical characteristic of **resonance** (and not the amplitude maximum).

2.7.1.3 Supercritical Area

If the excitation frequency is much larger than the natural frequency, i.e. $\Omega > \omega_e$, in the denominator of Eq. 2.41, the 1 can be neglected with respect to $(\Omega/\omega_e)^2$:

$$\frac{\hat{x}}{\frac{ur}{m}} = -1; \qquad \text{dimensionless} \tag{2.43}$$

or, dissolved for $\hat{x}$:

$$\hat{x} = -\frac{ur}{m} \qquad \text{m} \tag{2.44}$$

The system then oscillates with a constant amplitude, independent of the speed. Transformed and extended with Ω^2 to mass forces one obtains

$$\hat{x} m \Omega^2 = -u r \Omega^2 \qquad \text{N} \tag{2.45}$$

It can be seen that in the supercritical range the mass force of the moving system and the centrifugal force balance each other out.

When calculating the phase shift angle—if D is small—the arctan of a very small negative value is sought, so $\varphi \approx 180°$.

The **supercritical state** can also be described as follows: The common centre of gravity of the masses m and u remains at rest; seen from it, the centres of gravity of the masses m and u lie in opposite directions—hence the phase shift angles of 180° between the movements of the masses m and u.

2.7.2 Degrees of Freedom

The mass *m*, Fig. 2.7, has only one possibility of motion, it can be moved in the direction of the x-axis. It is said to have **one degree of freedom.** If, for example, a rotation is possible in addition to the displacement, two degrees of freedom are available.

A rigid body can be moved in space in a maximum of three independent directions and rotated about three independent axes; it therefore has six degrees of freedom. If the body is not rigid, but consists of several masses connected by spring elements, the number of degrees of freedom increases accordingly. If it is even a continuous structure (distribution of masses and stiffnesses), the number of degrees of freedom becomes infinite.

The question of the number of degrees of freedom is important mainly because it makes it relatively easy to determine the number of natural frequencies. Every system always has as many natural frequencies as there are degrees of freedom. In general, however, only *those* natural frequencies are of interests which are sufficiently excited in practice.

2.7.3 Dynamic Stiffness

To calculate the dynamic stiffness in Eq. 2.41 for the mass is used $m=c/\omega_e^2$ from Eq. 2.38. With $c=c_{\text{stat}}$ and transformed accordingly it becomes:

$$\frac{\hat{x}}{ur\frac{\omega_e^2}{c_{stat}}}=\frac{\left(\frac{\Omega}{\omega_e}\right)^2}{1-\left(\frac{\Omega}{\omega_e}\right)^2};\quad \frac{\hat{x}c_{stat}}{ur\Omega^2}=\frac{1}{1-\left(\frac{\Omega}{\omega_e}\right)^2}\qquad \text{dimensionless}\qquad (2.46)$$

Analogous to the static stiffness the dynamic stiffness is the quotient of the amplitude of the oscillating force ($ur\Omega^2$) and the amplitude of the vibration displacement ($\hat{x}$) caused.

From the expression on the right hand side in Eq. 2.46, the dynamic stiffness can be derived by forming the reciprocal and multiplying both sides by c_{stat} to:

$$c_{dyn}=\frac{ur\Omega^2}{\hat{x}}=c_{stat}\left[1-\left(\frac{\Omega}{\omega_e}\right)^2\right]\qquad \text{N/m}\qquad (2.47)$$

It is obvious that:

- In the subcritical area, i.e. for $\Omega<\omega_e$ the dynamic stiffness is constant and approximately equal to the static stiffness.
- In the resonance area, i.e. for $\Omega=\omega_e$ the dynamic stiffness becomes zero (without damping).

Fig. 2.11 Course of the dynamic stiffness—related to the static stiffness—over the speed ratio Ω/ω_e

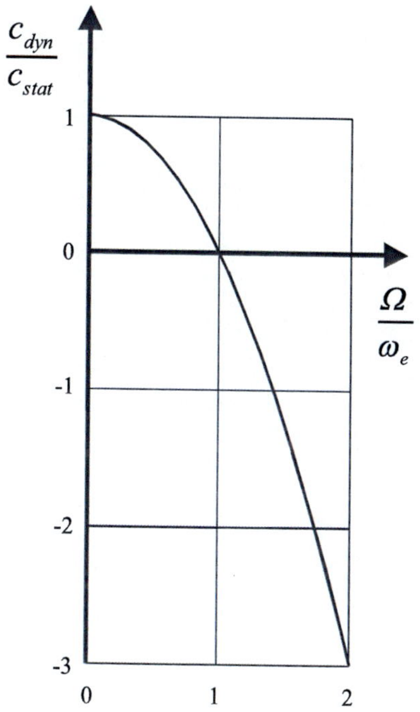

- In the supercritical area, i.e. for $\Omega > \omega_e$, the dynamic stiffness increases approximately quadratically with the angular frequency, i.e. it quickly becomes greater than the static stiffness (negative sign).

Forming the ratio of dynamic stiffness to static stiffness, we obtain Eq. 2.48.

$$\frac{c_{dyn}}{c_{stat}} = 1 - \left(\frac{\Omega}{\omega_e}\right)^2 \qquad \text{dimensionless} \tag{2.48}$$

The course of the dynamic stiffness—related to the static stiffness—over the speed ratio Ω/ω_e is shown in Fig. 2.11.

Terms and Explanations

3

Contents

3.1 Preliminary Note

Before theory and practice of rotor balancing technology are described in the next chapters, some very central terms should be explained in more detail to reduce problems of understanding.

A comprehensive overview of all terms and definitions important for balancing technology is given in Sect. 19.3 together with German and French technical terms.

H. Schneider, *Rotor Balancing*, https://doi.org/10.1007/978-3-662-66049-2_3

3.2 Rotor Balancing

Rotor balancing is an operation to check the mass distribution of a rotor (or assembly, component) and, if necessary, to correct it to ensure that the unbalance tolerances are kept.

3.3 Balancing Task

The balancing task for a rotor is defined by three criteria:

- The behaviour of the unbalances over the speed,
- The types of unbalance to be corrected or checked,
- The ability of the rotor to maintain or change the position of its mass elements and their centres of mass relative to each other over the speed range (see Table 4.3).

3.4 Rotor

A body rotating in service is a rotor.[1]

3.5 Unbalance

Every rotor has an unbalance. This can have many causes, such as: Design, material, manufacturing and assembly. Changes in unbalance condition during operation can also occur: Wear, deposits or changes in position or even loss of parts. e.g. blades.

Due to the inhomogeneity of the material and the individual actual data during production and assembly, it is inevitable that ***each*** rotor has its ***own*** unbalance distribution, even in series production.

Whether an unbalance is disturbing and must be corrected depends on its size and the unbalance tolerance.

[1]The designation "rotor" also includes bodies that do not rotate during service, but are rotatably mounted and are subjected to linear or alternating accelerations, e.g. the pointer of an analogue meter in a car.

3.6 Unbalance Condition

Based on principle the real unbalance state of a rotor cannot be determined but only the ***effects of*** the unbalance condition: During balancing, a set of unbalances is determined in 1, 2, or more radial planes which corresponds as closely as possible to this effect, and the necessary unbalance correction is then derived from this set.

It makes sense to describe the ***unbalance condition*** by the three types of unbalance:

- Resulting Unbalance $\vec{U}_r$,
- Moment unbalance $\vec{P}_r$ and
- Modal unbalance(s) $\vec{U}_{n,r}$.

3.7 Unbalance Behaviour

Unbalances can remain constant over the speed range of the rotor up to service speed, or they can change.

Modal unbalances are considered to be independent of speed—however, with shaft elastic behaviour the response of the rotor changes over the speed range.

3.8 Unbalance Tolerances

The permissible unbalance U_{per} is the unbalance tolerance for the entire rotor.

If correction is performed in several planes, the unbalance tolerance is divided, either between two or more planes, and/or between the different types of unbalance. The unbalance tolerances then indicate separately for each plane, or for each relevant type of unbalance, the permissible unbalance.

3.9 Correction

Unbalance correction is a process by which the mass distribution of the rotor is corrected: mostly by adding or removing masses, sometimes by shifting masses, rarely by shifting the shaft axis of the rotor.

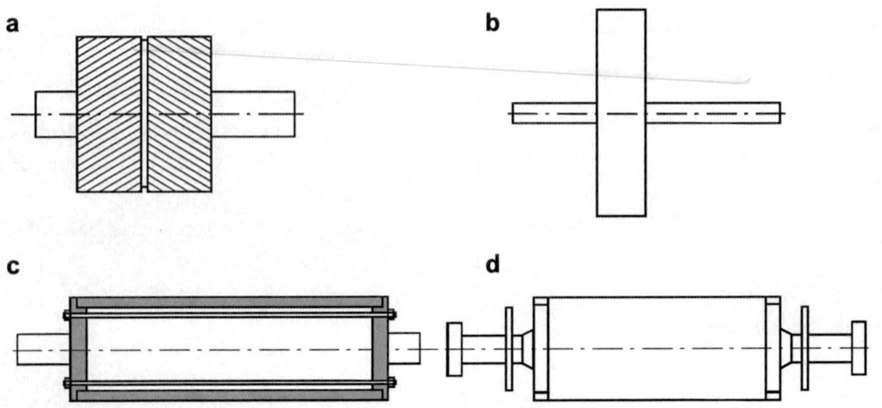

Fig. 3.1 Examples of rotor types that often exhibit certain rotor behaviour: **(a)** Solid gear wheel—***rigid behaviour.*** **(b)** Laval rotor (disk on elastic shaft)—***shaft elastic behaviour.*** **(c)** Drum with multiple tie rods (which deflect differently under centrifugal force)—***body elastic behaviour.*** **(d)** Generator with windings (which settle under centrifugal force)—***settling behaviour***

3.10 Correction Plane

This is a plane perpendicular to the shaft axis of the rotor, where an unbalance correction is performed. Depending on the balancing task, one, two or more correction planes are used.

3.11 Shaft Axis

The axis to which all unbalances are related to must be precisely defined; it is called shaft axis.

The shaft axis is the line between the journal centres and follows the deflected shape of the rotor due to gravity or any other constant force.

For rotors with rigid behaviour it is a rotor-fixed axis, i.e. it follows all movements of the rotor.

3.12 Rotor Behaviour

ISO 19499 describes the rotor behaviour by means of some typical examples, Fig. 3.1[2].

3.12.1 Rotors with Rigid Behaviour

Most rotors are designed and constructed in such a way that the unbalance condition and their shape change only negligibly up to the service speed of the rotor. They are referred to as rotors with ***rigid behaviour.***

3.12.2 Rotors with Flexible Behaviour

For some rotors, the changes in unbalance and/or shape are not negligible up to the operating speed. ***Flexible behaviour*** is a collective term, a distinction is made between *shaft-elastic behaviour, component-elastic behaviour* and *settling behaviour.*

3.12.2.1 Rotors with Shaft Elastic Behaviour

The changes of the unbalance condition are negligible up to the service speed, but in addition to the low-speed unbalances—resultant unbalance and moment unbalance—***modal unbalances*** also play a role, which excite flexural modes near to resonances.

3.12.2.2 Rotors with Component-Elastic Behaviour

The changes of the unbalance condition are ***not*** negligible up to the service speed, but are reversible, i.e. they disappear again with decreasing speed. Modal unbalances do not play a role here.

3.12.2.3 Rotors with Settling Behaviour

The changes of the unbalance condition are ***not*** negligible up to the service speed and are ***not*** reversible, i.e. they remain. After ***settling,*** the rotors show one of the other behaviours—from rigid behaviour to component-elastic behaviour.

[2] ISO 19499 points out that all these examples may also show other behaviour, i.e. they only show general tendencies.

3.13 Rotor Concept

With the conception of a rotor, certain machine-dynamic goals are pursued—they also may have an effect on the balancing task. The conceptions of *rigid rotor behaviour* and *shaft-elastic behaviour rotor are* mutually exclusive.

Component elastic behaviour and *settling behaviour* normally are not intended, but sometimes cannot be avoided.

4 Theory of Balancing

Contents

H. Schneider, *Rotor Balancing*, https://doi.org/10.1007/978-3-662-66049-2_4

4.1 Preliminary Note

In the theory of rotor balancing, important areas have been revised in the past decade with the aim of describing the states of the rotors and their behaviour with a comprehensive system in such a way that all possible balancing tasks can be derived from it.

4.2 General

The basic results were published as an introduction to balancing technology in ISO 19499[1] and later—especially regarding aspects of shaft-elastic rotors—in more detail in Beiblatt 1 of DIN ISO 21940-12.[2]

The new perceptions and interpretations particularly concern:

- The description of the unbalance state, Sect. 4.2.1.
- The rotor concept, Sect. 4.2.2.
- The rotor behaviour, Sect. 4.2.3.
- The determination of the unbalance tolerances, Sect. 4.2.4.
- The systematics of the rotor behaviour determining the feasible balancing methods Sect. 4.2.5.

Important details are outlined in this chapter and explained in detail in other chapters.

[1] ISO 19499: Mechanical vibration—Balancing—Guidance on the use and application of balancing standards. Will be replaced by ISO 21940: Mechanical vibration—Balancing of rotors, Part 1: Introduction.

[2] DIN ISO 21940: Mechanische Schwingungen—Auswuchten von Rotoren—Teil 12: Verfahren und Toleranzen für Rotoren mit nachgiebigem Verhalten, Beiblatt 1 (2015): Verfahren zum Auswuchten bei mehreren Drehzahlen. (English: DIN ISO 21940—Part 12: Procedures and tolerances for rotors with flexible behaviour, Supplement 1 (2015): Methods for balancing at multiple speeds).

4.2.1 Unbalance State

The unbalance state of a rotor was ***traditionally*** described by the unbalances along the rotor axis, which were then summarised and interpreted differently for the areas of the "rigid rotor" and the "shaft-elastic rotor".

The ***new system*** always starts with the three ***types of unbalance:***

- Resultant unbalance $\vec{U}_r$.
- Moment unbalance $\vec{P}_r$.
- Equivalent modal unbalances $\vec{U}_{ne,r}$.

Each of these types of unbalance[3] describes an independent aspect of the unbalance state.

Each of these aspects is also expressed in the different equations for calculation (see Sects. 4.5.1 to 4.5.4)—all together, however, describe the unbalance state completely.

If subsequently some unbalances do not play a role—i.e. do not need to be treated or checked—they are omitted for the selection of the balancing method.

Notes

1. This approach is sometimes not appreciated, as more details have to be measured and assessed than are finally necessary for the balancing process.
2. However, only if a certain detail was ***measured and assessed*** one can decide if it is relevant or if it can be neglected.
3. The check can be very different: in case of single rotors, a decision must be made in *each individual case,* in the case of repeat rotors, e.g. during the *type test,* and in the case of a rotor series after measuring an appropriate *collective.*

The amounts of these unbalances can be represented in a joint diagram, Fig. 4.1.

The resultant unbalance and the moment unbalance each have only one bar, the equivalent modal unbalances can have several bars on the speed axis, corresponding to the number of bending flexural modes of interest.

In order to generalise the representation, each unbalance has been normalised by the permissible unbalance U_{per}.[4] The operating speed n_{s} is fixed in the diagram, the flexural resonance speeds—here only the 1st speed $n_{\text{crit, 1}}$—are dependent on the rotor system and thus variable.

[3] In this scheme, all types of unbalance are actually in form of a resultant, i.e. formed from *all* unbalances on the rotor, hence the index r. In the following, however, only the resultant unbalance is always called "resultant" and designated with this index (to avoid confusion), for the moment unbalance and the modal unbalances the "resultant" is only sometimes used.

[4] The permissible unbalance U_{per} is the total permissible unbalance for the rotor, defined e.g. according to ISO 21940-11.

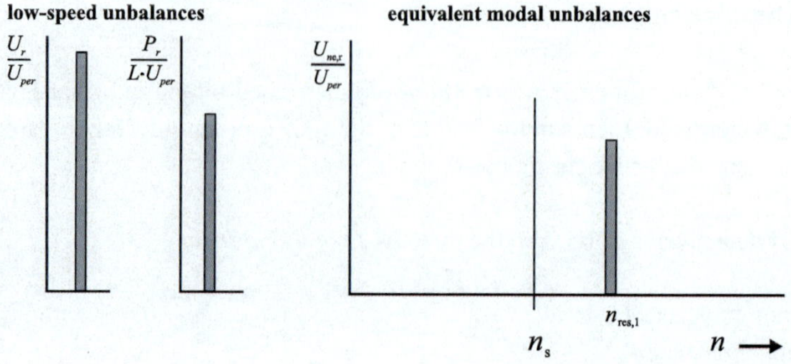

Fig. 4.1 Illustration of the different unbalance components. Each unbalance is related to the total permissible unbalance U_{per}. In order to obtain the same dimensions as U_{per}, the *moment unbalance is* represented as a *couple unbalance* with the distance L (see Sect. 4.5.3), the *modal unbalance* as *equivalent modal unbalances* (see Sect. 4.5.5)

4.2.2 Rotor Concept

The *concept of rotors* pursues certain machine-dynamic objectives in the development, production and operation of machines.

Rotors are typically designed as rigid *or* shaft elastic. These two concepts are mutually exclusive.

However, if the design of the respective concept does not sufficiently consider all conditions that influence the rotor behaviour, the objective can easily be missed. A sufficient balancing procedure for a rotor can therefore not be reliably derived from its *concept*—its *rotor behaviour* (see Sect. 4.2.3) is a better basis.

The other two rotor behaviours—body-elastic behaviour and settling behaviour—are not intentionally targeted, but they are sometimes unavoidable, e.g. due to the design and the materials used.

4.2.3 Rotor Behaviour

4.2.3.1 General

Whereas in the past rotors were labelled with terms such as *rigid rotor* or *shaft-elastic rotor,* today one speaks of the ***rotor behaviour*** and thus one can also describe a smooth transition from *rigid behaviour* to *shaft-elastic behaviour.*

This smooth transition in behaviour can be caused by very different influences, and appear in both directions.

Typical influences on the rotor behaviour are, e.g.—intentional or unintentional—changes in:

1. The operating speed.
2. The bearings and support.
3. The initial unbalance, e.g. depending on design, material, manufacturing process and assembly.
4. The permissible residual unbalances of the unbalance components.
5. The balancing methods and balancing processes.

While the first point is generally known, the second point is usually less considered. The other influences are largely unknown and are only now being systematised in the above-mentioned standards.

4.2.3.2 Effects of Rotor Behaviour

The following explanations are intended to illustrate the significance of these influences on planning and processes for rotor balancing:

1. Balancing always means reducing the initial unbalances to the permissible residual unbalances, or checking all relevant unbalances for compliance with the tolerances.
2. If only the so-called low-speed unbalances—resultant unbalance and/or moment unbalance—have to be taken into account (and if they are sufficiently unchanged over the speed range of interest), it is a *rigid* rotor behaviour.
   If modal unbalances also are relevant, a *shaft-elastic* rotor behaviour is present.
3. Any change in the *initial unbalances* or the *permissible residual unbalances* changes their relation and can therefore:
   - Allow a balanced rotor to become an out-of-tolerance rotor.
   - Turn a rotor where single-plane balancing is sufficient into a problem with two balancing planes.
   - Turn a rotor with *rigid behaviour* into one with *shaft-elastic behaviour.*
   - Or vice versa in all these cases.
4. A change in the balancing procedure can make additional unbalances relevant, e.g.:
   - Single-plane balancing, with changing to a different correction plane, can cause the moment unbalance to be out of tolerance: this makes two-plane balancing necessary.
   - Modal unbalances may become out of tolerance by changing the correction plane, making multi-plane balancing necessary.

4.2.3.3 Principle of Order

ISO 19499 describes not only the ***rigid behaviour*** and the ***shaft-elastic behaviour,*** but also two other forms, the ***body-elastic behaviour*** and the ***settling behaviour.***

Beiblatt 1 of DIN ISO 21940-12, systematises this area even further, see Table 4.3 in Sect. 4.2.5.

Rotor behaviour is defined by (quotation from Beiblatt 1 of DIN ISO 21940-12, up to and including Note 6, but modified):

- The behaviour of the unbalances over the speed.
- The types of unbalance to be corrected or checked.
- The ability of the rotor to maintain or change the position of its mass elements and their mass centres relative to each other over the speed range.

Notes

1. Unbalances usually do not change significantly with the speed. Modal unbalances are also not speed-dependent.[5] Only in special cases unbalances do change significantly with speed, see *body-elastic rotor behaviour* and *settling behaviour.*
2. Mass elements are useful to describe the mass distribution of the rotor and possible changes with speed. Mass elements can preferably be finite elements, parts, components or axial sections.
3. Rotor behaviour is also influenced by the design, manufacture and assembly of the rotor.
4. The response of the rotor to unbalances can change over the speed range, also due to the bearing conditions. Whether the response is acceptable or not is determined by the unbalance tolerances.
5. The speed range covers all speeds from standstill to maximum service speed, but may also include an overspeed range, e.g. to imitate service loads—temperature, pressure, flow, or similar.
6. Only changes in the position of mass elements that do *not* occur symmetrically to the shaft axis are of importance for balancing.

4.2.4 Unbalance Tolerances

The ***permissible unbalances*** thus play a very important role in addition to the initial unbalances, but they can also be represented in a similar diagram as the unbalances in Fig. 4.1 (see Fig. 4.2), so that the information can be superimposed later.[6]

This diagram is based on the effect of the various unbalances on the *rotor situation at service speed.* The low-speed unbalances are assumed to be fully effective (constant over the speed), while for the equivalent modal unbalances the resonance peaks must be taken into account.

The limits shown in Fig. 4.2 solely apply if only **one** unbalance has to be considered at a time. The limits for the *individual* low-speed unbalances referred to U_{per} are

[5] The *modal unbalances* are considered to be independent of speed: Although the response changes when passing the resonances, the excitation by the modal unbalances does not.

[6] The service speed n_s is fixed in the diagram, the flexural resonance speeds are variable depending on the rotor.

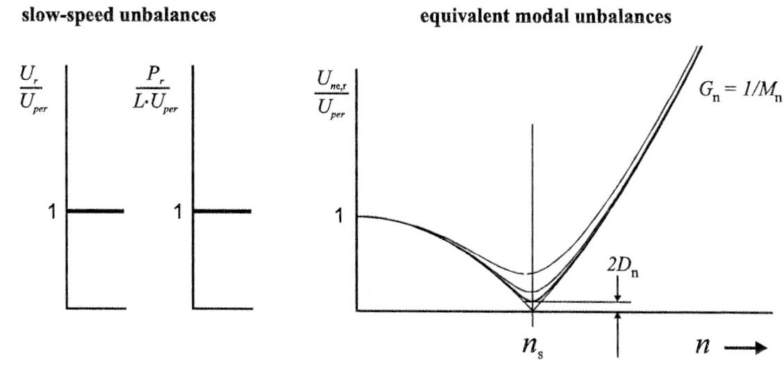

Fig. 4.2 Representation of the tolerance limits of the unbalance types, related to U_{per}, if they only occur individually. Limit curves G_n for different modal dampings D_n=0; 0.05; 0.1; 0.2

horizontal lines with the value 1. For the equivalent modal unbalances referred to U_{per}, the limit is given by the curves G_n, which reflects the resonance behaviour at different dampings. G_n is the reciprocal value of the modal transfer function M_n:

$$G_n = \frac{1}{M_n} = \frac{\sqrt{[1-(n_s/n_{crit,n})^2]^2 + 4D_n^2(n_s/n_{crit,n})^2}}{(n_s/n_{crit,n})^2}$$

or simplified:

$$G_n = \sqrt{\left[\left(\frac{n_{crit,n}}{n_s}\right)^2 - 1\right]^2 + 4D_n^2\left(\frac{n_{crit,n}}{n_s}\right)^2} \quad \text{dimensionless} \tag{4.1}$$

The curve starts with a value of 1 for low speeds—resonances far below the service speed. At service speed it drops to a minimum (which is determined by the modal damping) and then increases sharply with higher speeds, approaching a second degree polynomial curve.

If ***several*** unbalances are to be considered at the same time, the tolerances for each relevant unbalance must be reduced accordingly. Assuming that the unbalances are not (or not significantly) dependent on each other, Beiblatt 1 of DIN ISO 21940-12 recommends a calculation of the correction according to the least squares method for this purpose:

$$\sqrt{U_{r,per}^2 + \left(\frac{P_{r,per}}{L}\right)^2 + \sum_{n=1}^{N}(U_{ne,per}M_n)^2} = U_{per} \quad \text{g} \cdot \text{mm} \tag{4.2}$$

For a ***uniform*** distribution this means: If two unbalances (U_r, P_r) and additionally N modal unbalances are to be considered for the low-speed balancing, the calculated factor per unbalance is:

Table 4.1 Factors per unbalance depending on the number of unbalances (with uniform distribution)

Number of unbalances	1	2	3	4	5	6
Factor calculated	1	0.71	0.58	0.50	0.45	0.41
Recommended factor	1	0.70	0.60	0.50	0.45	0.40

$$\frac{1}{\sqrt{2+N}} \quad \text{dimensionless} \tag{4.3}$$

Depending on the total number of unbalances to be taken into account, the factors are as follows, see Table 4.1. For each unbalance, the tolerance is determined by multiplying U_{per} by this factor.

For ***uneven*** distribution, see Beiblatt 1 of DIN ISO 21940-12.

4.2.5 Balancing Task

The rotor in Fig. 4.1 runs below the 1st flexural resonance speed.

Assumption

In addition to the two low-speed unbalances, the 1st equivalent modal unbalance can have an influence here.

- A total of three unbalances are to be considered. Table 4.1 results in a factor of 0.60 per unbalance (with an even distribution).
- From the speed ratio $n_{\text{crit},n}/n_{\text{s}}$ (assuming the modal damping is negligible) the value on the modal weighting curve G_n and finally the ratio of the permissible unbalances to the total permissible unbalance can be derived, see Table 4.2.

The values in the last column indicate the factor by which the permissible total unbalance U_{per} must be multiplied to obtain the permissible value for each individual unbalance.

Figures 4.1 and 4.2 can be summarized in a single diagram: Fig. 4.3. If, in addition, the results of the tolerance calculation from Table 4.2 are entered—thick horizontal lines—one can immediately see which unbalances are out of tolerance and need correction: the two low-speed unbalances and the 1st equivalent modal unbalance.

With this system, all balancing tasks can be described precisely—from the rotor, which does not need to be balanced at all, through single-plane balancing, two-plane balancing, and to all levels of shaft-elastic rotor behaviour, see Sect. 4.6.

Table 4.2 Calculation of the permissible unbalances

Unbalance component U_x	U_{per} from ISO 21940-11	$n_{crit,n}/n_s$	G_n	Factor from Table 4.1	$U_{x,per}/U_{per}$
U_r	U_{per}	–	–	0.60	0.60
P_r/L		–	–		0.60
$U_{1e,r}$		1.40	0.96		0.58

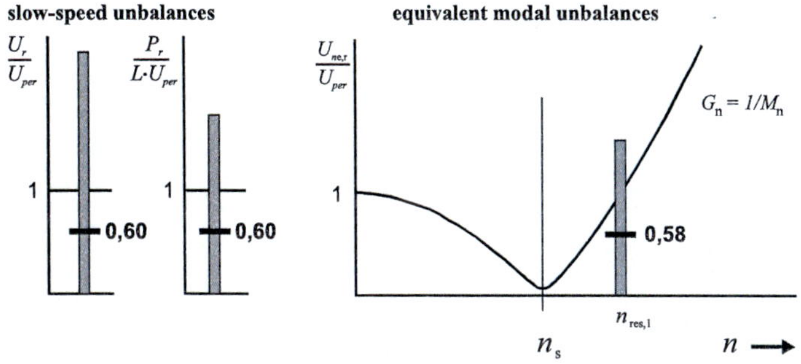

Fig. 4.3 Clarification of the balancing task by showing the unbalance amounts and the permissible limits. Conclusion: All 3 unbalances are out of tolerance and need correction

Beiblatt 1 of DIN ISO 21940-12, also provides a good overview in tabular form of the various rotor behaviours, the corresponding balancing standards and the recommended balancing procedures, Table 4.3.

In Table 4.3 three criteria are important:

- How do the respective unbalances behave over speed—constant or variable?
- Which unbalances need to be corrected or checked—resultant unbalance, moment unbalance, modal unbalances?
- Is the position of the mass elements of the rotor—kept, or does it change? And if it changes—according to what principle?

This allows a correct assignment of the balancing task within the two standards ISO 21940-11 as well as ISO 21940-12. Within ISO 21940-12 there is further division into different (9) balancing methods.

As can be seen from Table 4.3 that *shaft-elastic behaviour, body-elastic behaviour* and *settling behaviour* are summarised under the generic term *flexible behaviour.*

Table 4.3 Overview of the different rotor behaviours, the appropriate balancing standards, as well as the recommended balancing procedures, depending on the three criteria mentioned in the text. (Excerpt from Beiblatt 1 of DIN ISO 21940-12)

<table>
<tr><th>unbalance behaviour over speed / unbalances, to be corrected or to be checked</th><th colspan="5">constant</th><th colspan="2">variable</th></tr>
<tr><td>resulting unbalance</td><td>—</td><td>X</td><td>—</td><td rowspan="2">X</td><td rowspan="2">X</td><td rowspan="2">X</td><td rowspan="2">X</td></tr>
<tr><td>resulting moment unbalance</td><td>—</td><td>—</td><td>X</td></tr>
<tr><td>modal unbalance</td><td>—</td><td>—</td><td>—</td><td>—</td><td>X</td><td>—</td><td>—</td></tr>
<tr><td>position of mass elements</td><td colspan="4">is kept</td><td colspan="3">is changing</td></tr>
<tr><td>position is changed as</td><td colspan="4">—</td><td>elastic bending</td><td>elastic deformation</td><td>component setting</td></tr>
<tr><td>balancing necessary</td><td>no</td><td colspan="6">yes</td></tr>
<tr><td rowspan="2">rotor behaviour</td><td rowspan="2">—</td><td colspan="3" rowspan="2">rigid</td><td colspan="3">flexible</td></tr>
<tr><td>shaft-elastic</td><td>body-elastic</td><td>setting behaviour</td></tr>
<tr><td>related standard ISO 21940</td><td>—</td><td colspan="3">part 11</td><td colspan="3">part 12</td></tr>
<tr><td>balancing procedure recommended</td><td>—</td><td>single-plane balancing</td><td>two-plane balancing</td><td>three-(two-)plane balancing</td><td>procedure A to H</td><td>procedure I</td><td>allow setting at high speed before balancing</td></tr>
</table>

4.3 Unbalances and Correction

Many people, even those who have had little or no contact with rotor balancing technology, have already heard the word ***unbalance***, but it is often not interpreted correctly. Yet the physical phenomenon is known to most, but rather under the term ***centrifugal force***.

According to the ISO definitions[7] (for terms and definitions see Chap. 3 and Sect. 19.3), an unbalance is present in a rotating system (rotor) "when vibration forces or motion is imparted to it and its bearings from centrifugal forces of mass eccentricities".

In Sect. 2.6.9 it is recapitulated that a mass u, which rotates on a radius $\vec{r}$ with the angular frequency $\vec{\Omega}$, generates a centrifugal force $\vec{F}$. The product $u\,\vec{r}$ determines the cen-

[7] ISO 21940: Mechanical vibration—Rotor balancing—Part 2: Vocabulary.

trifugal force according to size and direction. In rotor balancing technology this is called unbalance $\vec{U}$. One can therefore write:

$$\vec{U} = u\,\vec{r} \qquad \text{kg} \cdot \text{m} \tag{4.4}$$

With:

- $\vec{U}$ Unbalance, a vector, SI unit: kg · m, frequently used unit: g · mm,
- u Unbalanced mass, a scalar, SI unit: kg, often used: g,
- $\vec{r}$ Radius, distance of the mass centre of the unbalance mass from the shaft axis (see Chap. 3), a vector, SI unit: m, often used: mm

The unbalance therefore is a vector and always has the same direction as the radius vector of the unbalance mass.

The unbalance is independent of the speed (n or $\vec{\Omega}$ are not included in the equation) provided that the radius $\vec{r}$ is constant (rigid rotor behaviour).

Task

What is the unbalance U generated by a mass $u=24$ g on a radius $r=500$ mm?

Solution

$U=ur=24 \cdot 500$ g · mm $=12{,}000$ g · mm (or 12 kg · mm).

Unbalance correction is usually done by adding or removing material in one or more correction planes (see Chap. 14). Ideally, the sum of the unbalances in each of these planes should be zero, i.e. the sum of the unbalance $\vec{U}$ and the unbalance correction $u_c\,\vec{r}_c$.

For a single correction plane, one can write:

$$\vec{U} + u_c\,\vec{r}_c = 0 \tag{4.5}$$

If we first look at the amounts of the two unbalances, we see that the ***product*** of the correction mass u_c with the correction radius $\vec{r}_c$ must be equal to the unbalance $\vec{U}$ of the rotor (not the correction mass equal to the unbalance mass).

If the direction is also taken into account, it becomes clear that the correction can take place at the same angular direction as the unbalance or in the opposite direction, see Fig. 4.4.

Therefore one can conclude:

1. The correction radius can be selected as desired, the correction mass is then calculated according to:

$$u_c = \frac{U}{r_c} \qquad \text{g} \tag{4.6}$$

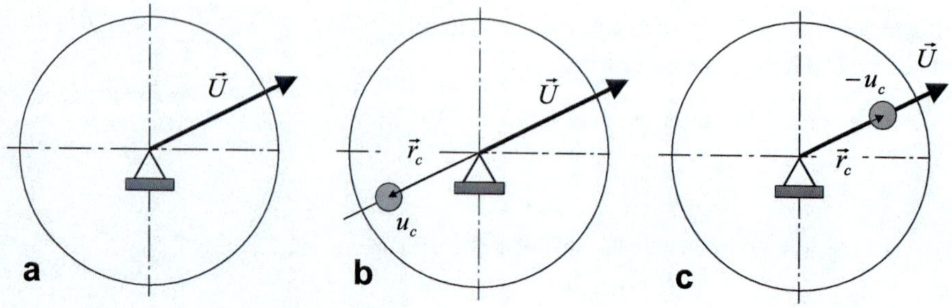

Fig. 4.4 Correcting an unbalance: (**a**) unbalance $\vec{U}$; (**b**) placing material on the opposite side: correction mass positive, (**c**) removing material on the same side: correction mass negative

or vice versa:

$$r_c = \frac{U}{u_c} \quad \text{mm} \tag{4.7}$$

2. The unbalance $\vec{U}$ can be corrected by

- Adding material on the opposite side,
- Removing material on the same side.

Task

An unbalance $U = 12{,}000\ \text{g} \cdot \text{mm}$ is to be corrected on the correction radius $r = 300$ mm. How large must the correction mass be?

Solution

With Eq. 4.7, rearranged: $u_c = \frac{U}{r_c} = \frac{12{,}000}{300} \frac{\text{g} \cdot \text{mm}}{\text{mm}} = 40\,\text{g}$.

This type of correction, where any angle can be used according to the respective position of the unbalance, is called ***polar* correction** (derived from polar coordinates, a specification of angle and radius).

If, due to the specific nature of the rotor or the type of correction, correction is only possible in certain directions (at fixed locations), this is referred to as ***fixed-location* correction.** The unbalance of a correction plane is divided into ***components*** according to the possible correction directions and each component is corrected separately, see Fig. 4.5.

4.4 Unbalance of the Disc-Shaped Rotor

Until now, the unbalance has been related to a theoretical case where only the unbalance has a mass, but not the radius and the axis. What is the situation with a real rotor with mass m?

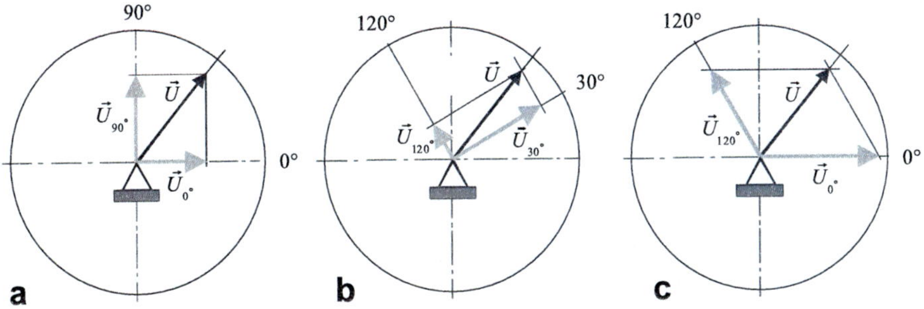

Fig. 4.5 Splitting the unbalance $\vec{U}$ into two components, e.g. for a fixed location correction. **a** into 90° components, **b** into shifted 90° components, **c** into 120° components

The simplest case is a ***disc-shaped rotor*** (axial extension negligible), which sits perpendicular to the shaft axis.[8] If the rotor rotates with the angular frequency $\vec{\Omega}$, each mass particle m_i on its radius $\vec{r}_i$ generates a centrifugal force $\vec{F}_i$ in Newton, see Fig. 4.6:

$$\vec{F}_i = m_i \vec{r}_i \Omega^2 \quad \text{N} \tag{4.8}$$

The vector sum of the centrifugal forces of all individual elements is the resulting centrifugal force from the rotor (Eq. 4.9):

$$\vec{F}_r = \sum_{i=1}^{n} m_i \vec{r}_i \Omega^2 \quad \text{N} \tag{4.9}$$

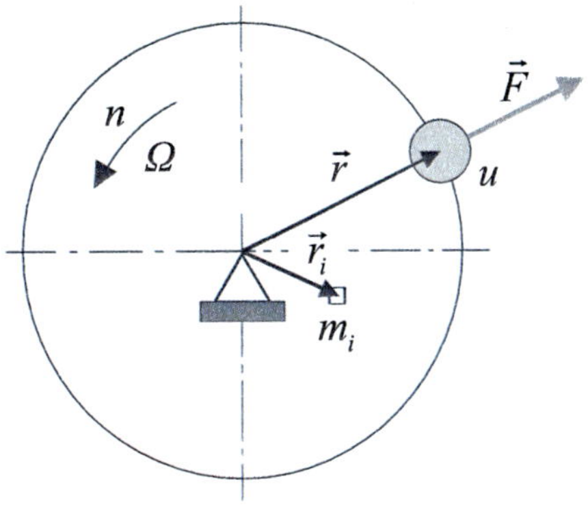

Fig. 4.6 Description of the unbalance of a mass particle, as well as the unbalance of the disc-shaped rotor

[8] If a disc is not perpendicular to the shaft axis, a moment unbalance occurs, see Sect. 5.7.

There are two possibilities:

- $\vec{F}_r = 0$: There is no centrifugal force, the rotor is without unbalance, a *completely balanced rotor,*
- $\vec{F}_r \neq 0$: The rotor has an *unbalance.*

The question now arises as to how best to specify the unbalance state.

The remaining centrifugal force ($\vec{F}_r \neq 0$) can be thought of as arising from a single unbalance $u\vec{r}$ or $\vec{U}$ (Eq. 4.10) and then cancel the speed influence on both sides (Eq. 4.11):

$$\vec{F}_r = \sum_{i=1}^{n} m_i \vec{r}_i \Omega^2 = u\vec{r}\Omega^2 \quad \text{N} \tag{4.10}$$

$$\sum_{i=1}^{n} m_i \vec{r}_i = u\vec{r} = \vec{U} \quad \text{g} \cdot \text{mm} \tag{4.11}$$

Therefore

- The unbalance state of a disc-shaped rotor (perpendicular to the shaft axis) can be fully described by *one* unbalance vector.
- The unbalance correction only requires *a* correction in *one correction* plane
- This correction plane must of course lie in the plane of the disc-shaped rotor.[9]

4.5 Unbalance of a General Rotor

In a rotor with a larger axial extension, unbalance can occur everywhere. The unbalance state of a ***general rotor*** arises from an infinite number of unbalance vectors distributed along the shaft axis of the rotor.

If all unbalance vectors were corrected in their respective planes, the rotor would be completely balanced. In practice, however, it is neither possible to measure all these unbalances nor to correct them—but fortunately this is also not necessary.

A summary description of the unbalance state is required, which is as unambiguous as possible and—in conjunction with other criteria—leads to the balancing method to be used in each case, as shown in Table 4.3.

[9] When correcting in a different plane, an additional moment unbalance occurs, see Sect. 4.5.2.

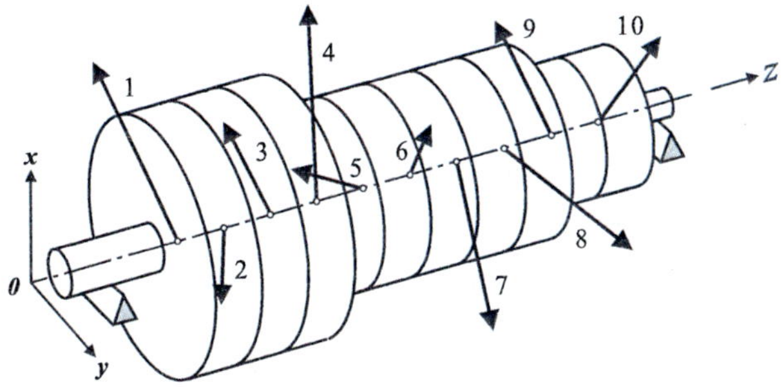

Fig. 4.7 General rotor modelled with ten disc elements (whose respective moment unbalance is negligible), each with an unbalance vector

For this description, one can use the disc-shaped rotor and the knowledge gained there: One mentally divides the general rotor into rotor elements of small axial extension, each of which corresponds to a disc-shaped rotor.

The unbalance of *one* rotor element can then be expressed by *one* unbalance vector (as long as its moment unbalance is negligible).

Due to the unbalances of *all imaginary* rotor elements, the unbalance state of the general rotor can thus be described sufficiently precise with a finite number of unbalances Fig. 4.7.

Note

Unfortunately, the real ***unbalance state*** of a rotor cannot be determined. In the balancing process, the ***effect*** of the unbalance condition is measured and a correction is determined in one, two or more planes that adequately correct this unbalance state. Specifying the unbalance state of a rotor by a limited number of unbalance vectors therefore corresponds to practical requirements.

Due to the perspective view, the position and size of the unbalances cannot be seen well. However, if one looks axially at the individual rotor elements, the unbalance vector can be correctly represented in size and angle for each element, Fig. 4.8. This division into ten rotor elements (in this case of equal width) is only for an easier overview and serves as a basis for the following graphical evaluations.

For a general description and mathematical calculation of the unbalance state, this scheme can be abandoned: Every conceivable unbalance vector can be represented by an unbalance vector $\vec{U}_k$ in a plane displaced from the origin *0* by the distance $\vec{z}_k$, Fig. 4.9.

As noted in Sect. 4.2.1 the unbalance state of a ***general rotor*** is fully described by three unbalance types:

- Resultant unbalance $\vec{U}_r$
- Moment unbalance $\vec{P}_r$
- Equivalent modal unbalances $\vec{U}_{ne,r}$

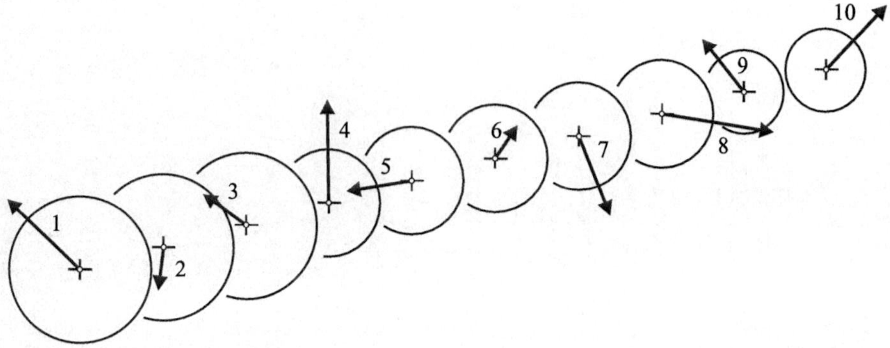

Fig. 4.8 The ten elements of the rotor of Fig. 4.7 seen axially, with their unbalance vectors correctly reproduced in size and angle

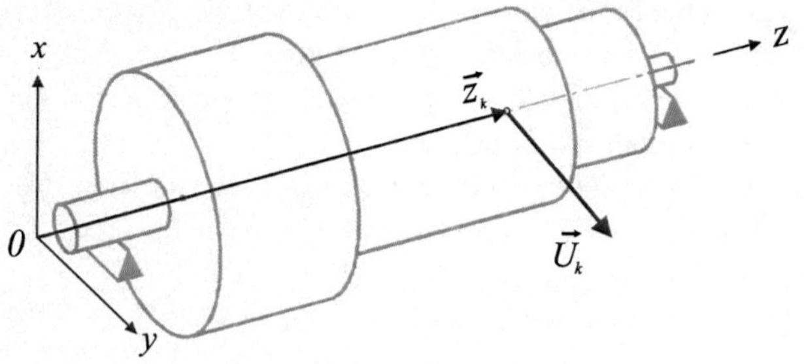

Fig. 4.9 Principle representation for the calculation of unbalances—unbalance vector $\vec{U}_k$ at distance $\vec{z}_k$ from origin *0*

The next sections explain how these different types of unbalance are formed from the unbalance state of the rotor.

4.5.1 Resultant Unbalance

The ***resultant unbalance*** $\vec{U}_r$ is the vector sum of all unbalances $\vec{U}_k$:

$$\vec{U}_r = \sum_{k=1}^{K} \vec{U}_k \qquad \text{g} \cdot \text{mm} \tag{4.12}$$

Here, k is the counter for unbalance planes, from 1 to K.

On the basis of Fig. 4.7 describes Fig. 4.10 the graphical determination of $\vec{U}_r$ as the vector sum of the ten individual unbalances.

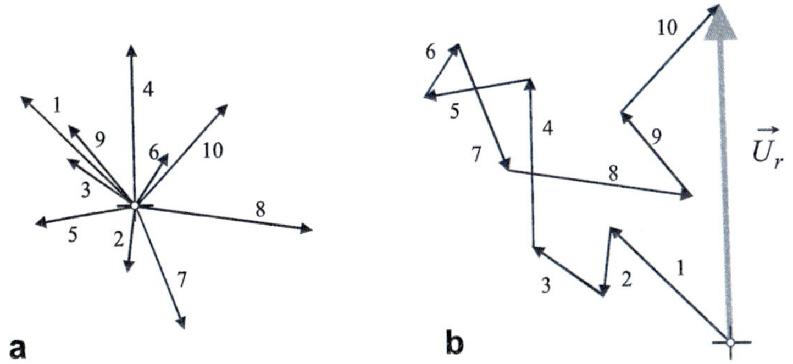

Fig. 4.10 Resultant unbalance: View in axial direction of the rotor (**a**), i.e. all unbalance vectors have the same origin. The vector sum of the unbalances in planes 1 to 10 (**b**) corresponds to the resultant unbalance $\vec{U}_r$

Notes

1. The ***resultant unbalance*** is always identical (in size and angular position), no matter which plane of the rotor is chosen for the representation.
2. However, the choice of this plane influences the size and angular position of the ***moment unbalance.***
3. If the resultant unbalance is stated in the mass centre plane, it becomes the ***static unbalance.***

4.5.2 Moment Unbalance

A ***moment unbalance*** can only be described correctly if a plane is defined for the resultant unbalance—for the representation, or for the correction, see Fig. 4.11.

The *individual moment unbalance* $\vec{P}_k$ is the vector product of the distance $\vec{l}_k$ (of the plane R for the resultant unbalance $\vec{U}_r$ to the plane of the unbalance $\vec{U}_k$) with the unbalance $\vec{U}_k$:

$$\vec{P}_k = \vec{l}_k \times \vec{U}_k \qquad \mathrm{g \cdot mm^2} \tag{4.13}$$

The moment unbalance is a rotational vector perpendicular to the rotor axis (see also Sect. 2.6), i.e. it is a different physical quantity than the unbalance, with a different unit ($\mathrm{g \cdot mm^2}$) and is denoted by P.

Sometimes it is clearer to relate all axial lengths to a common reference point (e.g. the origin 0 in Fig. 4.11), then the equation is:

$$\vec{P}_k = (\vec{z}_k - \vec{z}_R) \times \vec{U}_k \qquad \mathrm{g \cdot mm^2} \tag{4.14}$$

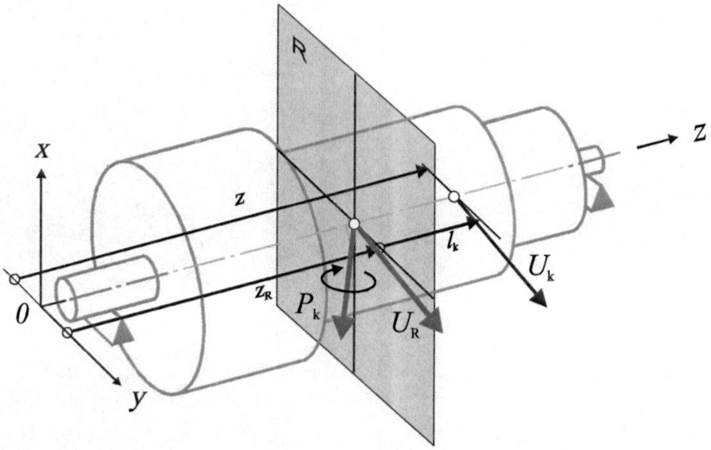

Fig. 4.11 Principle diagram for the calculation of an *individual moment unbalance*—the unbalance vector $\vec{U}_k$ forms the resultant unbalance $\vec{U}_r$ and the moment unbalance $\vec{P}_k$ in plane *R*

Here $\vec{z}_k$ is the distance from *0* to the plane of the unbalance vector $\vec{U}_k$ and $\vec{z}_R$ is the distance from *0* to the plane *R of* the resultant unbalance $\vec{U}_r$.

The ***resultant moment unbalance*** $\vec{P}_r$[10] is the vector sum of all *individual moment unbalances* $\vec{P}_k$, i.e.:

$$\vec{P}_r = \sum_{k=1}^{K} (\vec{z}_k - \vec{z}_{U_r}) \times \vec{U}_k \qquad \text{g} \cdot \text{mm}^2 \tag{4.15}$$

Here, *k* is the counter for the unbalance planes, from 1 to *K*.

Notes

1. The moment unbalance changes depending on the plane for the resultant unbalance.
2. In many cases it is useful for theory and practice to express the moment unbalance as a *couple unbalance.*

[10]The *individual moment unbalance* (Eq. 4.14) is formed by an individual unbalance vector $\vec{U}_k$. It is important for systematics, but does not play a major role in practice. In contrast, the *resultant moment unbalance is* formed by all unbalance vectors of the rotor (Eq. 4.15). In practice, this is a very important factor and is therefore simplified and referred to as *moment unbalance* in the remainder of this book.

4.5.3 Couple Unbalance

A *couple unbalance* corresponds in its effect to the *moment unbalance* it is intended to replace.

The couple unbalance is formed by two unbalance vectors of equal size but opposite angular position $-\vec{C}_k$ and $\vec{C}_k$ (unit e.g. g · mm) in two different axial planes. These planes can be chosen arbitrarily, e.g. at a distance $\vec{z}_{-C}$ and $\vec{z}_C$ of *0*, Fig. 4.12. The distance between the two planes is denoted here by the length $\vec{b}$.

To express an *individual moment unbalance* by an *individual couple unbalance*, the vector products must be the same. Using the lengths $\vec{z}_{-C}$ and $\vec{z}_C$ for the planes with the unbalance vectors $-\vec{C}_k$ and $\vec{C}_k$ results in:

$$\vec{P}_k = (\vec{z}_C \times \vec{C}_k) + (\vec{z}_{-C} \times (-\vec{C}_k)) \qquad \text{g} \cdot \text{mm}^2 \tag{4.16}$$

Or simplified:

$$\vec{P}_k = (\vec{z}_C - \vec{z}_{-C}) \times \vec{C}_k \qquad \text{g} \cdot \text{mm}^2 \tag{4.17}$$

Furthermore, the difference in lengths $\vec{z}_C - \vec{z}_{-C} = \vec{b}$ is the length $\vec{b}$ between the two planes, thus:

$$\vec{P}_k = \vec{b} \times \vec{C}_k \qquad \text{g} \cdot \text{mm}^2 \tag{4.18}$$

If Eqs. 3.18 and 3.14 are combined:

$$\vec{b} \times \vec{C}_k = (\vec{z}_k - \vec{z}_R) \times \vec{U}_k \qquad \text{g} \cdot \text{mm}^2 \tag{4.19}$$

Because of $\vec{C}_k = -(-\vec{C}_k)$ only the unbalance vector of one plane needs to be calculated and stated, the unbalance vector for the other plane results accordingly.

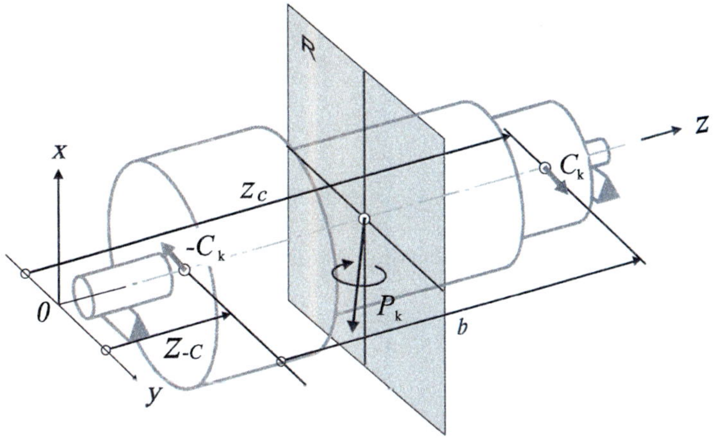

Fig. 4.12 Principle representation for the calculation of unbalances—the *individual moment unbalance* $\vec{P}_k$ can be expressed by the *individual couple unbalance* $-\vec{C}_k$ and $\vec{C}_k$

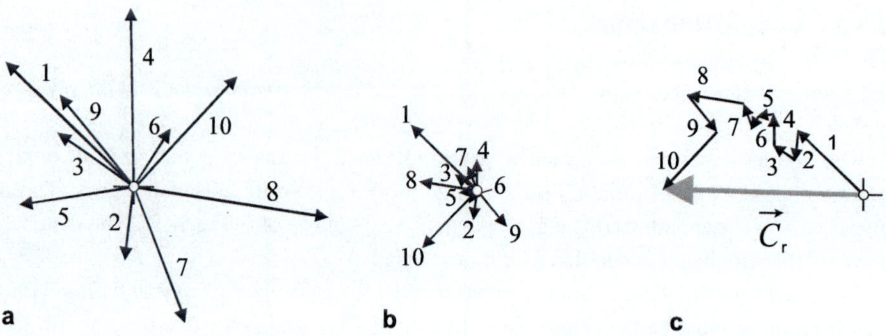

Fig. 4.13 Couple unbalance: (**a**) The individual unbalance vectors 1–10. (**b**) From these, the *individual couple unbalances* (Eq. 3.18) $\vec{C}_k$ are formed. (**c**) These unbalances are added vectorially to $\vec{C}_r$ the *resultant couple unbalance*

The ***resultant couple unbalance***[11] is the vectorial sum of the *individual couple unbalances* from Eq. 3.19, see Eq. 4.20.

$$\vec{b} \times \vec{C}_r = \sum_{k=1}^{K} (\vec{z}_k - \vec{z}_{U_r}) \times \vec{U}_k \qquad g \cdot mm^2 \tag{4.20}$$

Here too, because of $\vec{C}_r = -(-\vec{C}_r)$, only the unbalance vector of one plane needs to be calculated and specified.

Figure 4.13 shows the graphical determination of the couple unbalance $\vec{C}_r$, based on the unbalances of the rotor in Fig. 4.7 and on Fig. 4.12.

4.5.4 Modal Unbalance

Modal unbalances are based on the unbalance distribution and the deflections in the individual flexural modes, see Fig. 4.14.

Figure 4.14 shows modal deflections of an idealised rotor for the first three flexural modes (modal function Φ_n (z); $n=1, 2, 3$).[12] For this idealisation, a uniform mass and stiffness distribution as well as rigid bearings at the rotor ends were assumed.

[11] Analogous to moment unbalance (footnote 10), the *individual couple unbalance* is rarely used, but mostly the *resultant couple unbalance*. To simplify matters, this is usually called a *couple unbalance*.

[12] In the case of two dimensional flexural modes, the ordinate values are real numbers, but for three-dimensional flexural modes they are complex numbers.

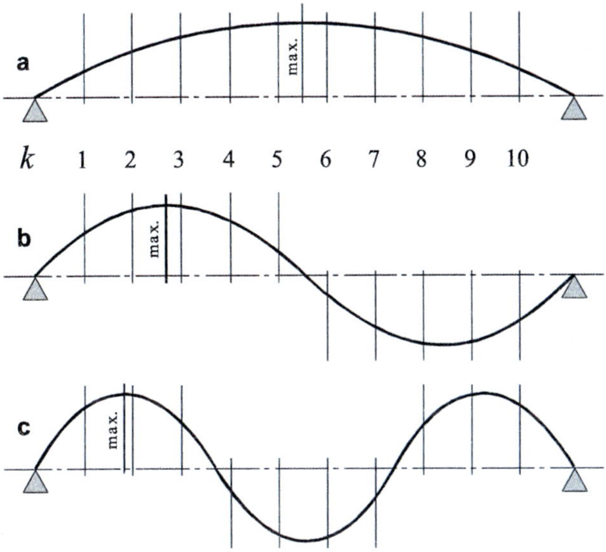

Fig. 4.14 The first three flexural modes of an idealised rotor (uniformly distributed mass and stiffness, rigid bearing at the ends): **(a)** 1st flexural mode $\Phi_1\ (z)$, **(b)** 2nd flexural mode $\Phi_2\ (z)$, **(c)** 3rd flexural mode $\Phi_3\ (z)$

Again we assume a rotor with several unbalances as shown in Fig. 4.7 and use the principle in Fig. 4.9.

If a single unbalance vector $\vec{U}_k$ is weighted with the ordinate value of the relevant flexural mode $\Phi_n\ (z_k)$ associated with this plane, the *individual modal unbalance* $\vec{U}_{n,k}$ is obtained:

$$\vec{U}_{n,k} = \vec{U}_k \Phi_n(z_k), \quad k = 1,\ 2,\ 3,\ \ldots, 10 \qquad \mathrm{g \cdot mm} \tag{4.21}$$

All these unbalances of a certain flexural mode $\vec{U}_{n,k}$ together form the *modal unbalance,* an unbalance distribution along the rotor. The vectorial sum of these unbalances gives the ***resultant modal unbalance*** in the *nth* flexural mode $\vec{U}_{n,r}$:

$$\vec{U}_{n,r} = \sum_{k=1}^{K} \vec{U}_k \Phi_n(z_k), \quad k = 1,\ 2,\ 3,\ \ldots, 10 \qquad \mathrm{g \cdot mm} \tag{4.22}$$

Notes

1. The *resultant modal unbalance* formed this way is correct in magnitude and angle, but cannot express the distribution of the *modal unbalance* along the rotor. However, this is not necessary because the *resultant modal unbalance* is only used as a basis for the calculation of the *equivalent modal unbalance* (see Sect. 4.5.5).

2. For these two unbalances—*modal unbalance* and *resultant modal unbalance*—a commonly agreed upon statement is difficult because there are different possibilities to normalise the ordinate values. The *equivalent modal unbalance* (see Sect. 4.5.5), on the other hand, is independent of the normalisation and therefore easier to handle in theory and practice.

4.5.5 Equivalent Modal Unbalance

If an *individual modal unbalance* $\vec{U}_{n,k}$ (Eq. 4.20) is divided by the *maximum ordinate value of the flexural mode* $\Phi_{n,\,\max}$, it becomes the *individual equivalent modal unbalance* in the *nth* flexural mode $\vec{U}_{ne,\,k}$. It acts in the plane of the maximum ordinate value (in the case of several identical maxima in a chosen plane). One can write:

$$\vec{U}_{ne,\,k} = \vec{U}_k \frac{\Phi_n(z_k)}{\Phi_{n,\,\max}}, \quad k = 1,\,2,\,3,\,\ldots,10 \qquad \mathrm{g \cdot mm} \tag{4.23}$$

The vectorial sum of all individual $\vec{U}_{ne,k}$ is the ***resultant equivalent modal unbalance*** in the *nth* flexural mode $\vec{U}_{ne,\,r}$:

$$\vec{U}_{ne,\,r} = \sum_{k=1}^{K} \vec{U}_k \frac{\Phi_n(z_k)}{\Phi_{n,\,\max}}, \quad k = 1,\,2,\,3,\,\ldots,10 \qquad \mathrm{g \cdot mm} \tag{4.24}$$

This division by $\Phi_{n,\,\max}$ and the vectorial summation can also be represented graphically, e.g. for the first flexural mode from Fig. 4.14 in Fig. 4.15:

Figure 4.15 shows the formation of the resultant equivalent modal unbalance in four steps:

a. Describes the ordinate values of the different planes.
b. Shows the unbalance vectors $\vec{U}_k$ from Fig. 4.7 when viewed along the axis.
c. Represents the *individual equivalent modal unbalances* $\vec{U}_{1e,\,k}$ (Eq. 4.23).
d. Shows how the vector sum of all $\vec{U}_{1e,\,k}$ forms the *resultant equivalent modal unbalance* $\vec{U}_{1e,\,r}$, Eq. 4.24.

For all relevant flexural modes, the ***equivalent modal unbalances***[13] are formed analogously to this example. In simple cases, this formation can be done graphically, but in general a computational evaluation with appropriate software is preferable.

Equivalent modal unbalances are used in addition to low-speed unbalances to specify unbalance states and unbalance tolerances of rotors. However, equivalent modal unbalance cannot be used for unbalance correction.

[13] The *individual equivalent modal unbalance* is actually only needed for calculations. In statements about the unbalance state of a rotor, the *resultant equivalent modal unbalance* is usually used, but then simplified and called *equivalent modal unbalance*.

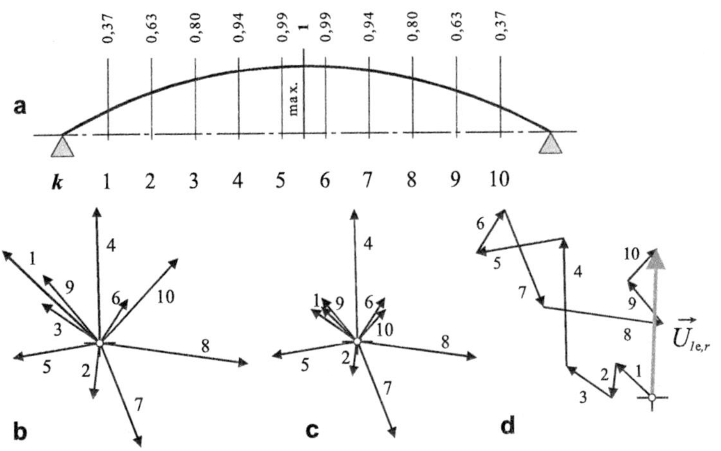

Fig. 4.15 Resultant equivalent modal unbalance in the 1st flexural mode: (**a**) describes the ordinate values of the different planes. From the unbalance vectors 1 to 10 (**b**) the *individual equivalent modal unbalances* $\vec{U}_{1e,k}$ are formed in (**c**). These unbalances are vectorially added to the *resultant equivalent modal unbalance* $\vec{U}_{1e,r}$ (**d**)

Notes

1. The *equivalent modal unbalance* $\vec{U}_{ne,r}$ is an individual unbalance vector in the axial position of $\Phi_{n,\ \max}$ with the same effect on the deflection of the rotor in this flexural mode as *all* the individual modal unbalances of this flexural mode combined.
2. The *equivalent modal unbalances* are mathematical quantities that can be used to make statements about unbalance state and unbalance tolerances for the shaft-elastic behaviour of rotors.
3. An *equivalent modal unbalance* **cannot** be used to correct a *modal unbalance.* Although the correction would (if set against the relevant *modal unbalance*) be able to correct the deflection in the associated flexural mode, but as an individual unbalance it would also change the low-speed unbalances and all other modal unbalances.

4.6 Overview of the Balancing Tasks

4.6.1 General

The following typical balancing tasks are described and explained using the above principles and forms of representation. Only the unbalances relevant to the task in question are considered, but it must be possible to exclude the others as irrelevant with sufficient certainty.

Relevant are all unbalances that are:

- Out of tolerance and therefore need to be *corrected.*
- Within tolerance, but still need to be *checked* for the size.

Unbalance corrections specifically change the desired unbalance components, but (usually) also influences others unintentionally.

Therefore, the correction must be well considered—which correction planes are used for which unbalances—and also the resulting change in the other unbalance components must be taken into account.

As a result of this influence, the balancing task may become more difficult, so that more unbalances have to be corrected than would correspond to the initial situation.

However, the balancing task can also be simplified, e.g. if the relevant modal unbalances become so small by skilfully correcting the low-speed unbalances, that they lie within the tolerances.

4.6.2 The Balanced Rotor

A balanced rotor exists when ***all*** relevant unbalance components are within the permissible values.

4.6.3 Single-Plane Balancing

Figure 4.16: the measured values for the different unbalance components show that **only** the resultant unbalance is above the permissible value. Single-plane balancing.

Assumptions

The moment unbalance is within tolerance and does not need to be controlled, the equivalent modal unbalance is also negligible.

Therefore

This means that single-plane balancing is sufficient (see Sect. 7.4.1), but the following should be noted:

- Correcting the resultant unbalance *in the selected* plane (for determining the moment unbalance) does not change the resultant moment unbalance, but can change the modal unbalances.
- Correcting the resultant unbalance in *another* plane changes the moment unbalance *and* can change the modal unbalances.
- Even *after the* resultant unbalance has been corrected, this unbalance must be the only significant one; otherwise one of the examples explained below will arise.

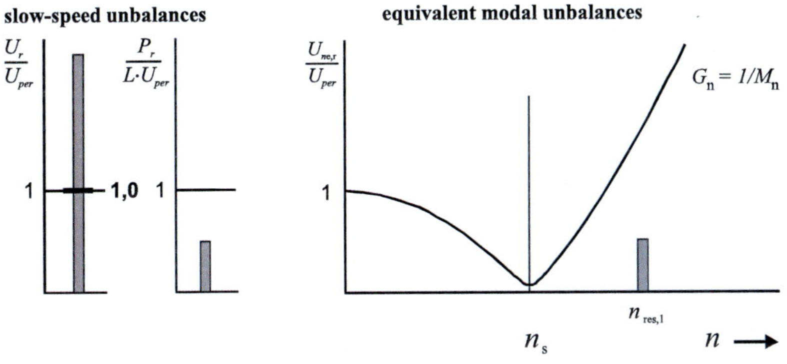

Fig. 4.16 Only the *resultant unbalance* needs to be corrected: Single-plane balancing is sufficient—Tolerance factor for this unbalance: 1.0

4.6.4 Two-Plane Balancing

Figure 4.17: the measured values for the different unbalance components show that the resultant unbalance **and** the moment unbalance are above the permissible values. The equivalent modal unbalance is negligible.

Therefore

Two-plane balancing is required (see Sect. 7.4.2). In addition to the instructions in Sect. 4.6.3 note:

- Also correcting the moment unbalance can change the modal unbalances.
- Even after the correction of the resultant unbalance and the moment unbalance, these unbalances must be the only ones to be taken into account; otherwise one of the examples explained below will arise.

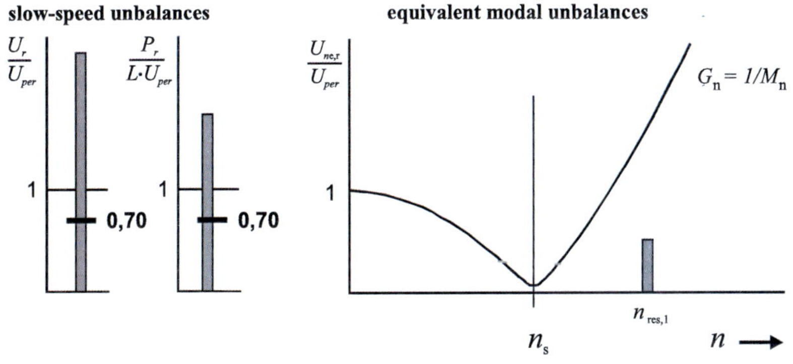

Fig. 4.17 The *resultant unbalance* and the *moment unbalance* must be corrected: Two-plane balancing—Tolerance factors for these unbalance components: 0.70

4.6.5 Multi-Plane Balancing

If, in addition to the low-speed unbalances, equivalent modal unbalances are also above the permissible values, multi-plane balancing is required (see Sect. 4.2.5). The most important different configurations will be described using four examples:

4.6.5.1 Example 1

Figure 4.18 shows a rotor running at subcritical speed (corresponding to Sect. 4.2.5).

More Details

- Resultant unbalance and moment unbalance are out of tolerance.
- The 1st equivalent modal unbalance—here above the service speed—is also out of tolerance.

Therefore

- Three unbalances are to be taken into account when determining the tolerances.
- At least three correction planes are required.
- Correction of the three relevant unbalances can change the higher modal unbalances.
- Even *after* the three unbalances have been corrected, these unbalances must be the only ones to be observed; otherwise one of the examples explained below will arise.

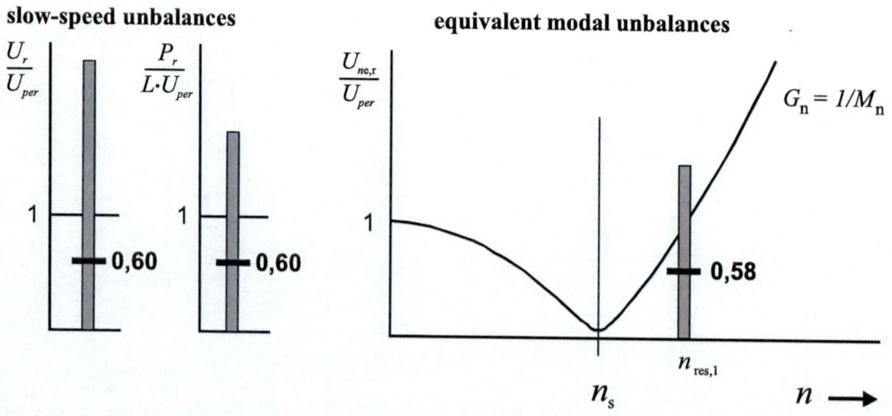

Fig. 4.18 Rotor running below the 1st flexural resonance speed: Both *low-speed unbalances* and the *1st equivalent modal unbalance* are outside the tolerance—tolerance factors for the unbalance components in this case: 0.58 to 0.60 (see Example 1)

4.6.5.2 Example 2

Figure 4.19 shows a rotor running at overcritical speed: service speed between the 1st and 2nd flexural resonance speed.

More Details

- Resultant unbalance and moment unbalance are out of tolerance.
- 1st and 2nd equivalent modal unbalance are out of tolerance (one below, one above service speed).[14]

Therefore

- Four unbalances are to be taken into account when determining the tolerances.
- At least four correction planes are required.
- Correcting the four unbalances can change the higher modal unbalances.
- Even *after* correcting the four unbalances, these unbalances must be the only significant ones, otherwise one of the examples explained below will arise.

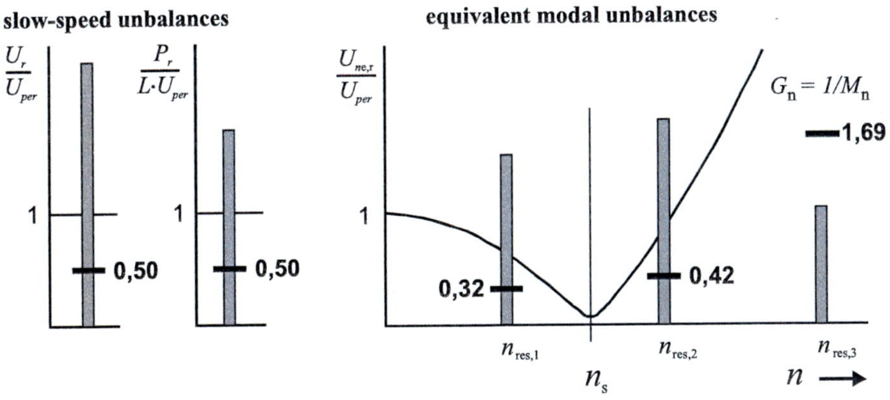

Fig. 4.19 Rotor operating above the 1st flexural resonance speed: Both *low-speed unbalances* and two *equivalent modal unbalances* are out of tolerance, one below the service speed, the other above—tolerance factors for the relevant unbalance components in this case: 0.32 to 0.50 (see Example 2)

[14] Even if flexural resonance speeds have already been passed before reaching service speed, one or even two flexural resonance speeds above the operating speed can be significant. This important detail is often overlooked even today and is only clearly described in DIN ISO 21940-12, Beiblatt 1.

4.6.5.3 Example 3

Figure 4.20 shows a rotor running at overcritical speed: service speed between the 2nd and 3rd flexural resonance speed.

For such complicated cases, where many modal unbalances have to be considered, it is best to compile the data in the form of a table, as explained in Sect. 4.2.5: see Tab. 4.4.

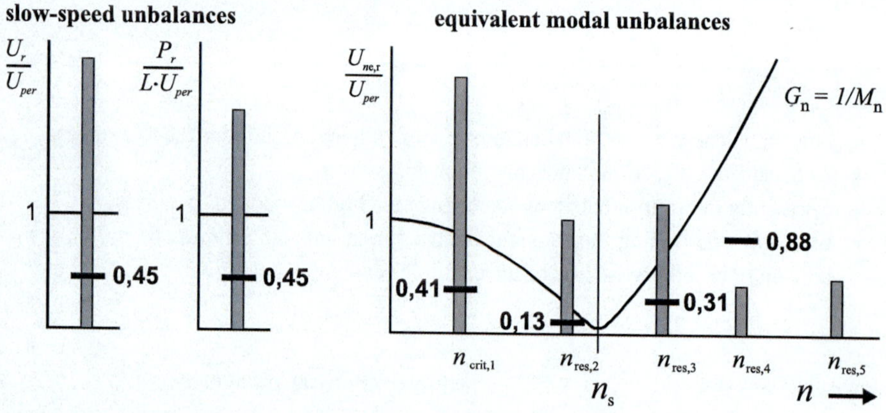

Fig. 4.20 Rotor operating above the 1st flexural resonance speed: Both *low-speed unbalances* and three *equivalent modal unbalances* are out of tolerance; two below the operating speed, one above—tolerance factors for the relevant unbalance components in this case: 0.13 to 0.45 (see Example 3)

Table 4.4 Calculation of permissible unbalances, Example 3[a]

Unbalance component U_x	U_{per} from ISO 21940-11	$n_{crit,n}/n_s$	G_n	Factor from Table 4.1	$U_{x,per}/U_{per}$
U_r	U_{per}	–	–	0.45	0.45
P_r/L		–	–		0.45
$U_{1e,r}$		0.29	0.92		0.41
$U_{2e,r}$		0.85	0.28		0.13
$U_{3e,r}$		1.30	0.69		0.31
$U_{4e,r}$		1.72	1.96		0.88

[a] As an conservative approach, the calculations were performed up to the 4th equivalent modal unbalance. The factor from Table 4.1 would then be 0.4 for a total of six relevant unbalances. However, since the 4th equivalent modal unbalance is far from the tolerance, the value 0.45 was used as the factor for ***five*** relevant unbalances

More Details

- Resultant unbalance and moment unbalance are out of tolerance.
- 1st to 3rd equivalent modal unbalances are out of tolerance (two below, one above service speed).[15]

Therefore

- Five unbalances are to be taken into account when determining the tolerances.
- At least five correction planes are required.
- Correcting the five unbalances can change the higher modal unbalances.
- Even *after* the five unbalances have been corrected, these unbalances must be the only significant ones; otherwise one of the examples explained below will arise.

[15] Here, two flexural resonance speeds have already been passed before service speed is reached, but one flexural resonance speed above service speed is still significant.

4.6.5.4 Example 4

Figure 4.21 shows a rotor running at overcritical speed: service speed between the 2nd and 3rd flexural resonance speed.

As already mentioned for Example 3, it is appropriate to compile the data for the calculations of the permissible unbalances in a table, see Table 4.5.

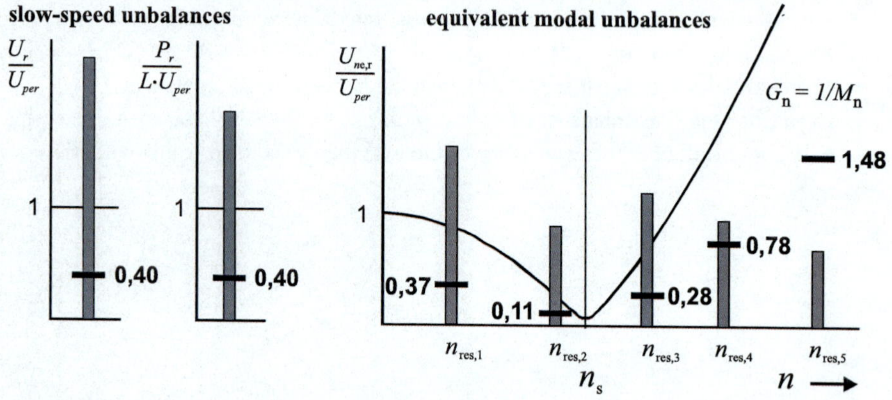

Fig. 4.21 Rotor operating above the 2nd flexural resonance speed: Both *low-speed unbalances* and four *equivalent modal unbalances* are out of tolerance, two below the service speed, two above—tolerance factors for the unbalance components in this example: 0.11 to 0.78 (see Example 4)

Table 4.5 Calculation of permissible unbalances, Example 4[a]

Unbalance component U_x	U_{per} from ISO 21940-11	$n_{crit,n}/n_s$	G_n	Factor from Table 4.1	$U_{x,per}/U_{per}$
U_r	U_{per}	–	–	0.40	0.40
P_r/L		–	–		0.40
$U_{1e,r}$		0.29	0.92		0.37
$U_{2e,r}$		0.85	0.28		0.11
$U_{3e,r}$		1.30	0.69		0.28
$U_{4e,r}$		1.72	1.96		0.78
$U_{5e,r}$		2.17	3.71		1.48

[a] As an conservative approach the calculation was carried out up to the 5th equivalent modal unbalance. The factor according to Eq. 4.3 would then be 0.38 for a total of 7 relevant unbalances. However, since the 5th equivalent modal unbalance is far from the tolerance, the value 0.40 was used as the factor for 6 six relevant unbalances

More Details

- Resultant unbalance and moment unbalance are out of tolerance.
- 1st to 4th equivalent modal unbalances are out of tolerance (two below, two above service speed).[16]

Therefore

- Six unbalances are to be taken into account when determining the tolerances.
- At least six correction planes are required.
- Correcting the six unbalances can change the higher modal unbalances.
- Even *after the* six unbalances have been corrected, these unbalances must be the only significant ones, otherwise an even more complicated example will result.

4.7 Conclusion of the New Perspective

The new form of presentation not only leads to a better overview, but also to new insight views, e.g.: Conclusion of the new perspective.

4.7.1 Significance of Resonances

It used to be common practice to rate the significance of flexural resonance speeds primarily depending on their distance from the service speed. Unfortunately ISO 19499 still implies this in its tables (to be deleted during the transfer to ISO 21940-1).

But apart from the frequency ratio—which goes into the boundary curve for the equivalent modal unbalances—the *total permissible unbalance,* the *number of relevant unbalances* and the *unbalance state* are important. Only from *all* these details, a clear statement can be made about the respective balancing task.

4.7.2 Significance of Flexural Resonances above the Service Speed

One can also clearly see that equivalent modal unbalances of one or even two flexural resonance speeds that are *above* the service speed must be taken into account and, if necessary, corrected for.

[16] Two flexural resonance speeds were passed before service speed. Still two flexural resonance speeds *above* service speed are relevant.

If the 1st flexural resonance speed is above the service speed, it has always been taken into account. But a 2nd one has been completely overlooked, even though it can also play a role in special cases (e.g. extremely large unbalances).

However, the flexural resonance speeds *above the* service speed have so far been completely excluded if flexural resonance speeds had already occurred *below* service speed.

4.7.3 Treatment of Flexural Resonances above Service Speed

However, flexural resonance speeds above the service speed are also not easy to handle:

- For flexural resonances *below* the service speed, measurements are usually taken close to the flexural resonance speed in order to make the measurements more *selective* and *sensitive*; this increases the accuracy of the evaluation.
- However, the flexural resonance speeds *above* the service speed normally cannot be approached. The usual means of increasing measuring quality during balancing are therefore missing in this case.

This can mean—e.g. selection—that:

- Firstly: the influence of these higher flexural modes is only unpleasantly noticeable—and can possibly only be dealt with—when the other unbalances have been reduced far enough.
- Secondly, that these flexural modes can no longer be sufficiently selectively excited with individual unbalances, but only with *modal unbalance sets*, in order to obtain good influence coefficients.

Concerning—e.g. sensitivity—there is a systemic all-clear:

- The tolerances of the equivalent modal unbalances for the higher flexural modes increase rapidly with increasing distance from the service speed, so that these unbalances do not have to be corrected as precisely as those below the service speed. This means that relatively low measurement sensitivity is still sufficient.

4.8 Handling Unbalance Tolerances

4.8.1 Concept

The derivation of unbalance tolerances is explained in detail in ISO 21940-11, and quality levels G are also stated for many machines types, see Chap. 7.

Attention!
The permissible total unbalance U_{per} is derived from the quality class G with the aid of service speed and rotor mass.

This limit (or quality grade) describes the permissible unbalance for the ***delivery*** and is therefore an important contractual detail for the manufacturer and the buyer.

For the ***balancing*** process, the limits must be drawn more narrowly, on the one hand because the measuring equipment incorporates errors, and on the other hand because often a different situation exists during balancing than in the delivery condition—or is intentionally created for economical or practical reasons—e.g. by:

- A different support/bearing system.
- A different assembly state.
- Balancing accessories, e.g. mounts, drivers, keys etc.
- Physical effects, such as constraining forces, axial thrust, or magnetic fields.

How such deviations can be recognised and how they can be reduced is described in ISO 21940-14, as well as how the remaining deviations can be quantified and expressed as a combined error ΔU, see Chap. 16.

The permissible unbalance indication for the balancing process therefore is smaller than U_{per} by ΔU, see Fig. 4.22. This concept was published in 1997 after its formulation in ISO 1940-2.

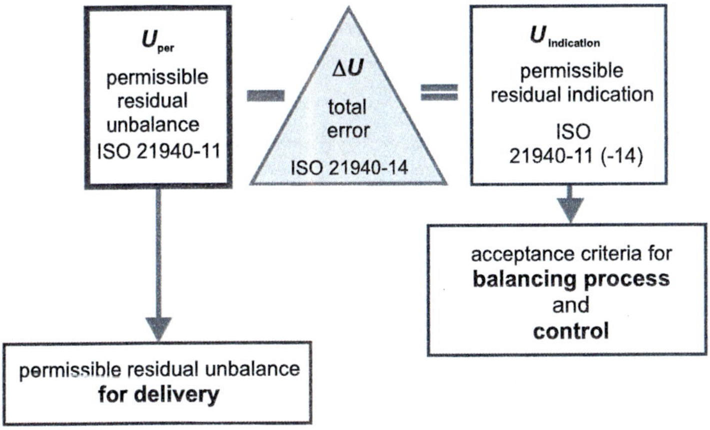

Fig. 4.22 The combined error of the balancing process is subtracted from the permissible unbalance for the delivery and results in the permissible indication for the balancing process and for balancing checks

Many users therefore reduce the permissible unbalance indication according to ISO 21940-11 by a certain percentage for the balancing process, but this is working "blindfolded", so to speak: with a small ΔU, the effort is unnecessarily large, with a large ΔU it may be far from sufficient.

4.8.2 Procedure

With the help of a careful determination of the possible balancing errors, the intended balancing process can be checked (and improved, if necessary) by determining whether, for example:

- The balancing machine is suitable.
- The balancing speed is correctly selected.
- The degree of assembly is sufficient.
- The errors are permissible for the parts assembled later.
- The intended drive and the drivers are sufficient.
- The adaptors used meet the requirements.

If all errors are expressed in multiples of U_{per}, the procedure can be presented most clearly. A few examples:

4.8.2.1 Example 1

Task

A small electric motor is to be balanced to G 6.3; the total permissible unbalance U_{per} results from the service speed and rotor mass. At the selected balancing speed, the balancing machine has a minimum achievable unbalance $U_{mar}=0.5\ U_{per}$ and for U_{per} an unbalance reduction ratio R_{UR} of 80% (see Chap. 13).

Solution

The error due to the R_{UR} is 0.2 U_{per}, the combined ΔU—sum of R_{UR} and U_{mar}—is therefore 0.7 U_{per}. This leaves only 0.3 U_{per} for the indication of the residual unbalance, an unfavourable situation.

4.8.2.2 Example 2

Task

A higher balancing speed is selected for the electric motor from example 1. In this case, the balancing machine has a minimum achievable unbalance $U_{mar}=0.2\ U_{per}$ and an unbalance reduction ratio R_{UR} of 90%.

Solution

The error due to the R_{UR} is 0.1 $U_{per,}$ the combined ΔU is therefore 0.3 U_{per}. This leaves 0.7 U_{per} for the indication of the residual unbalance: a favourable situation.

4.8.2.3 Example 3

Task

The electric motor in example 2 runs in antifriction bearings during operation, which can have an error of 0.5 U_{per}. Can it be balanced without the bearings?

Solution

The error due to the balancing machine again is 0.3 U_{per}, the error due to the antifriction bearings is added. The combined ΔU therefore is 0.8 U_{per}. This leaves only 0.2 U_{per} for the indication of the residual unbalance, an extremely unfavourable situation.

So either the rotor must be balanced *with* the bearings or service bearings with a significantly *lower error* must be used.

4.8.2.4 Example 4

Another example shows that considerations of errors in balancing can also interfere with the manufacturing or assembly process.

Task

A dual-flow low pressure turbine is to be balanced without the blades of the last stage (due to the large diameter). The error of the balancing machine is 0.2 U_{per}.

If the blades are mounted *arbitrarily*, it results in an error of 1.5 U_{per}.

If they are distributed according to their *masses*, the error is reduced to 0.5 U_{per}.

If a distribution corresponding to the *blade moments* is used (referred to the shaft axis), this results in an error of 0.3 U_{per}. Which distribution type should be chosen?

Solution

An arbitrary distribution (combined error 1.7 U_{per}) results in a negative permissible reading (−0.7 U_{per}), i.e. this method is to be excluded.

With a distribution according to the mass (combined error 0.7 U_{per}), only 0.3 U_{per} remains for the indication of the residual unbalance: a very unfavourable situation.

With a distribution according to the blade moments (combined error 0.5 U_{per}), 0.5 U_{per} remains for the indication of the residual unbalance: this procedure is therefore recommended.

Conclusion

The permissible indications for the balancing process are derived from the balancing target, but they are also process-dependent: from the *way* the balancing machine is used and sometimes from many other *boundary conditions* that can cause errors, see Chap. 16.

Theory of the Rotor with Rigid Behaviour

5

Contents

5.1 Preliminary Remark

An ideal **rotor with rigid behaviour** would show movements during rotation—on a slightly yielding bearing—but no deflection of the rotor. All displacements of the rotor would be caused by the movement of the bearings and the supporting structure.

H. Schneider, *Rotor Balancing*, https://doi.org/10.1007/978-3-662-66049-2_5

In reality, of course, no rotor is absolutely rigid up to the service speed (or overspeed) and will therefore also always show small deflections—in one or even several flexural modes.

5.2 Rotor Behaviour

The rotor behaviour can nevertheless be considered ***rigid as*** long as these deflections are negligible, i.e.:

- Are small compared to the unbalance tolerance.
- Do not need to be checked during balancing.

Figure 5.1 shows the unbalance display of a *rotor with rigid behaviour* on a typical hard-bearing balancing machine, calibrated to display unbalance units over the speed range. n_1 is a typical low balancing speed, n_2 the service speed of the rotor. The picture shows the change in the display between n_1 and n_2—on the left as a curve over the speed, on the right in a polar representation.

In addition, two unbalance states are shown: **(a)** unbalanced: out of tolerance, **(b)** corrected: within tolerance.

The change in the unbalance reading *I* over the speed range is negligible compared to the balancing tolerance (*tol.*) in both unbalance states.

By definition (see Chap. 3), only the low-speed unbalances—*resulting unbalance* and *moment unbalance*—are important for a rotor with rigid behaviour, i.e. the *modal unbalances* play no role in the description of the unbalance state.

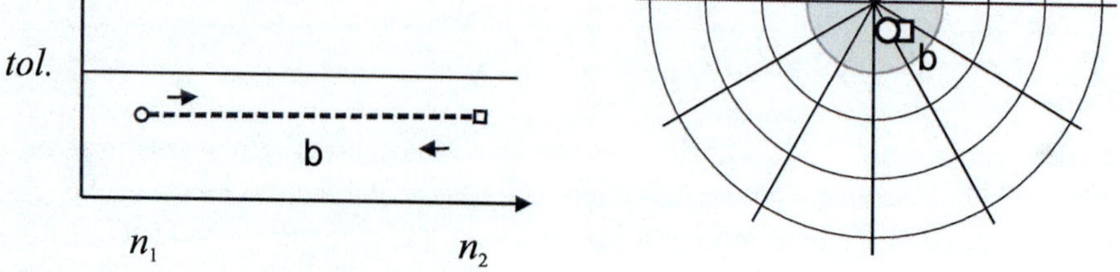

Fig. 5.1 Rotor with ***rigid behaviour,*** amount (left) and vector (right): Negligible change in unbalance indication *I* over the speed range, compared to *tol.*, the balance tolerance. Low speed n_1, service speed n_2; **a** unbalanced, **b** corrected

Note

However, the rotor behaviour can only be clearly described as *rigid* if the relevant unbalances do not change significantly with the speed (see Table 4.3), i.e. if *body-elastic behaviour* and *settlement behaviour* also are negligible.[1]

5.3 Unbalanced Condition

We start from the explanations on the general rotor in Sect. 4.5 and can obtain the *resulting unbalance* from distributed unbalance vectors according to Sect. 4.5.1 and the *moment unbalance according to* Sect. 4.5.2. From the moment unbalance *the couple unbalance* is obtained according to Sect. 4.5.3.

In practice, however, when balancing a rotor with rigid behaviour, unbalances in two arbitrarily selectable planes are mostly used, the *dynamic unbalance,* a superposition of the *resulting unbalance* with the *moment unbalance,* see Sect. 5.3.4.

Sometimes only the *resulting unbalance* needs to be taken into account, see Sect. 5.3.2. Although this type of unbalance is not bound to a specific plane, only ***after*** a plane has been assumed, a statement on the *moment unbalance* can be made, i.e. it can be determined whether it is really negligible or not.

A special case is the assumption of the *resulting unbalance* in the plane of the mass centre of the rotor, it is called *static unbalance,* see Sect. 5.3.1.

The best way to explain the principle and the effect of the various unbalances is to start from a *completely balanced rotor* and apply these unbalances or combinations of unbalances.

5.3.1 Static Unbalance

If a single unbalance is applied to a completely balanced rotor in *that* radial plane in which its mass centre lies (Fig. 5.2), it is called a **static unbalance** $\vec{U}_s$.

By means of a cross-section through the rotor at this point (Fig. 5.3), it is easy to see that:

- In a completely balanced rotor, the mass centre must be on the shaft axis (otherwise a centrifugal force would act, see Sect. 2.6.9).
- As a result of the added unbalanced mass, the mass centre moves away from the shaft axis.

[1] Reminder: The unbalance state can change by *body-elastic behaviour* and by *settling behaviour*, i.e. some unbalance vectors along the rotor axis change. This change can influence *all* three types of unbalance. The *modal unbalances*, on the other hand, are considered to be independent of rotational speed: although the response behaviour changes when passing a resonance, the excitation, i.e. the modal unbalance, does not.

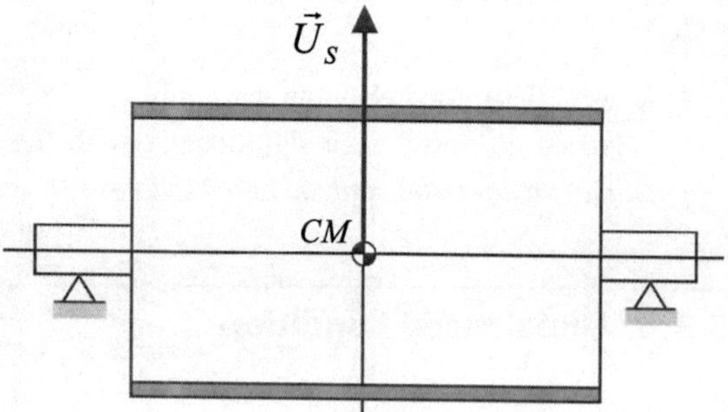

Fig. 5.2 *Static unbalance $\vec{U}_s$*: an unbalance acting at the mass centre *CM*

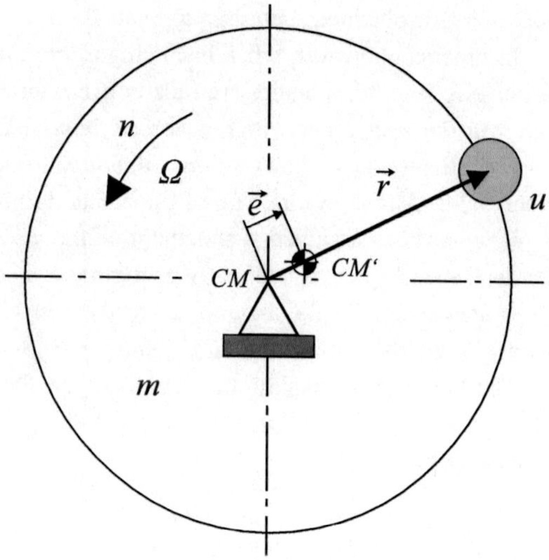

Fig. 5.3 The cross-section through the rotor in Fig. 5.1 shows (mass centre *CM* on the shaft axis) how, as a result of the unbalanced mass *u* on the radius $\vec{r}$, the new total mass centre *CM'* with the eccentricity $\vec{e}$ from the shaft axis is built

The equilibrium condition results in:

$$(m + u)\,\vec{e} = u\vec{r} \qquad \text{unbalance} \tag{5.1}$$

or

$$\vec{e} = \frac{u\vec{r}}{m + u} \qquad \text{length} \tag{5.2}$$

Since the unbalance mass *u* is in almost all cases much smaller than the rotor mass *m*, it is usually neglected in the denominator, resulting in

$$\vec{e} = \frac{u\vec{r}}{m} = \frac{\vec{U}_s}{m} \qquad \text{length} \tag{5.3}$$

$\vec{e}$ indicates how far and in which direction (angle) the mass centre is displaced from the shaft axis and is therefore called **mass centre eccentricity.** In general, $\vec{e}$ is much smaller than $\vec{r}$, the appropriate unit is therefore μm. If, for example, u is used in g, m in kg and r in mm, e is obtained in μm:

$$1\,\mu\text{m} = 1\,\frac{\text{g}\cdot\text{mm}}{\text{kg}} \qquad \text{length} \tag{5.4}$$

5.3.1.1 Example 1

Task

A rotor with $m = 600$ kg has a static unbalance $\vec{U}_s = 12{,}000$ g · mm. What is its mass centre eccentricity?

Solution

Mass centre eccentricity $e = \frac{U_s}{m} = \frac{12{,}000}{600} \frac{\text{g·mm}}{\text{kg}} = 20\,\mu\text{m}$.

The centrifugal force due to the static unbalance acts at the centre of gravity. In the case of a symmetrical and symmetrically mounted rotor, the forces in both bearings are equal in magnitude and direction.

5.3.1.2 Example 2

Task

What are the unbalanced bearing forces F_A and F_B on this rotor at a speed $n = 1000\ \text{min}^{-1}$?

Solution

Centrifugal force $F = U_s\,\Omega^2 \approx 0.012 \cdot 100^2\ \text{kg}\cdot\text{m}\ (\text{rad/s})^2 = 120$ N (Newton).

The unbalance U_s must be entered in kg · m (see Sect. 2.6.9). Furthermore, for a symmetrically mounted rotor $F_A = F_B$ and the sum of the bearing forces is equal to F:

$$F_A + F_B = F; \qquad F_A = F_B = \frac{F}{2} \approx 60\ \text{N}$$

Only *one correction* plane is required to correct the static unbalance: the mass centre plane. If correction is made in another plane, a moment unbalance will result as a side effect of the correction (see Sect. 5.3.3).

However, it is also possible to distribute the correction mass on two planes in such a way that the effect of a single mass in the plane of the mass centre occurs (i.e. the moment unbalance becomes zero), see Fig. 5.4.

From the equations:

$$u_{cI} + u_{cII} = u_c \quad \text{and} \quad u_{cI} f + u_{cII} g = 0 \quad \text{follows}$$

$$u_{cI} = u_c \frac{g}{g+f}; \qquad u_{cII} = u_c \frac{f}{g+f} \qquad g \qquad (5.5)$$

5.3.1.3 Example 3

Task

A rotor is to be balanced by a correction mass $u_c = 40g$. The distances of the correction planes I and II from the mass centre plane according to Fig. 5.4 are $f = 150$ mm, $g = 250$ mm. How large are the required correction masses u_{cI} and u_{cII} in planes I and II?

Solution

$$u_{cI} = u_c \frac{g}{f+g} = 40 \frac{250}{150+250} \text{ g} \frac{\text{mm}}{\text{mm}} = 25 \text{ g}$$

$$u_{cII} = u_c \frac{f}{f+g} = 40 \frac{150}{150+250} \text{ g} \frac{\text{mm}}{\text{mm}} = 15 \text{ g}$$

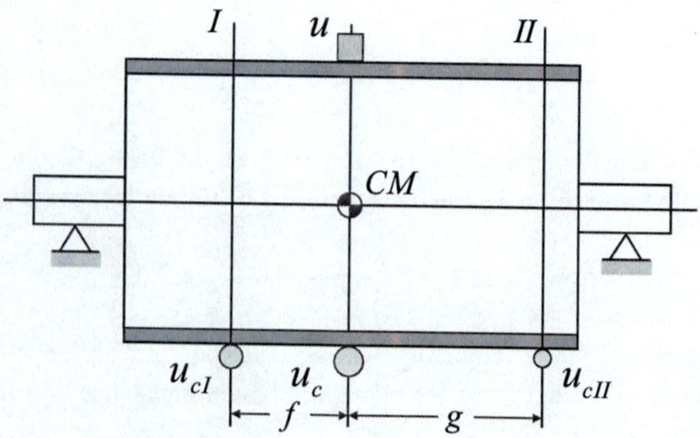

Fig. 5.4 Distribute the correction mass u_c between the correction planes I and II. It is assumed that the radii are equal. If not, the radius ratio must be considered

5.3.1.4 Example 4

Task

The same rotor is to be balanced so that both correction planes are on the same side, where $f = 200$ mm and $g = 600$ mm are fixed, see Fig. 5.5.

Solution

(Here f is used with negative sign)

$$u_{cI} = u_c \frac{g}{-f + g}; \quad u_{cII} = u_c \frac{-f}{-f + g},$$

so

$$u_{cI} = u_c \frac{g}{-f + g} = 40 \frac{600}{400} \text{ g} \frac{\text{mm}}{\text{mm}} = 60 \text{ g}$$

$$u_{cII} = u_c \frac{-f}{-f + g} = 40 \frac{-200}{400} \text{ g} \frac{\text{mm}}{\text{mm}} = -20 \text{ g}$$

For u_{cII}, material is either taken away in the same angular position as u_c or added on the opposite side.

Note

If the correction radii are not all the same, the correction masses are converted accordingly (inversely proportional to the radii).

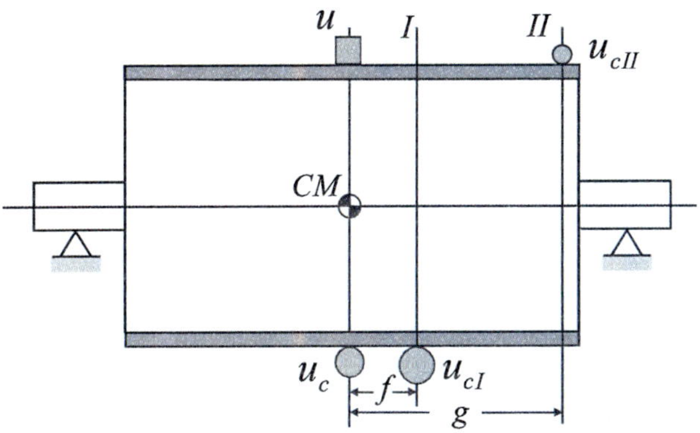

Fig. 5.5 Distribution of the correction mass to correction planes placed on one side. It is assumed that the radii are equal. If not, the radius ratio must be taken into account

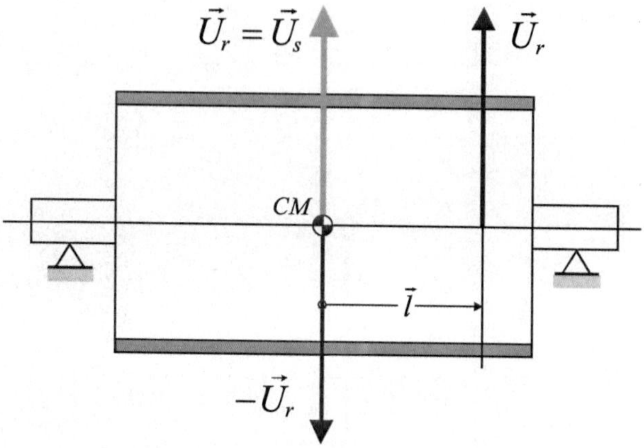

Fig. 5.6 A ***resulting unbalance*** acts like a *static unbalance* and a *couple unbalance* in the same longitudinal plane of the rotor

5.3.2 Resulting Unbalance

If a single unbalance is applied to a completely balanced rotor in any plane, it is called a *resulting unbalance.*[2]

It corresponds to a combination of a *static unbalance* with a *moment unbalance with* the characteristic that both lie in the same longitudinal plane of the rotor. The situation can best be understood by means of Fig. 5.6:

A *resulting unbalance* U_r is the distance l away from the mass centre. If the same unbalance vector is applied at the mass centre and also the same vector with a negative sign (opposite direction), the newly inserted unbalances cancel each other out—nothing has changed compared to the initial situation.

The system of unbalances can now be interpreted as follows: The unbalance vector at the mass centre (same direction as U_r) is a *static unbalance* U_s, the two remaining unbalances form a *couple unbalance* $-U_r, U_r$, thus representing a *moment unbalance.*

5.3.2.1 Example 1

In a plane $l = 200$ mm away from the plane of the mass centre, there is an unbalance $U_r = 400$ g · mm.

Task

How big are the static unbalance and the moment unbalance?

[2] *Static unbalance* is therefore a special case of *resultant unbalance*.

Solution

- Static unbalance $U_s = U_r = 400$ g · mm.
- Moment unbalance $P_r = l \cdot U_r = 200 \cdot 400$ g · mm^2 = 80,000 g · mm^2.

Such an unbalance condition can—if the correction plane is freely selectable—be completely corrected by a correction in *a single* plane.

However, some important boundary conditions must be fulfilled: single-plane balancing requires that:

- The situation is correctly assessed.
- The plane is found where the *resulting unbalance* acts.

If this is successful, the static unbalance and the moment unbalance are corrected simultaneously, although two (or even three) correction planes would normally be necessary for both unbalance types.

5.3.2.2 Example 2

The unbalance of a rotor is measured on a balancing machine, a static unbalance $U_s = 1000$ g · mm and a moment unbalance $P_r = l\ C_r = 350{,}000$ g · mm^2 located in the *same longitudinal plane* are determined (Fig. 5.7).

Task

Position of the correction plane where both types of unbalance can be eliminated with a single correction (i.e. how large is l and in which direction is it, seen from the mass centre plane)?

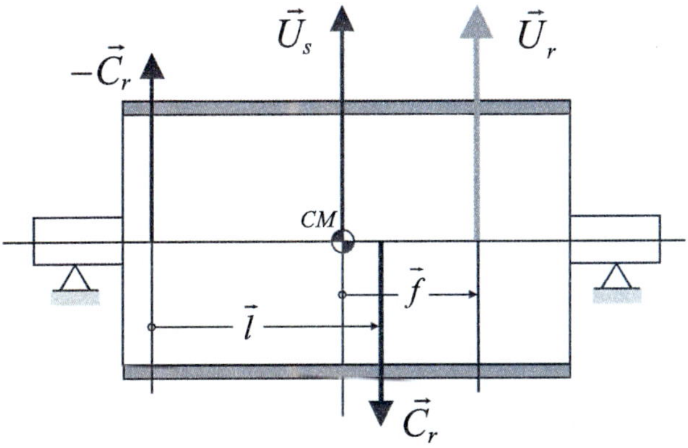

Fig. 5.7 Determining the right correction plane for a *resulting unbalance* U_r—unbalance condition given by a *static unbalance* U_s and a *couple unbalance* $-C_r$, C_r

Solution

Resulting unbalance $U_r = U_s = 1000$ g · mm.

Distance of the plane of the resulting unbalance from the plane of the mass centre (the moment unbalance $f\,U_r$ must be used with a negative sign because its direction of rotation is opposite to the moment formed by P_r):

$$-f\,U_r = P_r; \quad f = -\frac{P_r}{U_r} = -\frac{350{,}000}{1000}\,\frac{\text{g}\cdot\text{mm}^2}{\text{g}\cdot\text{mm}} = -350\ \text{mm}$$

The correction plane is therefore 350 mm to the left of the mass centre plane.

5.3.3 Moment Unbalance

If two unbalances of the same size are attached to a completely balance rotor in two different radial planes so that they are exactly opposite each other—a couple unbalance—they form a *moment unbalance* (Fig. 5.8).

If the two planes are separated by the length l and the two unbalances are $-\vec{C}$ and $\vec{C}$, then the moment unbalance is:

$$\vec{P} = \vec{l} \times \vec{C} \qquad \text{g}\cdot\text{mm}^2 \tag{5.6}$$

The vector $\vec{P}$ is perpendicular to the longitudinal plane in which the unbalances lie, similar to a torque vector. Simplified, one can write (i.e. without vector sign and without vector product):

$$P = lC = lur \qquad \text{g}\cdot\text{mm}^2 \tag{5.7}$$

In General

- *Moment unbalance* is the alternative to *static unbalance:* the mass centre of the rotor has no eccentricity.
- Due to the moment unbalance, the inertia axis is inclined, see Sect. 5.4.2.
- With the same lC it does not matter whether the two planes in which the unbalances act are symmetrical to the mass centre or asymmetrical.
- The mass centre does not even need to lie between the two unbalance planes—the moment unbalance and its effects are always the same.
- The unbalances cause an unbalance moment[3] (centrifugal moment), which *always* causes equal but opposite forces in both bearings.

---

[3] The difference between *moment unbalance* and *unbalance moment* should be noted: *Moment unbalance* is a special type of unbalance, while *unbalance moment* is the centrifugal torque due to a moment unbalance.

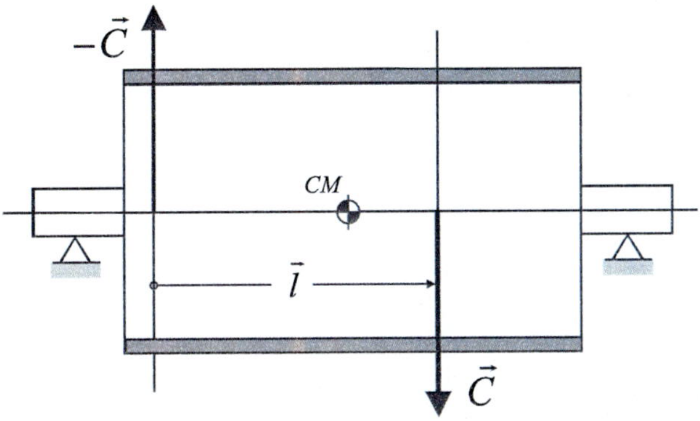

Fig. 5.8 A moment unbalance is created here by a couple unbalance: two opposite, equally sized unbalances $-\vec{C}$ and $\vec{C}$ with plane spacing l

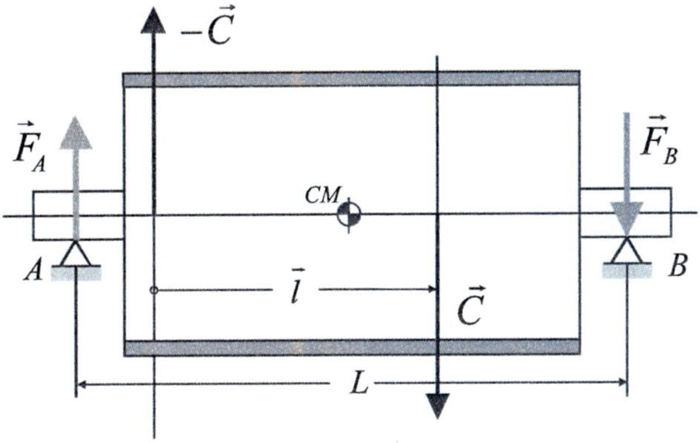

Fig. 5.9 Bearing reactions $\vec{F}_A$ and $\vec{F}_B$ due to a *moment unbalance* $\vec{l} \times \vec{C}$

5.3.3.1 Example

A rotor has two angularly opposed unbalances (couple unbalance $-C$, C) of 6000 g · mm each; the plane spacing is $l = 700$ mm, the bearing spacing is $L = 1000$ mm (Fig. 5.9).

Task

What is the magnitude of the moment unbalance and, at a speed $n = 1000\ \text{min}^{-1}$, the unbalance-related bearing forces F_A and F_B?

Solution

Centrifugal force F *of* unbalance C: $F = C \cdot \Omega^2 \approx 0.006 \cdot 100^2$ kg · m · (rad/s)$^2 = 60$ N (here C must be applied in kg · m; Ω was set to equal *n/10*, see Sect. 2.6.9).
Moment unbalance: $M_u = l \cdot F \approx 0.7 \cdot 60$ m · N $= 42$ N · m
Bearing load: from $L \cdot F_A = -M_u$ and $L \cdot F_B = M_u$ follows:
$F_A = -M_u/L \approx -42/1$ N · m/m $= -42$ N (load opposite to bearing B)
$F_B = M_u/L \approx 42/1$ N · m/m $= 42$ N

5.3.4 Dynamic Unbalance

The *general* unbalance condition of a rotor with rigid behaviour is the **dynamic unbalance**. One can imagine that it is formed from the superposition of the two possible unbalances—a *resultant unbalance* and a *moment unbalance.* In the general case, these unbalances have different angular positions.

However, the *dynamic unbalance* can also be expressed as *complementary unbalances* in two arbitrarily selectable planes.

Note

In practice, the *dynamic unbalance of* a rotor—depending on the task—is described as follows:

- **Mostly** by specifying the *complementary unbalance vectors* in two planes.
- **Sometimes** by specifying the *resulting unbalance* and the *moment unbalance.*

Each type of representation can of course be transferred to the other.

First, we derive the dynamic unbalance from the resultant unbalance and the *moment unbalance* (Fig. 5.10).

The resulting unbalance is assumed here to be *static unbalance in* the plane of the rotor mass centre. The *moment unbalance* is represented by the *couple unbalance,* i.e. the unbalances $-\vec{C}$ in the left-hand plane and $\vec{C}$ in the right-hand plane with a distance of l.

For a position of the planes *symmetrical to* the mass centre, the static unbalance is distributed in half to both planes, the couple unbalance is calculated with the plane ratio l/L and then the vector sum is formed in each plane.

$$\vec{U}_I = \frac{\vec{U}_s}{2} - \vec{C}\frac{l}{L} \qquad \text{g} \cdot \text{mm} \tag{5.8}$$

$$\vec{U}_{II} = \frac{\vec{U}_s}{2} + \vec{C}\frac{l}{L} \qquad \text{g} \cdot \text{mm} \tag{5.9}$$

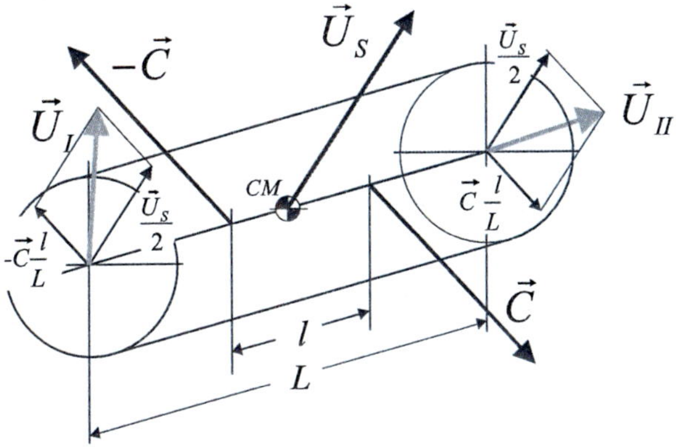

Fig. 5.10 *Dynamic unbalance*: conversion of a *static unbalance* $\vec{U}_s$ and a *moment unbalance* (represented by the unbalance pair $-\vec{C}, \vec{C}$) into two *complementary unbalances* $\vec{U}_I$ and $\vec{U}_{II}$

If the planes are *asymmetrical* with respect to the mass centre, the static unbalance is distributed according to the lever ratios, whereby Figs. 5.4 and 5.5 can be used as a basis. The conversion of the moment unbalance remains unchanged.

Vice versa, a dynamic unbalance can also be decomposed into a static unbalance and a moment unbalance, (geometry as in Fig. 5.10):

$$\vec{U}_s = \vec{U}_I + \vec{U}_{II} \qquad \text{g} \cdot \text{mm} \qquad (5.10)$$

$$\vec{P}_r = \vec{f} \times \vec{U}_I + \vec{g} \times \vec{U}_{II} \qquad \text{g} \cdot \text{mm}^2 \qquad (5.11)$$

There is another possibility to derive the *dynamic unbalance* directly from the general unbalance condition, Fig. 5.11:

According to the laws of statics, each individual unbalance can be converted into two arbitrarily selectable planes *I* and *II* (e.g. the end planes) and added up vectorially. The results are the complementary unbalances.

$$\vec{U}_I = \sum_{k=1}^{K} \vec{U}_k g_k / b \qquad \text{g} \cdot \text{mm} \qquad (5.12)$$

$$\vec{U}_{II} = \sum_{k=1}^{K} \vec{U}_k f_k / b \qquad \text{g} \cdot \text{mm} \qquad (5.13)$$

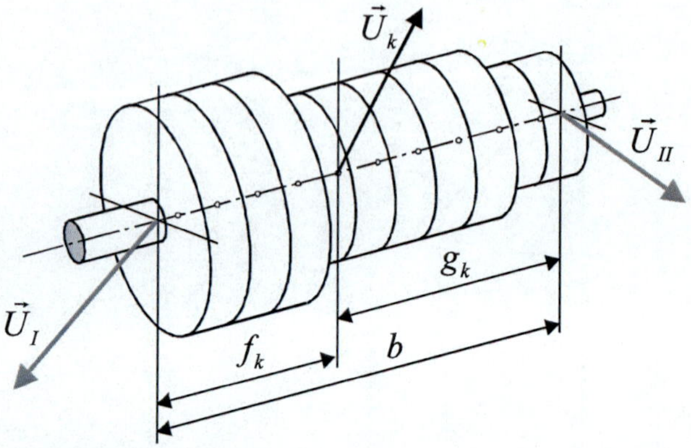

Fig. 5.11 All unbalances $\vec{U}_k$ of the *general rotor* divided into disks are converted into the end planes. The vector sums in these planes *I* and *II* are $\vec{U}_I$ and $\vec{U}_{II}$, they are called ***complementary unbalances*** and form the dynamic unbalance

Note

In general, the magnitude and angle of the two unbalance vectors depend on the position of the correction planes. It is particularly important that *both unbalance* vectors change, even if only *one* correction plane is chosen differently.

Therefore

- The unbalance state of a rotor with rigid behaviour can be completely described by *two complementary unbalances* in *two* arbitrarily chosen planes.
- The correction of the unbalance of such a rotor generally requires *a* correction in *each* plane.

Resultant unbalance, static unbalance and moment unbalance can be regarded as special cases of dynamic unbalance.

5.4 Display of the Unbalance Condition

5.4.1 Unbalances

In order to make it clear that the unbalance condition can and should be represented differently depending on the approach and the task, the unbalance state of a *rotor with rigid behaviour is* expressed in six variants in the guideline ISO 21940-11 Fig. 5.12.

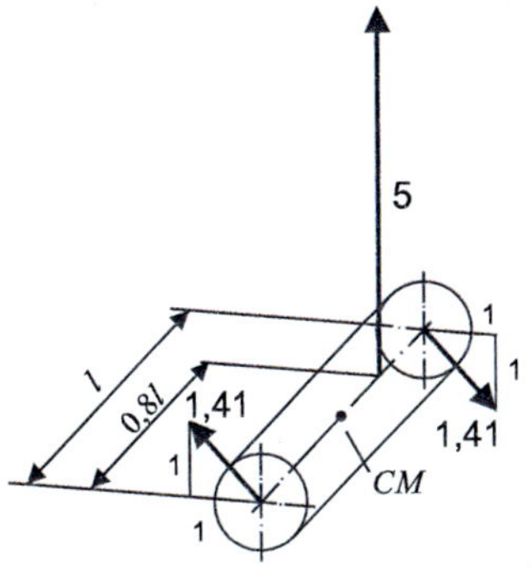

a) *Resultant unbalance vector* with an accociated *couple unbalance* in the end planes

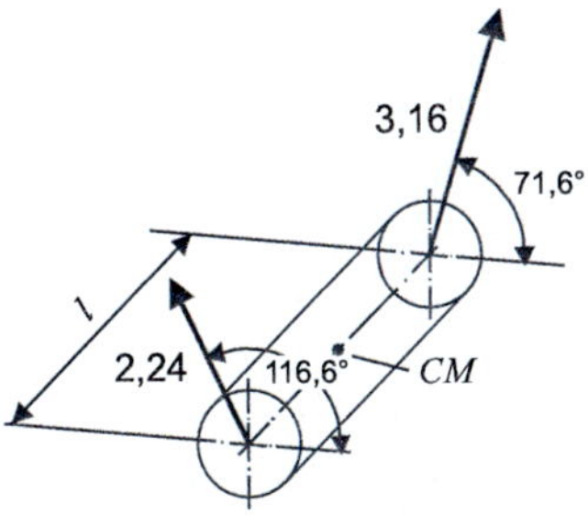

d) *Two unbalance vectors* in each of the end planes: a *dynamic unbalance*

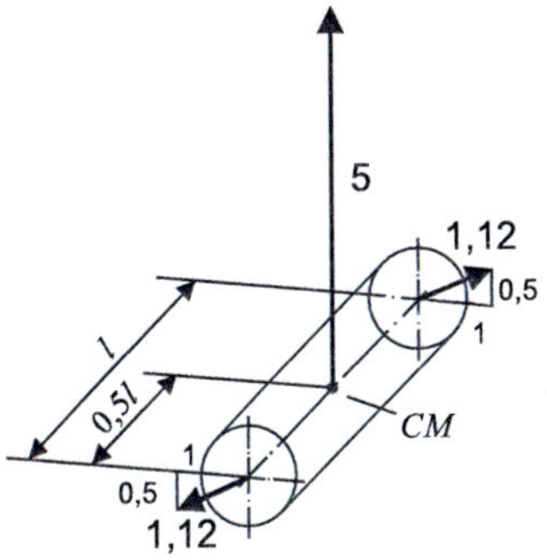

b) Special case of a): *resulting unbalance vector* located at the centre of mass *CM*, with an associated *couple unbalance* in the end planes

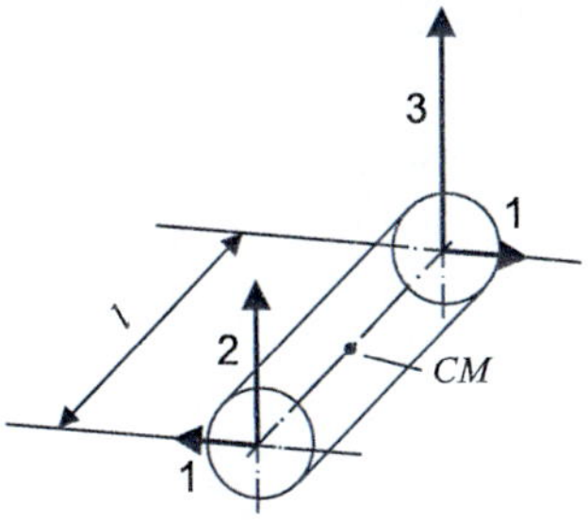

e) *Two 90° unbalance components* in each of the end planes

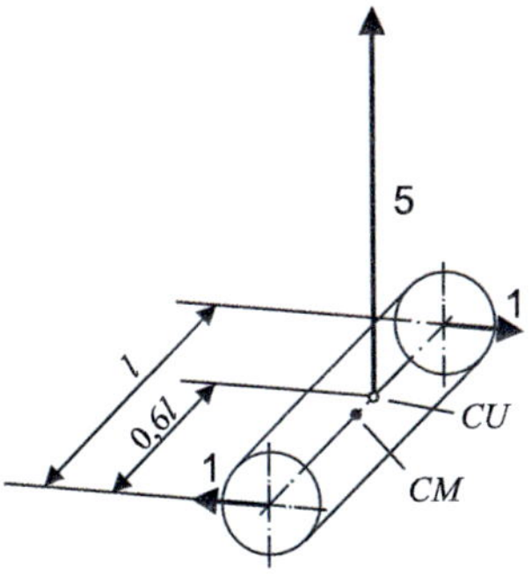

c) Special case of a): *resulting unbalance vector* located at the centre of unbalance *CU*.
The associated *couple unbalance* is a minimum, in a plane perpendicular to the *resulting unbalance*

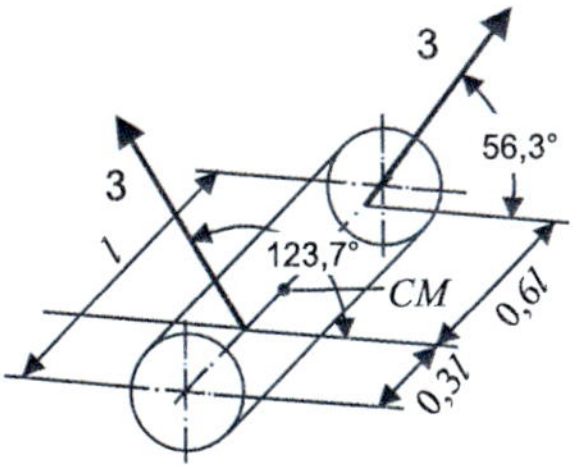

f) *Two unbalance vectors* in each of any two other planes. Another *dynamic unbalance*

Fig. 5.12 Different representations of one and the same unbalance state according to ISO 21940-11 (representation modified). *Left: resulting unbalance* and *couple unbalance* (**a–c**), *right complementary unbalances* in two planes (**d–f**). All unbalances in g·mm, lengths in mm. *CM* mass centre, *CU* centre of unbalance in **c**

Some basic relationships can be better clarified if we start from the two types of unbalance: *resultant unbalance* and *moment unbalance* (Fig. 5.12, left). In contrary, *dynamic unbalance* (Fig. 5.12, right) is mostly used for balancing considerations.

In both types of representation, it is clearly visible how unbalances change when planes are altered.

The expression used in Fig. 5.12 *centre of unbalance*[4] is rarely used. However, if the resulting unbalance is assumed to be in this plane, the result is the smallest possible moment unbalance, which can be advantageous.

The state of unbalance can be characterised by unbalances, apart from the representations discussed so far, but also by indicating the position of the central, adjacent mass axis of inertia to the shaft axis, as described below.

5.4.2 Position of the Axis of Inertia

In the case of a **completely balanced rotor,** the axis of inertia *I–I* coincides with the shaft axis *S–S:* There is mass symmetry, and thus no centrifugal force and no unbalance moment (Fig. 5.13).

If a **static unbalance** (resulting unbalance in *S*) is added, the inertia axis moves parallel from the shaft axis by the mass centre eccentricity *e* (Fig. 5.14).

The eccentricity of the mass centre can be calculated according to the equation derived in Sect. 5.3.1:

$$\vec{e} = \frac{\vec{U}_s}{m} \qquad \text{length} \tag{5.14}$$

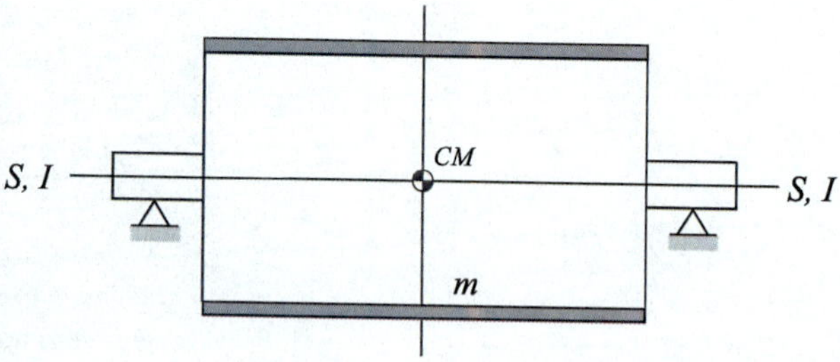

Fig. 5.13 With a completely unbalanced rotor, the inertia axis *I–I* and the shaft axis *S–S* coincide

[4]The centre of unbalance *CU* is that location of the *resulting unbalance* on the shaft axis for which the *moment unbalance* is a minimum. When considering whether a single-plane correction is sufficient, this aspect can be decisive.

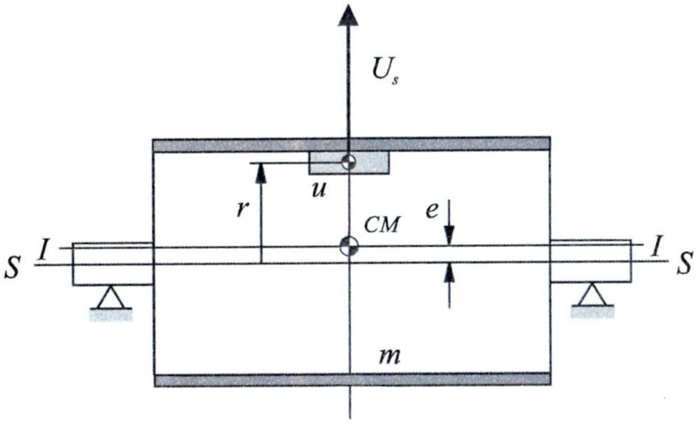

Fig. 5.14 Due to a *static unbalance*, the axis of inertia *I–I is* displaced parallel from the shaft axis *S–S* by the mass centre *eccentricity e*

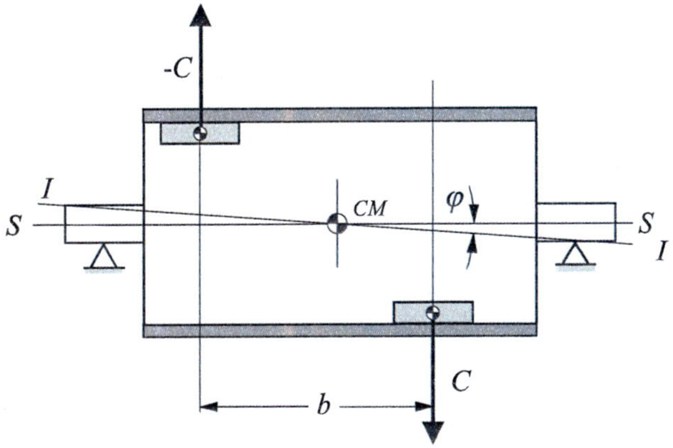

Fig. 5.15 A *moment unbalance*—represented by a *couple unbalance* (*−C and C*) with plane distance *b*—rotates the inertia axis *I–I out* of the shaft axis *S–S* by the angle $\vec{\varphi}$, the mass centre remains on the shaft axis

If a **moment unbalance** is applied, the inertia axis forms an angle with the shaft axis, but always intersects the shaft axis at the mass centre (Fig. 5.15).

This is independent of the plane in which the moment unbalance acts. The angle $\vec{\varphi}$ (in radians) can be calculated with:

$$\vec{\varphi} = \frac{1}{2} \arcsin \frac{2\vec{P}}{J_x - J_z} \qquad \text{rad} \tag{5.15}$$

For small angles, the equation simplifies to

$$\vec{\varphi} = \vec{P}/(J_x - J_z) \qquad \text{rad} \tag{5.16}$$

With:

- $\vec{P}$ Moment unbalance
- J_x Mass moment of inertia about the transverse axis through the mass centre (in practical cases mostly equal to J_y)
- J_z Mass moment of inertia about the longitudinal axis through the mass centre

Task

What is the angle $\vec{\varphi}$ due to moment unbalance $\vec{P} = 100\ \text{g} \cdot \text{m}^2$, when $J_x = 90\ \text{kg} \cdot \text{m}^2$ and $J_z = 20\ \text{kg} \cdot \text{m}^2$?

Solution

$$\vec{\varphi} = \frac{\vec{P}}{J_x - J_z} = \frac{0.1}{90 - 20} \ \frac{\text{kg} \cdot \text{m}^2}{\text{kg} \cdot \text{m}^2} \approx 0.0014\ \text{rad}; \quad \text{in angle degrees:} \quad \vec{\varphi} \approx 0.08°$$

Note that the difference of the moments of inertia (denominator of the equation) is positive for long rotors, so the angle turns *with* the moment unbalance.

For short (disc-shaped) rotors the difference is negative, i.e. in this case the angle turns in the *opposite direction* to the moment unbalance.

If the moments of inertia J_y and J_x are different, the moment unbalance is split into the directions of the *principal axes of inertia* x and y and calculated with the associated moments of inertia, i.e.

$$\vec{\varphi}_x = \vec{P}_x/(J_x - J_z); \quad \vec{\varphi}_y = \vec{P}_y/(J_y - J_z) \qquad \text{rad} \tag{5.17}$$

5.4.3 Overview

A general **resulting unbalance** displaces the inertia axis *and* angles it up. However, since both unbalances have the same angular position, there is necessarily an intersection with the shaft axis (Fig. 5.16).

To calculate the exact position, Eqs. 5.14 and 5.16 can be combined with $\vec{\varphi} \times \vec{h} = \vec{e}$:

$$\frac{\vec{f} \times \vec{U}_r}{J_x - J_z} \times \vec{h} = \frac{\vec{U}_s}{m}; \quad \frac{\vec{P}}{J_x - J_z} \times \vec{h} = \frac{\vec{U}_s}{m} \qquad \text{length} \tag{5.18}$$

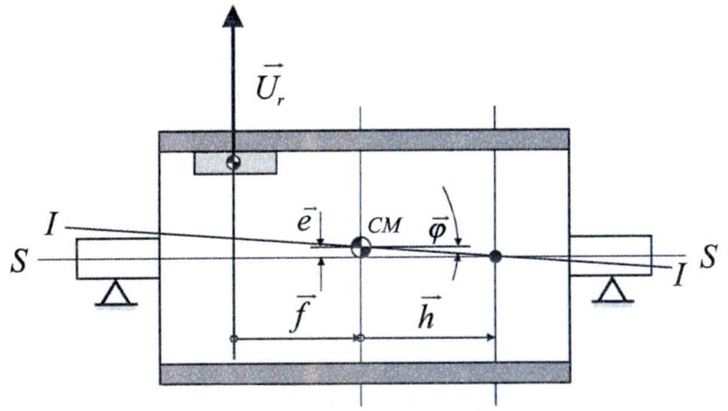

Fig. 5.16 A *resulting unbalance* at a distance *f* from the mass centre shifts the mass centre by the distance *e* and rotates the inertia axis *I–I* out of the shaft axis *S–S* by the angle $\vec{\varphi}$. There is always a point of intersection with the shaft axis, here at distance *h* from the mass centre

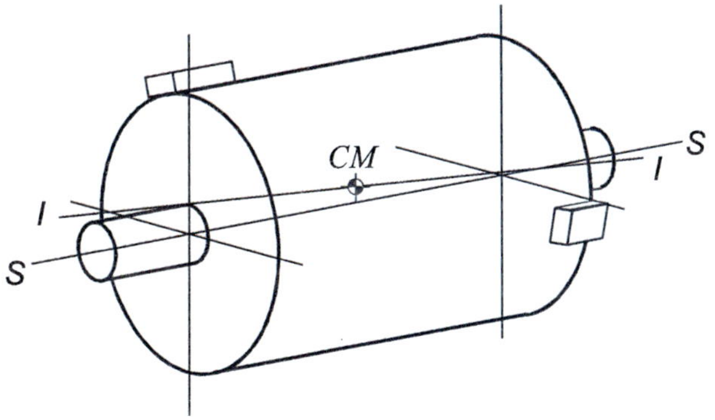

Fig. 5.17 In the case of a *dynamic unbalance,* the inertia axis *I–I* and the shaft axis *S–S* are skewed to each other: in general, they do not have a point of intersection

Seen from the plane of $\vec{U}_r$, for $J_x > J_z$ the intersection point is behind the plane of the mass centre, for $J_x < J_z$ it is in front of it, i.e. on the same side.

In the case of a **dynamic unbalance**, the axis of inertia and the shaft axis lie arbitrarily, generally the axes are skewed to each other, i.e. they have *no* point of intersection. Unfortunately, this situation cannot be accurately represented in a plane drawing (Fig. 5.17).

The other types of unbalance discussed here can be interpreted as special cases of dynamic unbalance:

- Intersection at infinity (parallel position of the axes): *static unbalance,*
- Intersection between the mass centre and the infinite: *resulting unbalance,*
- Intersection at the mass centre: *moment unbalance.*

Note

The axis position approach is particularly useful when the influence of geometries is assessed—e.g. fit and run-out errors.

Theory of the Rotor with Flexible Behaviour

6

Contents

H. Schneider, *Rotor Balancing*, https://doi.org/10.1007/978-3-662-66049-2_6

6.1 Preliminary Note

In Chap. 5, a rotor with rigid behaviour was assumed, whose unbalance and shape do not change with speed, see Fig. 5.1. What about a rotor whose behaviour is not rigid, but flexible, whose unbalance state changes with speed, or which shows deflections?

The underlying system was described in Chap. 4, a good overview is given in Table 4.3.

A distinction is made between ***plasticity*** (the deformation remains even after the load has been removed, usually called "settlement behaviour") and ***elasticity*** (the deformation recedes when the load is removed). In connection with rotor balancing, *elasticity* is conveniently subdivided into ***component-elasticity*** and ***shaft-elasticity.***

In the cases of *plasticity* and *component-elasticity,* the unbalance of the rotor is speed-dependent; in the case of *shaft elasticity, it is* not. The correct approach during balancing is very different in all three cases.

Today, especially rotors, which are highly loaded by the service speed, can exhibit quite considerable elastic and plastic deformations.

It should be clearly stated that only *those* deformations that occur asymmetrically to the shaft axis need to be considered.

These deformations can show very different appearances and require different measures. They are explained below.

6.2 Settling Behaviour

Rotors with plastic deformations often reach a steady-state condition at higher speeds, which is then maintained at lower speeds: settling behaviour (Fig. 6.1).

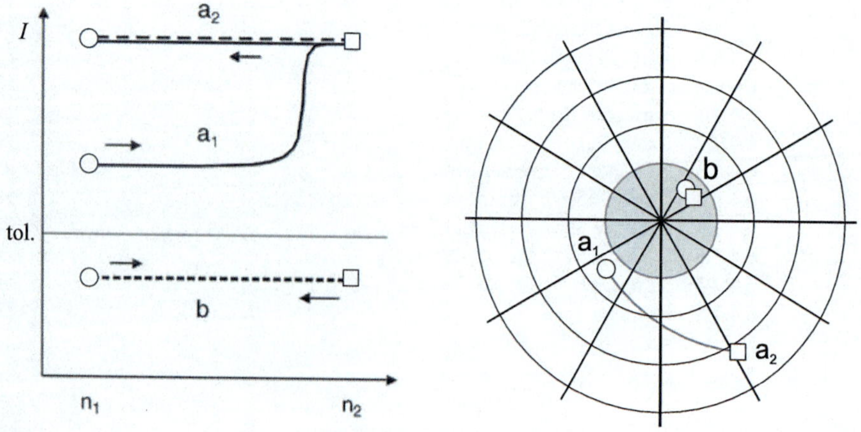

Fig. 6.1 Unbalance indication A with a ***settling behaviour***. Magnitude (left) and vector (right): **a** unbalanced, **b** balanced, n_1 balancing speed (low speed), n_2 service speed (a_1) *during*, (a_2) *after* deformation/settling

By running at a speed that experience has shown to be near service speed (or an amount below or above, depending on the type), a stable state of unbalance can usually be achieved for all speeds up to the service speed. Examples: settling of the windings of electric armatures or of the shrink fit of impellers in turbines.

If the rotor behaves rigid after the spin cycle, it can then be balanced at any speed (below service speed).

If, in addition to plasticity, some form of elasticity also occurs, proceed after the spin cycle as described in Sects. 6.3 and 6.4.

6.3 Component-Elastic Rotor Behaviour

If partial masses of the rotor, which have their centre of gravity neither on nor very close to the shaft axis, shift elastically as a result of the speed-dependent centrifugal forces, this is referred to as component-elastic behaviour.

The unbalance condition generally changes more and more rapidly with an increase in speed, the material loads can become very large and may lead to the fracture of the connecting elements (between these masses and the shaft axis), see Fig. 6.2.

Characteristically is, that no reversal of this tendency can be observed with a further increase in speed; i.e. the unbalance condition does not improve again. However, there are cases in which the displacement of the masses can only go to a stop, so that from then on a stable unbalance state prevails.

It is important to note that these eccentric masses *cannot* be symmetrised in themselves—i.e. the component elasticity cannot be eliminated by mass correction—as these masses are too far away from the shaft axis.

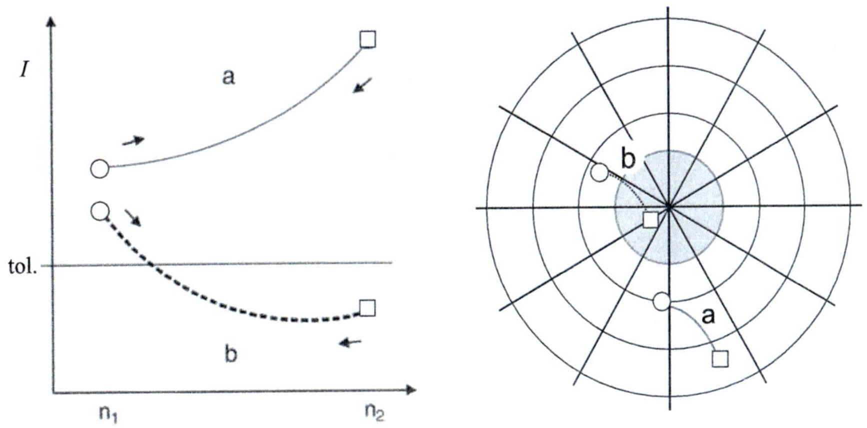

Fig. 6.2 Unbalance indication with ***component-elastic rotor behaviour***. **a** change in initial unbalance, **b** change after correction. For other designations see Fig. 6.1

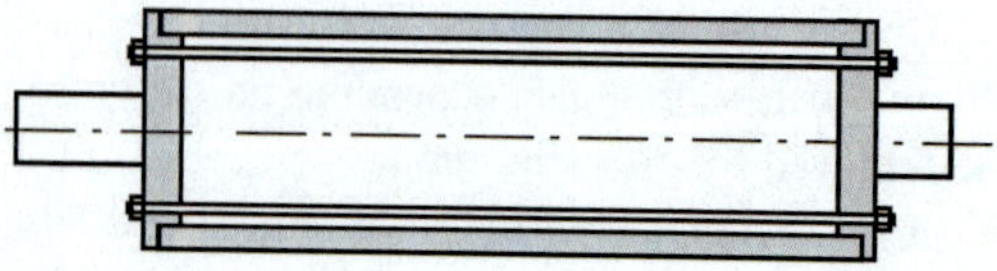

Fig. 6.3 Unevenly preloaded tie rods give this roll a component-elastic behaviour

If the reason for the component-elastic behaviour cannot be eliminated, such rotors must be balanced at service speed, or at a speed at which a stable state of unbalance is achieved. Afterwards, they are usually out of tolerance at low speeds.

If, for example, the forces and vibrations during run-up to service speed shall also be within certain limits, a compromise correction may be necessary, averaging run-up and service speed.

Example

In a hollow roll six tie rods are installed on equal radius between the end pieces, *one* of which was inadvertently not properly preloaded Fig. 6.3. This tie rod shifts more than the other five due to the centrifugal forces—this results in a component-elastic behaviour: the tie rod cannot (due to its design) be centred in the rotor, i.e. installed in the shaft axis. ◀

Apart from the method described above—balancing at service speed—there is the possibility here of reducing the asymmetry by precisely preloading the tie rods and thus the component-elasticity to such an extent that the assembly gains a rigid behaviour.

Conclusion

A better option (compared to compromise balancing) is therefore the elimination of component-elasticity, which requires different measures—in design or assembly—depending on the cause.

6.4 Shaft Elastic Rotor Behaviour

If (inherently rigid) masses whose mass centres lie on or near the shaft axis shift elastically as a result of centrifugal forces, the rotor behaviour is called shaft-elastic, Fig. 6.4

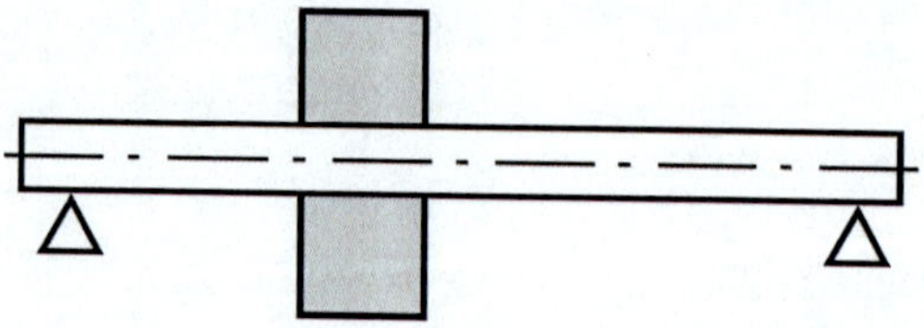

Fig. 6.4 Idealised rotor with shaft-elastic behaviour (Laval rotor)

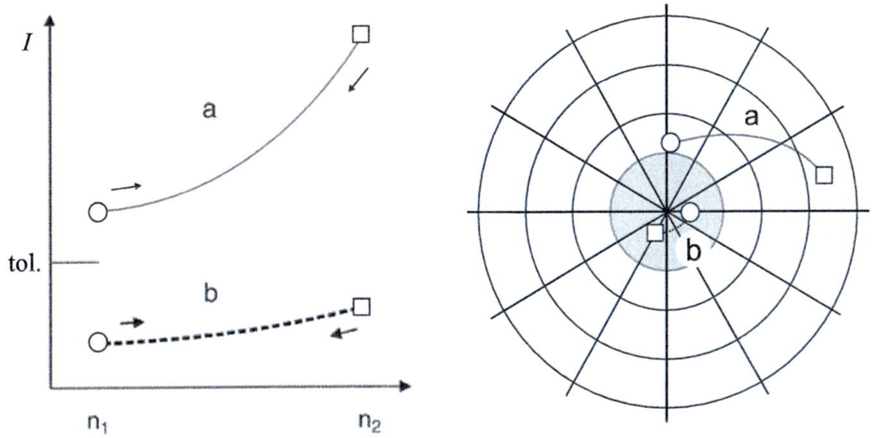

Fig. 6.5 Unbalance indication of shaft-elastic rotor behaviour below its 1st flexural resonance speed: **a** initial unbalance, **b** after balancing. For other designations see Fig. 6.1. The unbalance tolerances for higher speeds cannot be indicated in this representation because they follow a different systematic, see Chap. 8

If this rotor is operated below its 1st flexural resonance speed, the change in the unbalance indication initially looks similar to that of a component-elastic rotor behaviour (Fig. 6.2), see Fig. 6.5, run **(a)**: initial unbalance.

The essential difference only becomes apparent after the special balancing process required here: in contrast to the component-elastic rotor behaviour, the rotor with shaft-elastic behaviour is in tolerance over the entire speed range after a proper unbalance correction.

If the service speed is above the 1st flexural resonance speed, a completely different characteristic emerges: The unbalance indication changes more and more rapidly with an increase in speed, the deformation reaches a maximum and then falls again, Fig. 6.6.

This is exactly the appearance of a resonance as described for the single-mass spring system (see Sect. 2.7.1).

If the speed is increased further, more flexural resonances often follow. In contrast to plasticity and component elasticity, shaft elasticity is often intended by design, e.g. to keep the bearing forces and vibrations small via a supercritical speed.

Although shaft-elastic rotor behaviour are not as common as rigid rotor behaviour, it is often found, especially in high-quality rotor systems, e.g. in textile machines, paper machines, turbo pumps and compressors, turbochargers, turbines and turbo generators, in jet engines. The correct treatment of shaft-elastic rotor behaviour is therefore of significant economic importance.

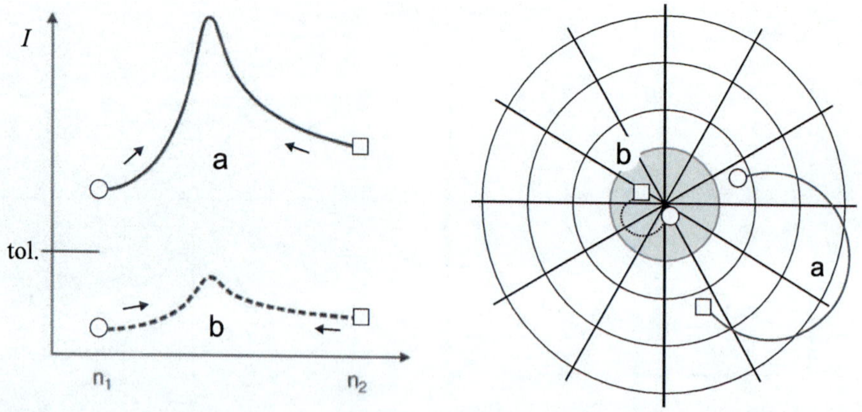

Fig. 6.6 Unbalance indication of a rotor with shaft-elastic behaviour, passing the 1st flexural resonance speed: **a** initial unbalance, **b** balanced. For other designations see Fig. 6.1. The unbalance tolerances for higher speeds cannot be given here (see Chap. 8)

Fig. 6.7 An idealised rotor with shaft-elastic behaviour, shown as a solid, long, thin roll with rigid bearings at both ends

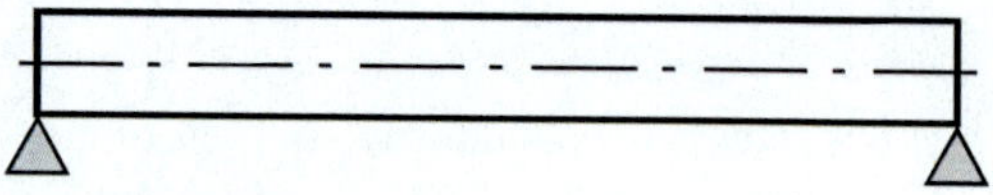

6.4.1 Idealised Rotor with Shaft-Elastic Behaviour

The simplest way to imagine a rotor with ***shaft-elastic behaviour is*** as a roll supported at both ends, Fig. 6.7

The rotor dynamics are also relatively easy to understand. It is essential that masses and flexibility (stiffness) are distributed over the entire length of the rotor: in the idealised rotor they are evenly distributed. It is therefore a system with an infinite number of degrees of freedom, thus also with an infinite number of flexural resonance speeds (see Sect. 2.7.2).

However, only *those* resonances that lie *below* the maximum service speed and *above* it in its vicinity are important. Moreover, only the vibrations *transverse* to the shaft axis are taken into account during balancing, i.e. in the case of the rotor with shaft-elastic behaviour, the different flexural modes.

6.4.2 Influence of Bearing Stiffness

The first three flexural modes for *absolutely rigid* bearings show Fig. 6.8.

There are vibration nodes in each bearing (and symmetrically in-between); the vibration modes are sinusoidal (because of uniformly distributed mass and stiffness).

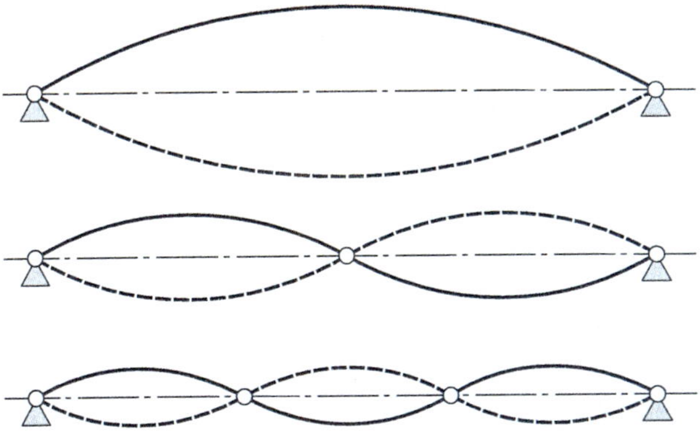

Fig. 6.8 The first three flexural modes of the rotor with shaft-elastic behaviour of Fig. 6.7 with ***absolutely rigid*** bearings

With *very soft* bearing support, the first two of the three modes are substantially different, Fig. 6.9

At low speeds, the rotor does not yet show any deflection, it oscillates in parallel in the 1st mode and with its ends in opposite directions in the 2nd mode. Only in the 3rd mode of the system the rotor is a flexural mode. It should be noted that its ends already oscillate in the opposite direction to the middle part, i.e. the oscillation nodes are not located at the ends but are shifted somewhat towards the middle.

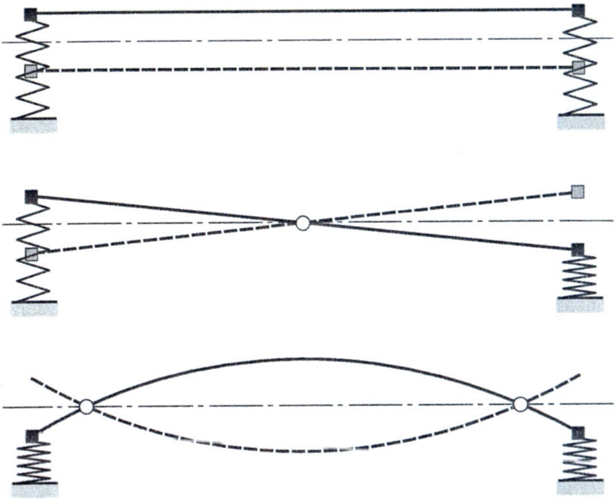

Fig. 6.9 The first three modes of the rotor with shaft-elastic behaviour of Fig. 6.7 with ***very soft*** bearing

In the load condition of the rotor (deflection), the flexural modes with the same number of nodes are always similar and therefore comparable. Therefore, the 1st mode of the rigid bearing must be assigned to the 3rd mode of the soft bearing if the rotor condition has priority.

Analogous to a vibrating string whose tone pitch (frequency) is raised by tapping (shortening the distance between nodes), the speed at which the 3rd flexural mode of the soft bearing occurs is higher than the speed of the 1st flexural mode with an absolutely rigid bearing.

In practice, the bearing support is very often only *slightly flexible*. In this case, the bearings always move slightly with the rotor, so that the nodes of the flexural mode lie outside the bearing planes, see Fig. 6.10.

The resonance speeds (speeds at which flexural modes occur) are somewhat lower than with the absolutely rigidly supported rotor, as can be seen from the larger node spacing.

These three bearing supports and flexural modes do not exist in isolation from each other. The continuous transition between the different bearing stiffnesses and their influence on the critical speeds of the rotor can best be shown by means of a diagram, see Fig. 6.11.

The critical speed n_e of the rotor system (or the rotor speed n) is plotted on the horizontal axis; the vertical axis indicates the bearing stiffness.

The dynamic stiffness as defined in Sect. 2.7.3 is used. The scale ranges from $-\infty$ (infinite mass stiffness) to $+\infty$ (infinite spring stiffness). The absolutely rigid bearing (Fig. 6.8) corresponds to the stiffness value $+\infty$, the soft bearing (Fig. 6.9) to positive values close to zero. The stiff bearing (slightly yielding), see Fig. 6.10 is then close to $+\infty$.

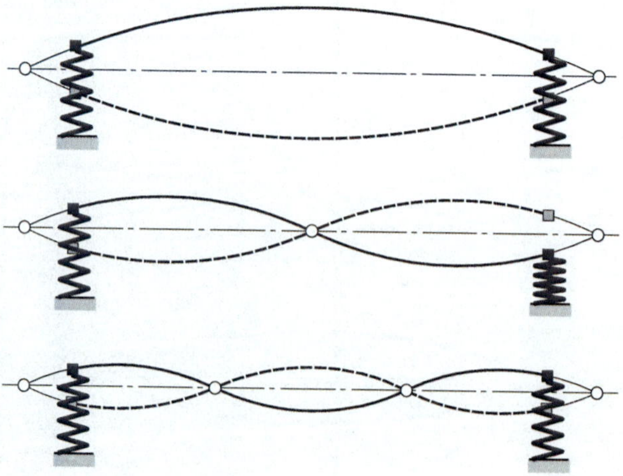

Fig. 6.10 The first three flexural modes of the rotor with shaft-elastic behaviour of Fig. 6.7 with ***slightly flexible*** bearing

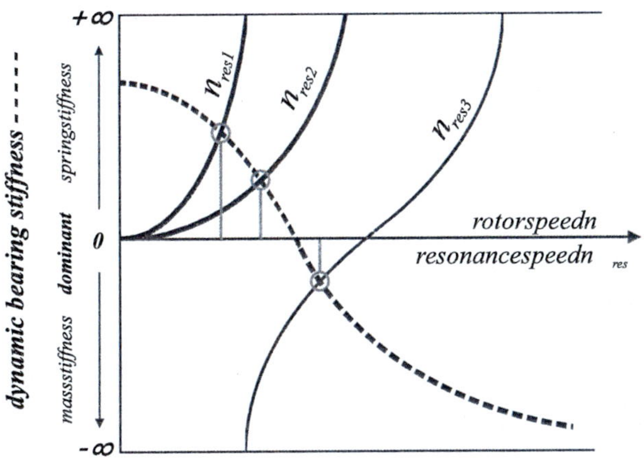

Fig. 6.11 Diagram showing the critical speeds of a rotor with shaft-elastic behaviour as a function of the dynamic bearing stiffness

The course of the critical speeds $n_{\text{res 1}}$ to $n_{\text{res 3}}$ as a function of the dynamic stiffness is marked by the corresponding curves. The curves for $n_{\text{res 1}}$ and $n_{\text{res 2}}$ start at the origin, i.e. at zero bearing stiffness the resonance speeds also become zero. The curves for $n_{\text{res 3}}$ and all higher critical speeds start at stiffness $-\infty$ and at a speed where the curve—for the respective 2 counter lower critical speed—ends at $+\infty$.

Static stiffnesses in this diagram are straight lines parallel to the speed axis. If a dynamic bearing stiffness is to be considered (the bearing mass vibrating with the rotor is the mass, the support is the spring stiffness), the corresponding curve (see Sect. 2.7.3) can be drawn directly into the diagram (dashed curve). The intersections of this curve with the curves of the critical speeds result in the rotor speeds at which the rotor-bearing system exhibits flexural resonance speeds.

From such a diagram, one can see at *which* speeds resonance occurs, but unfortunately not how *critical* these states are: This is decided by the damping in the system (which cannot be seen in the two-dimensional diagram) and the excitation.

Note

If the bearing support has different dynamic stiffnesses in radial directions, there are different critical speeds for the two main directions of stiffness, i.e. the resonances occur independently in each of the two main directions.

6.4.3 Flexural Resonance Speeds at Standstill

Even without the shaft-elastic rotor rotating, its flexural resonance speeds can be determined. For this purpose, exciters are used which either apply an alternating force of variable frequency to the rotor in a selectable direction perpendicular to the shaft axis, or which have a rotating effect. Instead, a single impulse can also be applied and the vibration response evaluated (impulse hammer).

If the support is not changed by the rotor at standing still—in the case of plain bearings, for example, the oil film is missing at standing still—the resonance frequency measured at standstill often agrees well with the flexural resonance speed measured under rotation. The prerequisite, however, is that the gyroscopic forces that would lead to a shift of the critical speeds to higher values during rotation are negligible. This condition is fulfilled in many such rotors—mostly long-stretched bodies.

6.4.4 General Rotor with Shaft-Elastic Behaviour

In general, masses and stiffnesses are not evenly distributed along the rotor length. The bearings are not located at the ends; there may be more or less large overhung masses. It follows that the flexure modes are no longer sinusoidal, but must be calculated (or measured) in each individual case. Nevertheless, the principles explained for the idealised shaft-elastic rotor also apply to the general case.[1]

If the nodes of a flexural mode coincide with *both* bearings, the bearing stiffness has no influence on this flexural resonance speed. In this (extremely rare) borderline case, *this* resonance cannot be observed by measuring in the bearings (forces or vibrations) because no measurable values occur. Measurements must then be taken in other planes on the rotor—usually contactless.

6.4.5 Unbalance Effects on the Rotor with Shaft-Elastic Behaviour

The flexural resonance speeds are excited by one or more unbalances, see Fig. 6.12.[2]

This always results in the same flexural mode (i.e. the flexure line is always similar, the position of the nodes is always the same)—in the ideal case, but also with moderate damping in the system—regardless of the radial plane where an individual unbalance lies or how the unbalances are distributed along the rotor.

---

[1] At least as long as the nodal spacing of the 1st flexural resonance mode in soft bearings is not larger than the bearing distance. If this case occurs, then some tendencies turn around. However, this extreme case occurs so rarely in practice that a description is not needed here.

[2] According to the definition, at flexural resonance speeds the flexure of the rotor dominates over the movement of the bearings.

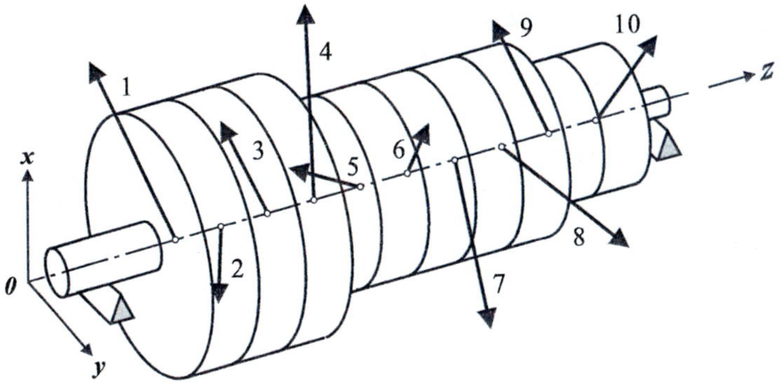

Fig. 6.12 Rotor modelled with 10 disc-shaped elements with a single unbalance vector each

In practical cases, the damping of the system is often so small that the flexural mode is flat, i.e. it lies in *a longitudinal plane* of the rotor, even if the unbalances have different angular positions.

The flexural mode is only dependent on the rotor and bearing data. Of course, the amplitude of the flexure depends on the size of the unbalance, but furthermore also on the plane in which the unbalance lies: In the nodal planes the unbalance cannot excite the flexure, but outside the nodes it does—the larger the bending arrow is in this plane, the stronger the impact, see Fig. 6.13.

6.4.5.1 Modal Unbalances

From the unbalance distribution and the respective bending line, the ***modal unbalance*** $\vec{U}_n$ can be calculated for each flexural mode. It is the sum of the products of a single unbalance $\vec{U}_k$ and the bending arrow of the flexural mode in this plane $\phi_n(z_k)$:

$$\vec{U}_n = \sum_{k=1}^{K} \vec{U}_k \phi_n(z_k) \tag{6.1}$$

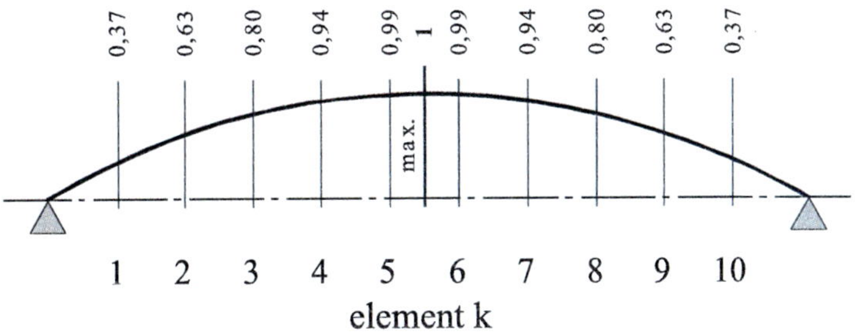

Fig. 6.13 Flexure line of the 1st flexural mode with bending arrows and numerical values to indicate the impact of unbalance on the flexure. For ease of verification, the ideal rotor in rigid bearings was used here

This modal unbalance is an *unbalance distribution* in the corresponding flexural mode and is therefore—and because of different normalisation possibilities—difficult to use in practice.

6.4.5.2 Equivalent Modal Unbalances

The ***equivalent modal unbalance*** in the nth flexural mode $\vec{U}_{ne}$ is *that* single unbalance in the *most sensitive* plane which corresponds to the modal unbalance in its effect on the nth flexural mode. The equation is:

$$\vec{U}_{ne} = \frac{\sum_{k=1}^{K} \vec{U}_k \, \phi_n(z_k)}{\phi_{\max}} = \sum_{k=1}^{K} \vec{U}_k \frac{\phi_n(z_k)}{\phi_{\max}} \qquad \text{g} \cdot \text{mm} \tag{6.2}$$

This calculation can also be represented graphically, using the second term of Eq. 6.2 as a basis. In this case, the local values of the flexural mode are first related to the maximum value, then multiplied by the local unbalance and summed up, see Fig. 6.14.

6.4.6 Balancing a Rotor with Shaft-Elastic Behaviour

In the following, a ***systematic step-by-step*** procedure is described to achieve a basic understanding of the balancing of such rotors. Later, possible simplifications will be mentioned.

For low-speed balancing of a rotor with shaft-elastic behaviour (i.e. at a speed at which it is still rigid), any unbalance condition can be handled by a correction in any two planes (see Sect. 5.3.4).

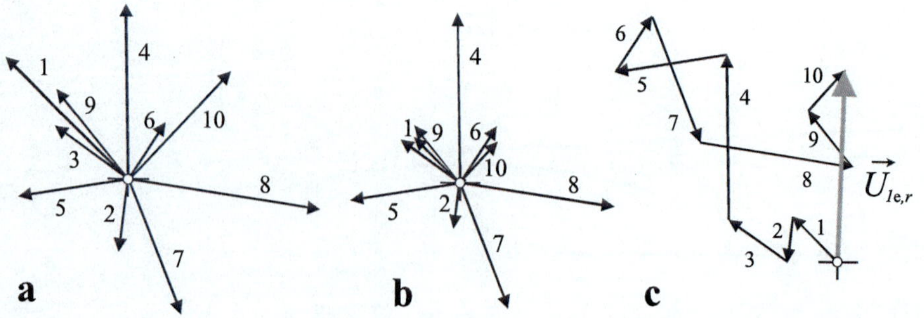

Fig. 6.14 Derivation of the equivalent unbalance $\vec{U}_{ne}$ of a rotor (view in the direction of the rotor axis). **a** the individual unbalances 1–10, corresponding to Fig. 4.14, **b** weighted individual unbalances (with the ratio of the respective bending arrow to the maximum value), and **c** forming vector sum $\vec{U}_{ne}$

If the rotor has e.g. the unbalance mass *u*, it is normally corrected at low speed by corresponding correction masses in correction planes I and II, see Fig. 6.15.

However, the unbalance mass and the correction masses have a fundamentally different effect on the modal unbalance of the shaft-elastic rotor (see Sect. 6.4.5). The result is a modal unbalance (which in unfavourable cases can even be larger than that caused by the unbalance mass alone).

In order to reduce the modal unbalance to the desired level, additional correction masses must be placed. This always requires more than two correction planes, because these masses must not worsen the correction achieved for low speed, i.e. they must not cause any dynamic unbalance of the rigid rotor. This means that the resulting unbalance (and thus the sum of the forces) and the moment unbalance (and thus the sum of the force moments) of the additional correction masses must be zero.

This group of correction masses for a flexural mode is called a *mass set*. The individual masses have a fixed ratio to each other—depending only on the plane distances and the correction radii—and are also fixed in their angular position to each other (same angle or 180° offset).

In resonance, this mass set also has an effect on the flexural mode. However, since the amount of the mass set can be selected at will and the whole mass set can be brought into any angular position, any deflection can be created or thus also eliminated.

For each flexural mode (where the rotor deflects, see Fig. 6.10), a different set of masses is required. The number of correction planes must exceed the number of nodes of the flexural mode by at least one. The minimum number for the first three flexural modes is therefore (in order): 3, 4 and 5 correction planes.

If the rotor is to be calibrated according to Fig. 6.16 for three flexural modes, (at least) the correction planes I to V must be present.

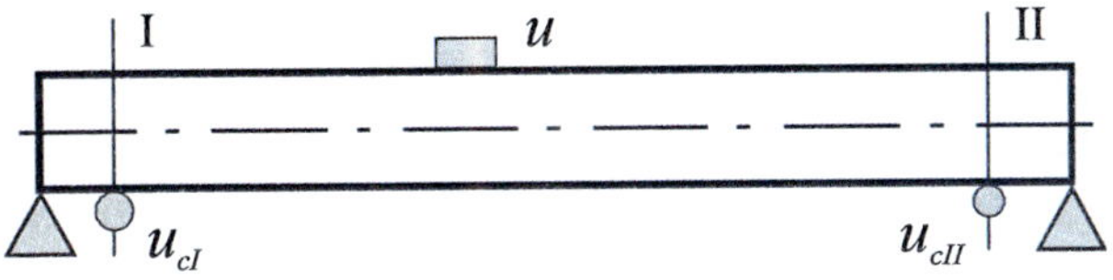

Fig. 6.15 Low-speed correction of the unbalance mass *u* by correction masses $u_{c\ I,II}$ in planes I and II

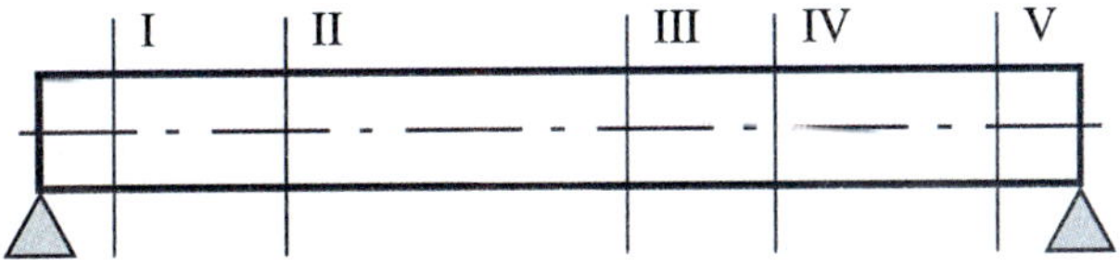

Fig. 6.16 An idealised shaft-elastic rotor with five correction planes

For each flexural mode, some correction planes are chosen so that the impact on the deflection is as large and unambiguous as possible. The masses, which are only to keep the influence on the balancing state achieved so far small, are placed as close as possible to the bearings or the respective nodes.

6.4.6.1 First Flexural Mode

For the 1st flexural mode (with two nodes and planes I, III and V), see Fig. 6.17, the determining equations for the unbalances of the mass set are:

$$U_I - U_{III} + U_V = 0; \qquad aU_I - bU_V = 0 \tag{6.3}$$

If *one* unbalance is assumed, e.g. U_{III} of the centre plane[3], the associated unbalances in the other planes are:

$$U_I = U_{III}\frac{b}{a+b}; \qquad U_V = U_{III}\frac{a}{a+b} \tag{6.4}$$

Instead of a calculation, the correct distribution can also be measured, e.g. as follows:

- One of the three unbalances, e.g. in the middle plane, is applied.
- Then the unbalances required in the other planes are measured (by an additional low-speed measuring run) and applied. This is a simple way to take even care of different balancing radii.

6.4.6.2 Second Flexural Mode

For the 2nd flexural mode (with three nodes and the planes I, II, IV and V), see Fig. 6.17—only two equations can be set up due to the equilibrium of forces and moments, which are not sufficient for a determination of the four unbalances, even if one unbalance is assumed. As an additional requirement, however, it is added here that this mass set must not disturb the *flexure* in the 1st flexural mode, i.e. with two nodes.

For the general case, the calculation is rather extensive. However, if the correction planes are *approximately symmetrical* and the masses and stiffnesses are approximately evenly distributed, the sensitivity of the rotor in its second bending mode is approximately equal (but opposite) in planes I and V, as well as in planes II and IV so that further conditions can be added:

$$U_I - U_V = 0; \qquad U_{II} - U_{IV} = 0 \tag{6.5}$$

The equation for the moment unbalance is thus simplified to

$$dU_I - bU_{II} = 0 \tag{6.6}$$

[3] Centre plane does not mean the centre of the rotor, but the middle of the 3 planes: it may well be asymmetrical, i.e. the flexural mode does not need to be known precisely—a good estimate is sufficient.

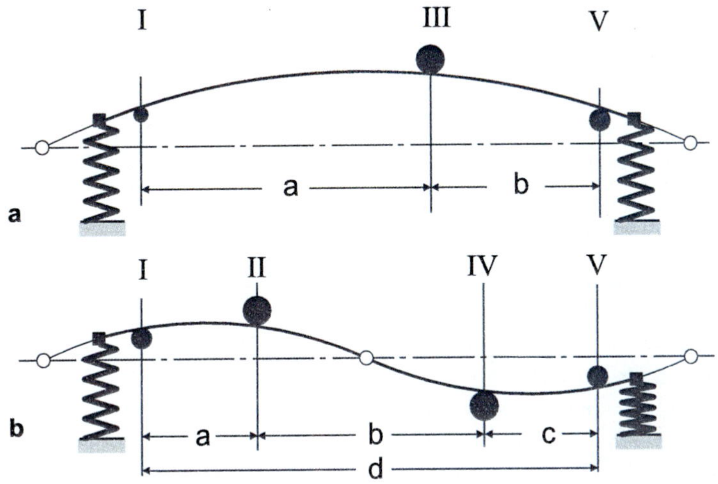

Fig. 6.17 Mass sets for different flexural modes: **a** for the 1st flexural mode, **b** for the 2nd flexural mode

or, if U_{II} is assumed, to

$$U_I = -U_{II}\frac{b}{d}; \qquad U_{IV} = -U_{II}; \qquad U_V = U_{II}\frac{b}{d} \tag{6.7}$$

If a calculation of the flexural modes for the general case is too time-consuming, estimation is sufficient to choose the correction planes sensibly. A suitable 4-mass set can then be determined experimentally, see Fig. 6.18.

Tuning of a 4-mass set that it does not interfere with low-speed balancing and a flexure with two nodes:

- In planes I, II and V a 3-mass set (see Sect. 6.4.6.1) is added, which does not disturb the balancing condition of the rigid rotor. But this mass set influences the flexures with *two* and *three* nodes.
- The influence on the flexure with *two* nodes is fully compensated by a second 3-mass set in planes I, IV and V.
  - This 3-mass set also does not influence the rigid rotor. Due to the two sets of 3 masses, only the flexure with *three* nodes has changed.
- The masses in planes I and V are each contracted to one mass and form together with the masses in planes II and IV the desired 4-mass set, which does not disturb neither the unbalanced state of the rigid rotor, nor the flexural mode with two nodes.
- The 4-mass set must now be adjusted in size and angular position so that the flexure with three nodes is corrected to the desired degree.

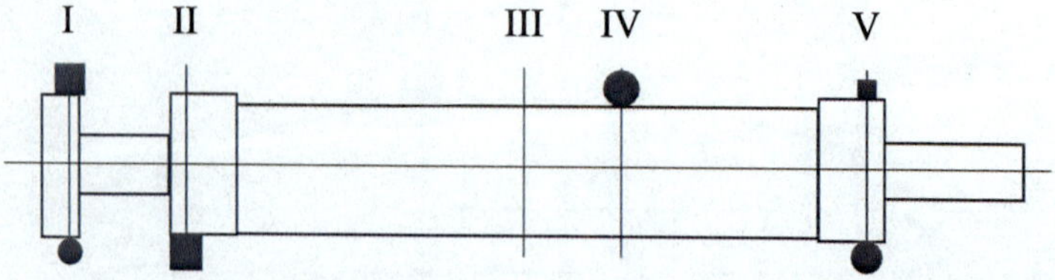

Fig. 6.18 Correct tuning of a 4-mass set (from two 3-mass sets) to correct for the 2nd flexural mode: ▪ first 3-mass set. • second 3-mass set to correct for the influence of the first set on the 1st flexural mode. Each set of 3 masses is tuned so that it does not interfere with the low-speed correction

6.4.6.3 Third Flexural Mode

To handle the 3rd flexural mode, the 5-mass set must be set so that it does not affect the unbalance condition of the rigid rotor and the deflection in the 1st and 2nd flexural mode.

The procedure explained above can be continued accordingly, but today a computer system would certainly be used for such a difficult task, so that the manual way does not need to be explained in detail here.

6.4.7 Choice of Correction Planes

6.4.7.1 Variety of Rotors

The variety of rotors with shaft-elastic behaviour is almost limitless, see Fig. 6.19—how to choose the right correction planes?

It is of course helpful to have calculated or measured flexural modes. But you do not need to know the exact flexural modes, because a good estimation is usually sufficient (see also Sect. 10.3), if some principles are also taken to heart.

Principles

- Each correction plane should be as effective as possible, i.e. the flexural mode under consideration should have a large deflection at this point.
- A clear curvature of the bending line must occur between *at least two* correction planes.
- At points of large deflection but low slope, single unbalances have the main effect.
- At points of large slope, with low deflection, moment unbalances have the main effect.

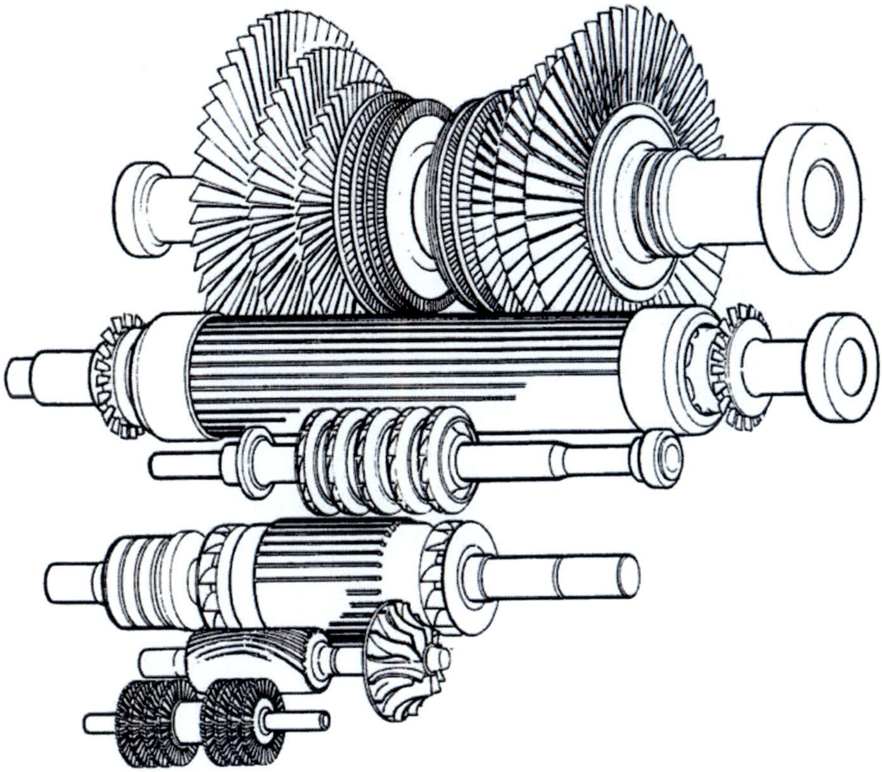

Fig. 6.19 Diversity of rotors with shaft-elastic behaviour—from a small vacuum pump to a large low-pressure turbine

- A combination of planes with a node between them is very effective.
- Test unbalances can be set individually or as a set in order to obtain IC (influence coefficients). What is useful or necessary depends on the balancing task and the chosen planes.
- The IC obtained should be checked for effectivity, i.e. they must adequately represent the flexural mode under consideration.
- It is useful to estimate the effectiveness of the correction planes for the flexural mode. This is best done by assuming modal sets and calculating their equivalent modal unbalance.

Different examples are illustrated below.

6.4.7.2 Example 1

The basis is an idealised rotor—uniform stiffness and mass distribution over the length, rigid bearing at the ends—with deflections normalised to the maximum value, for which a modal unbalance set is applied for the 1st flexural mode, see Fig. 6.20.

In the simplest case, this is a test unbalance U in the centre plane (II), and $-1/2U$ in each of the bearing planes (I and III): this set has no effect on the low-speed unbalances, so it is (backwards) orthogonal.

The equivalent modal unbalance of this unbalance set (see Sect. 6.4.5.2) is therefore $1U$. This value is used as a benchmark for the other examples.

Conclusion

In this case, the ICs could be obtained by single unbalances, or by a modal unbalance set.

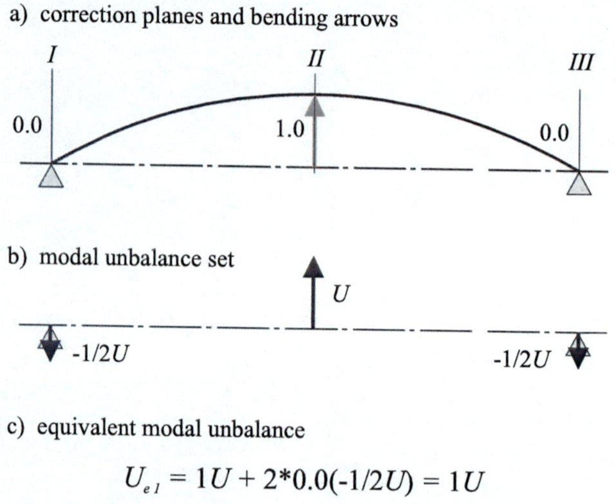

Fig. 6.20 An idealised rotor with deflections normalised to the maximum. And a modal unbalance set for the 1ˢt flexural mode in planes *I* to *III*: The equivalent modal unbalance is $1U$

6.4.7.3 Example 2

This example is more practical: The centre plane *II* is kept (for $1U$), but in addition 2 planes are used along the rotor (for $-1/2U$ each), here in the quarter planes, see Fig. 6.21.

The equivalent modal unbalance of this unbalance set is only $0.29U$[4]. This set is therefore much less effective than in example 1. In order to have the same flexure, the unbalances for the test set must be chosen 3 times as large.

> **Conclusion**
> Modal unbalance set would presumably have advantages in gaining ICs over single unbalances with this low effectiveness.

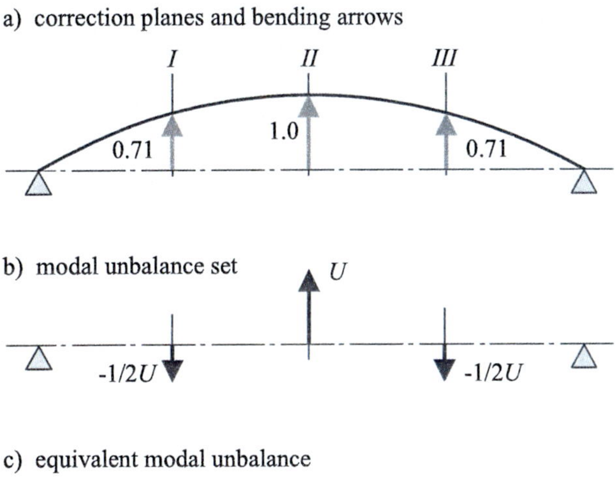

$$U_{e1} = 1U + 2{*}0.71(-1/2U) = 0.29U$$

Fig. 6.21 The idealised rotor of Fig. 6.20. Using the quarter planes, the equivalent modal unbalance is only $0.29U$

[4] Other planes can also be used for *I* and *III*. Here all the planes could be shifted laterally, i.e. they do not have to lie centrally on the bending line. The effectiveness of the set decreases only slowly with small displacement. If the planes *I* and *III are* not symmetrical to plane *II*, the lever arms must be taken into account as explained in footnote 3.

6.4.7.4 Example 3

If the rotor has overhung ends—e.g. couplings where unbalances can be set—a completely different situation arises, see Fig. 6.22:

The equivalent modal unbalance of this set is very large: 2.0U. The set is therefore twice as effective as in example 1. In order to have the same effectiveness, the unbalances here only need to be chosen 1/2 times as large.

> **Conclusion**
> In any case, individual unbalances for the formation of the ICs would work fine here.

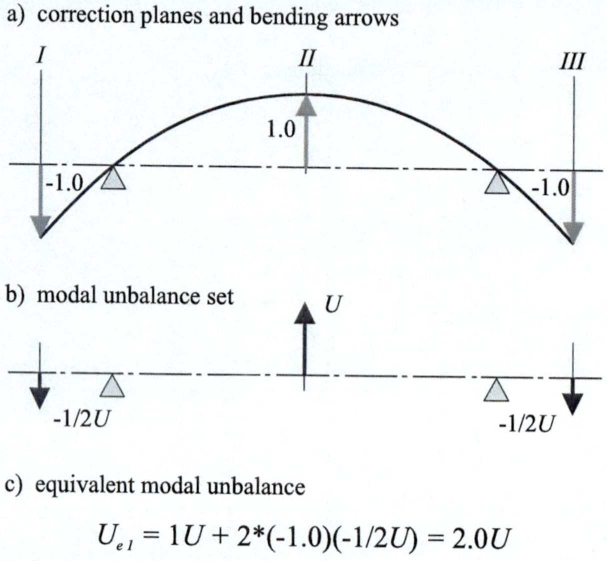

Fig. 6.22 The idealised rotor of Fig. 6.20, but with overhung ends. Using these planes, the equivalent modal unbalance is very good: 2.0U

6.4.7.5 Example 4

Serious mistakes are often made in rotors with ***rigid sections***: If all 3 planes are selected in *one* rigid section, the bending arrows may all be very large, but the condition that there should be a considerable curvature between at least 2 planes is not fulfilled, see Fig. 6.23

The equivalent modal unbalance of this set is really $0.0U$[5] .

It is therefore totally inoperative: even extremely large unbalances would not help here! If, despite this, the impression arises that the first flexural mode is influenced, it is probably due to an unprecise setting. But this becomes immediately apparent because then there are also influences on the low-speed unbalances (i.e. not a sufficiently modal set).

> **Conclusion**
> In this case, neither single unbalances nor modal sets would work for obtaining ICs.

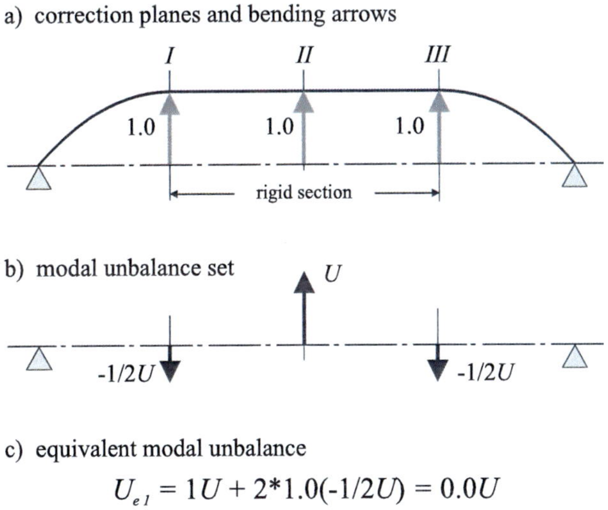

Fig. 6.23 A rotor with one ***rigid section***. With *all* planes in this section the equivalent modal unbalance is zero, i.e. the unbalance set is completely ineffective

[5]Even with any position of the 3 planes in the rigid section—and the correspondingly adjusted unbalances of the modal set (see footnote 3)—its effectiveness is zero. Even if the bending line in the rigid section is inclined to the left or to the right, the equivalent modal unbalance is *always* zero.

6.4.7.6 Example 5

For rotors where a section is ***almost rigid*** (i.e. shows little bending), the situation is as follows, see Fig. 6.24:

The bending arrows differ only minimally, i.e. there is an equivalent modal unbalance, but it is extremely small: $0.05U$. To have the same effectiveness as in example 1, the unbalances of the test set must be chosen approx. 20 times as large!

Conclusion

Such a situation should be avoided. But if it cannot be avoided, a modal set ***must*** be used here! Individual unbalances of the required size would generate far too large centrifugal forces and make it impossible to run up to service speed.

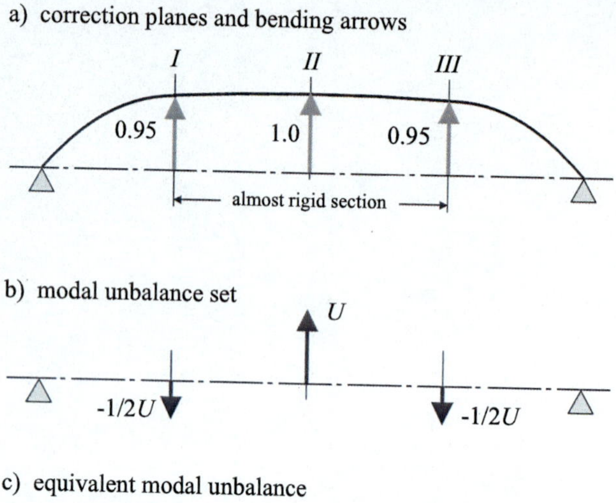

$$U_{e1} = 1U + 2{*}0.95(-1/2U) = 0.05U$$

Fig. 6.24 A rotor with an ***almost rigid section***. With *all* planes in this section, the equivalent modal unbalance is extremely small: $0.05U$, i.e. hardly usable. Such a situation should be avoided

6.4.7.7 Example 6

For rotors with *one* ***rigid section*** and an ***overhanging shaft end,*** the situation is quite different as soon as this shaft end can be used for a correction plane, see Fig. 6.25:

The equivalent modal unbalance of this set is $0.9U$. It is therefore approximately as effective as in example 1.

> **Conclusion**
> In this case, single unbalances or a modal set can be used to obtain the ICs.

The situation would improve even more if, instead of equal plane distances, see Fig. 6.25, plane *II* was shifted to the right for a distance ratio of 2/3 to 1/3. Result: The equivalent modal unbalance of this set would be approx. 30 % better, i.e. about $1.2U$.

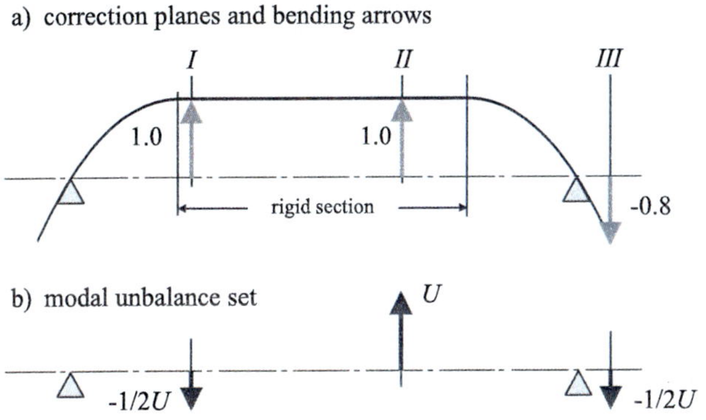

c) equivalent modal unbalance

$$U_{e\,I} = 1U + (1.0)(-1/2U) + (-0.8)(-1/2U) = 0.9U$$

Fig. 6.25 A rotor with *one* rigid section and with *an* overhung shaft end that can be used as a correction plane. The equivalent modal unbalance has an approximately normal size: $0.9U$

6.4.7.8 Example 7

However, the situation would improve even more if there are two overhung shaft ends and *both* can be used, see Fig. 6.26:

The equivalent modal unbalance of this set is $1.8U$. It is therefore almost twice as effective as in example 1.

> **Conclusion**
> In this case, it is always possible to work with individual unbalances to obtain ICs.

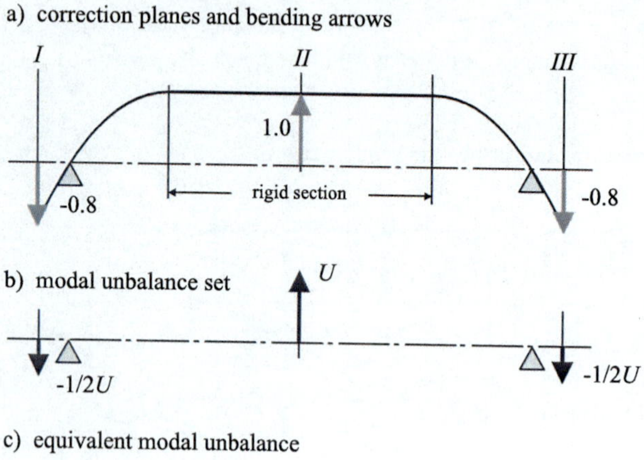

$$U_{e1} = 1U + 2*(-0.8)(-1/2U) = 1.8U$$

Fig. 6.26 A rotor with rigid section and with ***two*** overhung shaft ends that can be used for two correction planes. The equivalent modal unbalance is extremely good: $1.8U$

Tolerances for Rotors with Rigid Behaviour

7

Contents

H. Schneider, *Rotor Balancing*, https://doi.org/10.1007/978-3-662-66049-2_7

7.1 Preliminary Note

Balancing a rotor with rigid behaviour usually aims for a certain running smoothness in operation. This smoothness is often defined as the root mean square value of the vibration velocity at service speed (see Sect. 18.1).

On a balancing machine, however, these rotors are usually balanced at lower speeds, and the bearing support does not match the service condition. The above-mentioned vibration limits are therefore not applicable here.

A property of the rotor with rigid behaviour, which is independent from the above-mentioned boundary conditions, is its unbalance (see Chap. 5).

7.2 Basics

When balancing, the aim is not a "perfectly balanced rotor", but from a technical point of view a certain tolerance is permissible, which should not be undercut for economic reasons.

For the standardisation of the unbalance tolerance, a major task arose: A suitable scale was sought with which rotors of less than 1 g mass (e.g. balance wheel of a clock) up to 320 t (low-pressure turbine of a nuclear power plant) can be assessed just as easily as slow-running machine tool spindles with 100 min^{-1} and dental drill turbines, which can have a speed of up to 1,000,000 min^{-1}.

ISO 1940-1 for decades had provided guidance on the appropriate determination of unbalance tolerances and their verification. In the 2004 edition,[1] a very significant

[1] ISO 1940-1:2004 Mechanical vibration—Requirements for the balance quality of rotors in a constant (rigid) condition—Part 1: Determination and verification of the unbalance tolerance.

change was added: special reference planes were used for the tolerances, not the correction planes as before. This led to a more precise statement on the unbalance condition of the rotor, but also had serious effects on the balancing process and even on the balancing machines, which in future must better support the achievements with the tolerance target.

The latest edition of this standard has now been published in the 21940 series of standards as Part 11[2] and also formulates fundamental innovations in the area of acceptance criteria:

- In case of balance errors, the differences between the balancing condition and the delivery condition are emphasised more strongly.
- For checking the residual unbalance 3 areas are defined: Measured values are in tolerance, range of uncertainty, measured values are outside tolerance.

These criteria always apply, no matter by whom, where and how measurement take place—but the errors to be taken into account can be different in each case.

7.2.1 Tolerance Planes

Traditionally, unbalance tolerances have been specified in correction planes. This is actually wrong, because unbalance tolerances should, if possible, be specified in planes in which they do not influence each other. More precisely: in planes in which there is no significant difference, regardless of whether the residual unbalances dominantly form a resulting unbalance or a couple unbalance.

In a rotor with rigid behaviour, there are always two ideal planes for the unbalance tolerances. Usually these planes are close to the bearing planes. To simplify matters, the bearing planes can therefore be used as reference planes for the unbalance tolerances. These planes are then referred to as tolerance planes.

Notes

- For unbalance tolerances, only the magnitude is generally considered, the angle is arbitrary, so there are two tolerance fields (two-plane task).
- While the *vectors* of a dynamic unbalance can be converted to other planes, this is not possible with tolerance fields.

[2] ISO 21940 Mechanical vibration—Rotor balancing, part 11 (2016): Procedures and tolerances for rotors with rigid behaviour.

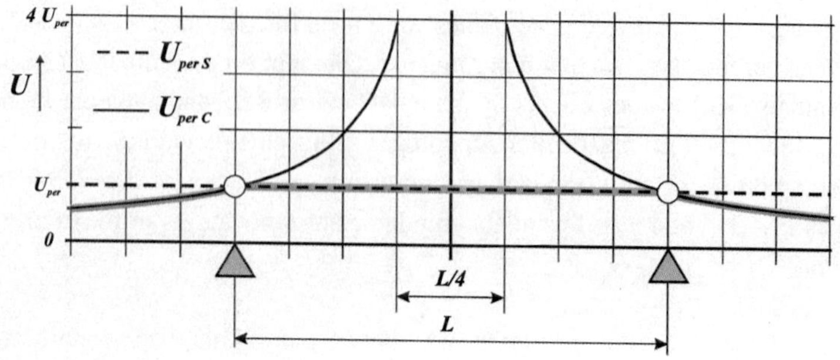

Fig. 7.1 Course of the permissible static unbalance U_{perS} and the permissible couple unbalance $U_{per\ C}$ via the rotor length (and beyond). There is equality in the tolerance planes (the bearing planes). The course of the permissible dynamic unbalance for the traditional procedure is shown in grey

However, if we look at tolerances for a static unbalance and a couple unbalance, we can see for different plane distances[3], Fig. 7.1:

The permissible static unbalance $U_{per\ S}$ remains constant; the permissible couple unbalance $U_{per\ C}$ is inversely proportional to the plane distance. The values are the same in the tolerance planes.

Traditional Approach

Up to now, when defining tolerance zones for a dynamic unbalance, only the smaller tolerance value was allowed in the correction planes, i.e.: between the bearings the value for the static unbalance, outside the bearings the value of the couple unbalance (grey thick line in Fig. 7.1). For all angles between the residual unbalances you are then on the safe side.

This Should Change

With many balancing machines, it is already possible to measure the unbalance in the bearing planes as well and thus check whether the rotor is in tolerance.

However, so far only a few measuring devices can check this tolerance in the tolerance planes (e.g. bearing planes) *and* at the same time give the right correction instructions for the correction planes. Innovations are still necessary here.

[3] For simplicity, symmetrically positioned planes are assumed here, but similar tendencies apply to asymmetrical planes also.

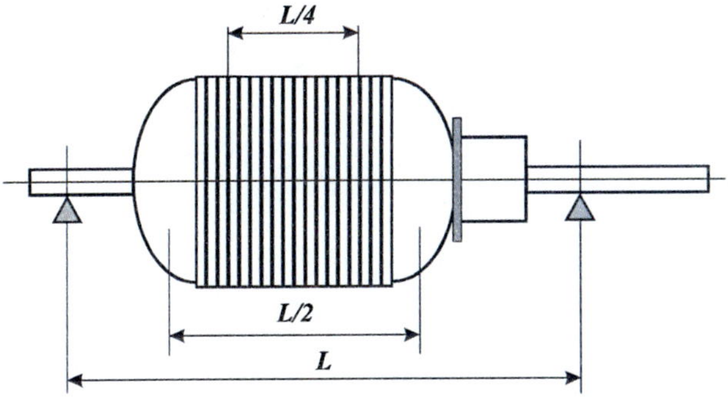

Fig. 7.2 Typical situation with small electric armatures: The distance between the correction planes is much smaller than the bearing distance. If a dynamic unbalance tolerance is specified in the correction planes, a couple unbalance must be corrected to 1/2 to 1/4 of the value actually required, i.e. unnecessarily accurate, see Fig. 7.1. Therefore, ISO 21940-11 recommends the use of special tolerance planes (e.g. bearing planes)

7.2.2 Correction Planes

For rotors that are out of tolerance, a correction must be made. This correction can only be done in planes where material can be added, removed or relocated. These are usually not the tolerance planes, at least not if the bearing planes are used for this purpose.

In case of small wound electric armatures, for example, either correction masses are usually added to the winding heads, or milling is carried out in the stack. The spacing of the correction planes varies drastically in these cases and is sometimes only 1/2 to 1/4 of the bearing spacing, see Fig. 7.2.

For these rotors, too, the tolerances for a dynamic unbalance have so far mostly been specified for the correction planes. If the residual unbalances dominantly form a static unbalance, the tolerance is well used. If they mainly form a couple unbalances, this unbalance must be balanced to 1/2 to 1/4 of the actually required value, see Fig. 7.1. This is not justifiable for economic reasons.

The ***number*** of correction planes required depends on the size and distribution of the unbalance, as well as on the design of the rotor and the position of the correction planes.

Generally, as many correction planes are needed as there are tolerance planes, i.e. one or two.

Although theoretically any rotor with rigid behaviour can be balanced in two planes, in practice sometimes more than two balancing planes are used, e.g. if:

- The resulting unbalance and the couple unbalance are corrected independently of each other, i.e. if the planes of the couple unbalance are not also used to correct the resulting unbalance.
- The correction is distributed over the rotor length.[4]

7.2.3 Limitation of the Permissible Residual Unbalance

For a disc-shaped rotor where the moment unbalance is negligible, the unbalance condition can be described by a single vector, the unbalance $\vec{U}$.

The rotor is in tolerance if the amount of this unbalance is not larger than the permissible unbalance U_{per}, i.e.

$$U \leq U_{per} \qquad \text{g} \cdot \text{mm} \tag{7.1}$$

This statement applies to every rotor shape, i.e. also to a general rotor.

The permissible unbalance U_{per} is also in the general case the total tolerance in the plane of the mass centre. In the case of two-plane correction, this value must be allocated to both tolerance planes, see Sect. 7.5.2.

As explained in Sect. 4.8, U_{per} is the value for ***delivery.*** For ***balancing,*** the permissible unbalance U_{per} must be reduced by the combined error ΔU that occurs during balancing compared to the condition at delivery: Errors when measuring the unbalance and due to a different constellation of the rotor. The following sections mainly describe the basis, i.e. the permissible unbalances for delivery (for the permissible unbalance indications during balancing see Chap. 16).

7.3 Similarity Considerations

Since unbalance tolerances are to be defined for a wide range of rotors, similarity considerations help to form benchmarks.

7.3.1 Rotor Mass and Permissible Residual Unbalance

In general, the larger the rotor mass, the larger the permissible unbalance may be. It is therefore appropriate to relate the permissible residual unbalance U_{per} to the rotor mass m.

[4] Sometimes the distribution of the correction over the rotor length is necessary to maintain the function or strength of the component, or because of limitations in the balancing planes, e.g. in crankshafts, if the correction is done in the counterweights.

The permissible specific unbalance is e_{per}:

$$e_{per} = \frac{U_{per}}{m} \qquad \text{unit m} \tag{7.2}$$

As explained in Sect. 4.8 for U_{per}, also e_{per} is the value for the delivery of the rotor. Sometimes the errors that occur during *balancing* compared to the *delivery are* also expressed as eccentricities: The eccentricity e_{per} permitted for the delivery must then be reduced by the error Δe for balancing.

ISO 21940-11 makes 5 comments on U_{per} and e_{per} that are very important, some of which also refer to the above-mentioned problem—conditions on delivery/on balancing machine:

Notes

(Cited from ISO 21940-11, but modified to cite sections of this book).

1. The SI unit for U_{per} /m is kg · m/kg (kilogram · metre per kilogram), or m (metres). But more practical is the unit g · mm/kg (gram · millimetre per kilogram), which corresponds to μm (micrometres), as many permissible residual specific unbalances are between 0.1 μm and 10 μm.
2. The term e_{per} is useful especially if geometric tolerances (e.g. runout, play) are related to unbalance tolerances.
3. In case of a rotor with only a resultant unbalance, e_{per} is the distance of the centre of mass from the shaft axis. However, in the case of a general rotor with both resultant unbalance and resultant moment unbalance present, e_{per} is an artificial quantity containing the effects of the resultant unbalance as well as of the resultant moment unbalance.
4. There are limits for achievable specific residual unbalance, e_{per}, depending on the set-up conditions in the balancing machine (e.g. centring, bearings, drive).
5. Small values of e_{per} can only be achieved in practice if the accuracy of shaft journals (roundness, straightness, etc.) is adequate. In some cases it can be necessary to balance the rotor in its own service bearings, using belt-, air- or self-drive. In other cases, balancing needs to be carried out with the rotor completely assembled in its own housing with bearings and self-drive, under service condition and temperature.

7.3.2 Service Speed and Permissible Residual Unbalance

Practical experience (statistical evaluations of damage cases) showed that for *similar rotors* the permissible residual specific unbalance e_{per} usually changes in inverse proportion to the rotor speed *n*. The relationship can be written: $e_{\text{per}}\, n = \text{constant}$, or better:

$$e_{per}\, \Omega = \text{const} \qquad \text{m/s} \tag{7.3}$$

The expression $e_{\text{per}}\ \Omega$ is the orbital speed of the rotor mass centre (see Sect. 2.6), usually expressed for balancing in mm/s.

The same dependence results from similarity considerations: In geometrically similar rotors (e.g. turbochargers) with the same—because material-related—circumferential speed, the same stresses are generated in the rotor and the same surface pressures in the bearings if the characteristic value $e_{\text{per}}\ \Omega$ is kept constant (assuming rigid bearings).

A simpler approach: $e_{\text{per}}\ \Omega$ is a velocity like the circumferential velocity of the rotor, and if *this* is kept constant, $e_{\text{per}}\ \Omega$ must also be kept constant from a similarity point of view.

7.3.2.1 Special Cases

There are special cases where the geometric similarity is not given and accordingly the determination of $e_{\text{per}}\ \Omega$ does not fit:

Example

Rotors whose service speed is significantly below the maximum speed provided for in the design: The rotor of a three-phase motor is designed for 3000 min^{-1}, but this time it runs in a stator for 1000 min^{-1}. In such a case the determination according to Eq. 7.3 may be too restrictive. ◀

Solution

Even if e_{per} was calculated for a service speed of 1000 min^{-1}, a higher value (corresponding to the speed ratio 3000/1000) can be permitted. This then leads to equal loads in the system again.

7.4 Determining the Permissible Residual Unbalance

7.4.1 General

Since there are also "permissible initial unbalances"—e.g. in the case of jet engine rotors assembled from many individual parts—one must correctly speak of "permissible residual unbalances"; however, most often only "permissible unbalance" is said in abbreviated form and thus the final state is meant.

The suitable requirement for the balancing quality can be found various ways:

- The assignment to ***quality grades*** is based on many years of worldwide experience with many rotor types (see Sect. 7.4.2). This method is used most frequently.

- An ***experimental determination*** is sometimes used for series products (see Sect. 7.4.3).
- Special cases are the determination on the basis of ***permissible bearing forces*** and on the basis of ***permissible vibrations*** caused by unbalances (see Sect. 7.4.4).
- The determination based on ***proven experience*** can be interesting for companies with a documented balancing history (see Sect. 7.4.5).

ISO 21940-11 recommends that the procedure on which the unbalance tolerance and the balancing quality are based be agreed between manufacturer and customer.

7.4.2 Balancing Grades G

7.4.2.1 Classification

For the first ISO 1940 standard, experience with unbalance tolerances was collected worldwide and classified on the basis of similarity considerations (see Sect. 7.3). From this, balance quality grades G were developed, which allow a classification into classes for typical machine types.

The quality grades are designated according to the magnitude of the product $e_{per}\ \Omega$, with the unit mm/s. The product $e_{per}\ \Omega$ could take on any value; however, a number of fixed values have been agreed upon.

These values differ by a factor of 2.5 for each step. In some cases, especially with high balancing quality (small G-values, small unbalance tolerance), a finer graduation may be necessary, e.g. with a factor of 1.6.

Quality grades from G 0.16 to G 4000 are used. If the amount of $e_{per}\ \Omega$ is e.g. 6.3 mm/s, the grade is called G 6.3. Over decades of years, many classifications were adjusted.

Table 7.1 reproduces Table 1 from ISO 21940-11:2016. Some machines are represented in several quality classes, e.g. electric motors in classes G 6.3, G 2.5 and G 1, according to their different design and use.

ISO 21940-11 gives a number of notes on this table, which of course refer to the various chapters of this standard. Here the notes are listed in the same order, but modified and related to the sections of this book:

Table 7.1 Guidance for balancing quality grades for rotors with rigid behaviour (Table 1 from ISO 21940-11:2016)

Machinery type: General examples	Balance quality grade G[5]	Magnitude $e_{per} \cdot \Omega$ mm/s
Crankshaft drives for large, slow marine diesel engines (piston speed below 9 m/s), inherently unbalanced	G 4000	4000
Crankshaft drives for large slow marine diesel engines (piston speed below 9 m/s), inherently balanced	G 1600	1600
Crankshaft drives, inherently unbalanced, elastically mounted	G 630	630
Crankshaft drives, inherently unbalanced, rigidly mounted	G 250	250
Complete reciprocating engines for cars, trucks and locomotives	G 100	100
Cars: wheels, wheel rims, wheel sets, drive shafts	G 40	40
Crankshaft drives, inherently balanced, elastically mounted		
Agricultural machinery	G 16	16
Crankshaft drives, inherently balanced, rigidly mounted		
Crushing machines		
Drive shafts (cardan shafts, propeller shafts)		
Aircraft gas turbines	G 6.3	6.3
Centrifuges (separators, decanters)		
Electric motors and generators (of at least 80 mm shaft height) of maximum rated speed up to 950 min^{-1}		
Electric motors of shaft heights below 80 mm		
Fans		
Gears		
Machinery general		
Machine tools		
Paper machines		
Process plant machines		
Pumps		
Turbo chargers		
Water turbines		
Compressors	G 2.5	2.5
Computer drives		
Electric motors and generators (of at least 80 mm shaft height) of maximum rated speeds above 950 min^{-1}		
Gas turbines and steam turbines		
Machine-tool drives		
Textile machines		

(continued)

[5] ISO 21940—Part 11 uses a comma to separate the integers from the decimals.

Table 7.1 (continued)

Machinery type: General examples	Balance quality grade G[5]	Magnitude $e_{per} \cdot \Omega$ mm/s
Audio and video drives	G 1	1
Grinding machine drives		
Gyroscopes	G 0.4	0.4
Spindles and drives of high-precision systems		

Notes

1. Typically, completely assembled rotors are classified here. Depending on the particular application, the next higher or lower grade may be used. For components see Sect. 7.7.
2. All item listed are rotating if not otherwise mentioned (reciprocating) or self-evident (e.g. crankshaft drives).
3. For some additional information on the chosen balance quality grade, see Fig. 7.3, which contains generally used areas (service speed and balance quality grade G) based on common experience.
4. For some machine-specific International Standards stating unbalance tolerances exist (Author's comment: they do not always base on ISO 21940-11).
5. The selection of a balancing quality grade G for a machine type requires due consideration of the expected duty of the rotor when installed in-situ, which typically reduces the grade to a lower level if lower vibration magnitudes are required in service.
6. The shaft height of a machine without feet, or a machine with raised feet, or any vertical machine, is to be taken as the shaft height of a machine in the same base frame, but of the horizontal shaft foot-mounted type. When the frame is unknown, half the machine diameter should be used.

This classification into balance quality grade G represents a recommendation based on previously obtained experience. If these standard values are complied with, satisfactory smooth running *in service* can be expected in all probability. It is conceivable that this list will be supplemented or amended as new rotor systems emerge or new aspects for classification arise.

Figure 7.3 shows the progression of the balance quality grade with the speed. The light field represents the range where general experience in the application is available (see note 3 in Sect. 7.4.2.1).

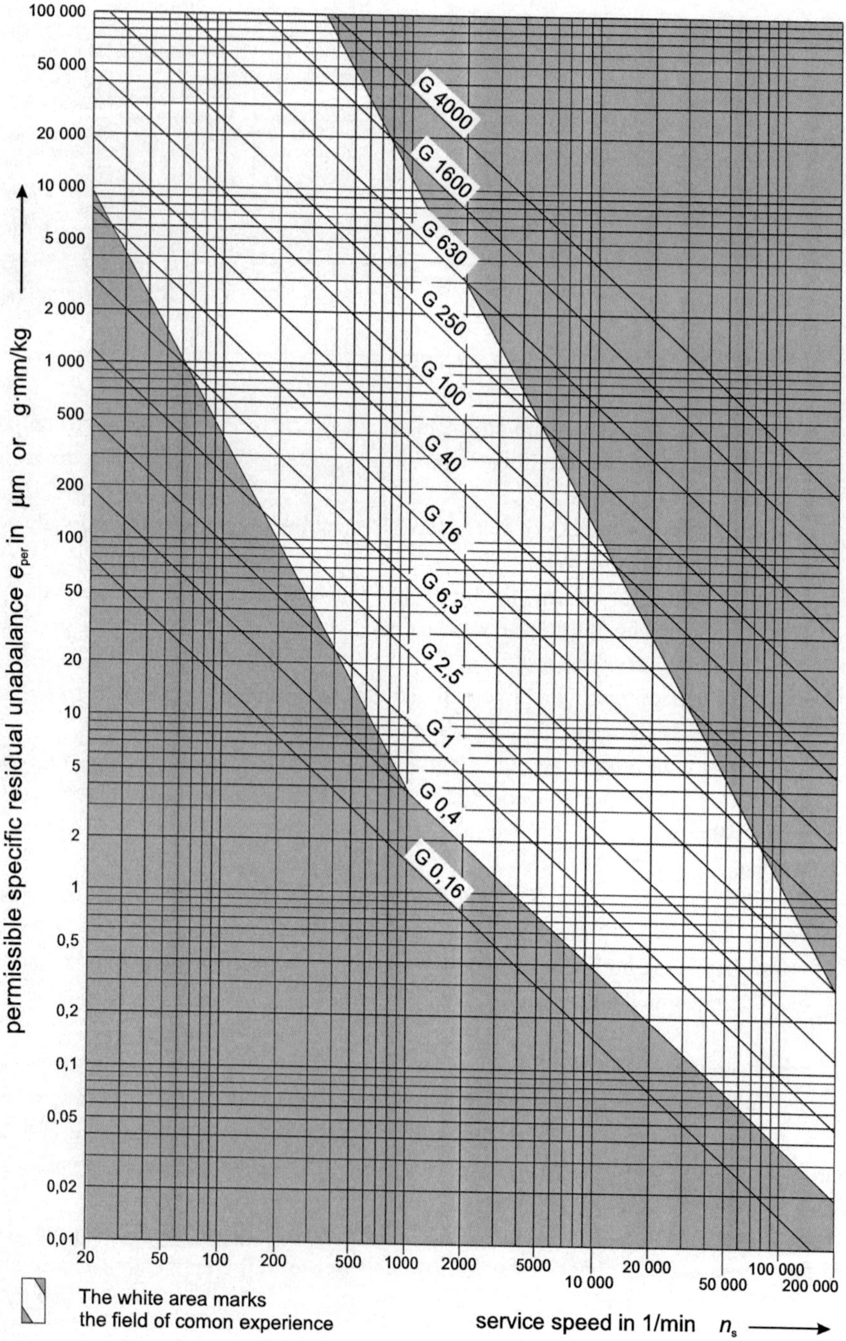

Fig. 7.3 Permissible residual specific unbalance as a function of the maximum service speed for different balance quality grades G (similar to ISO 21940-11, Fig. 2)

Question

What is the permissible residual unbalance e_{per} for quality grade G 6.3 at service speed $n = 3000\ \text{min}^{-1}$?

Solution

Find $3000\ \text{min}^{-1}$ on the speed axis (horizontal), go vertically up to the line G 6.3, from there horizontally to the left to the e_{per} axis and read there: $e_{\text{per}} \approx 20\ \mu\text{m}$ (or $20\ \text{g} \cdot \text{mm/kg}$).

7.4.2.2 Special Cases

The balance quality grade assumes a typical machine design where the rotor mass represents a certain part of the complete machine. If a design deviates clearly from this assumption, appropriate adjustments are necessary.

Practical Examples

Electric motors with a shaft height below 80 mm are classified in balance quality grade G 6.3; from there the permissible unbalance can be derived, see Sect. 7.4.2.1. This value of the permissible unbalance is based on the assumption that the rotor mass represents a typical percentage of the machine mass, e.g. 30%.

- With comparatively ***light*** rotors, e.g. ironless DC armatures, it can happen that the rotor mass reaches only a much smaller percentage. At 10% of the total mass, 3 times the value of the above-mentioned permissible unbalance may then be permitted (if the bearing arrangement is suitable for that load).
- In contrast, the proportion of rotor mass can be extremely ***large***, e.g. in the case of a motor with an external rotor. If this has e.g. 90% of the total mass, only 1/3 of the normal permissible unbalance may be permitted.

7.4.2.3 Permissible Residual Unbalance

The value e_{per} in the example for Sect. 7.4.2.1 naturally can also be calculated: G 6.3 means a permissible orbital velocity of the mass centre ($v_{\text{per}} = e_{\text{per}}\,\Omega$) of 6.3 mm/s. Then:

$$e_{per} = \frac{v_{per}}{\Omega} \approx \frac{6.3}{300} = 0.021\ \text{mm}, \qquad \text{or} \quad 21\ \mu\text{m} \tag{7.4}$$

The permissible unbalance U_{per} is then obtained (Eqs. 7.2 and 7.4) to:

$$U_{per} = e_{per}\, m = \frac{v_{per}}{\Omega} m \qquad \text{g} \cdot \text{mm} \tag{7.5}$$

Question

What is the permissible residual unbalance U_{per} for a rotor of $m = 125$ kg at $e_{per} = 21$ μm?

Solution

$$U_{per} = e_{per} m = 21 \cdot 125 \approx 2600 \text{ g} \cdot \text{mm}$$

(The result is rounded: a more precise specification of the unbalance tolerance would make no sense).

7.4.3 Experimental Determination

For large-scale products, it can be useful to determine the required balancing quality experimentally in order to optimise the balancing process. The measurements are usually carried out in the service condition.

In order to determine the actual permissible limit value for a particular rotor, this rotor is first balanced as well as possible (approximately to 1/10 to 1/20 of the recommended standard value). Then test unbalances of increasing magnitude are applied to the rotor until the influence of the unbalance begins to stand out from the level of the other disturbances in the service state, i.e. until this unbalance has a noticeable influence on the vibrational state, the running smoothness or the function of the machine.[6]

If the rotor is balanced in two planes, dynamic test unbalance in two planes or static unbalance and moment unbalance should be used for the test.

For dynamic test unbalances the tolerance planes (see Sect. 7.2.1) should be used. If this is not possible, the different effects of a static unbalance compared to a couple unbalance must be taken into account.

Furthermore, the limit value must be set in such a way that the changes in the unbalance condition to be expected during service can still be tolerated.

7.4.4 Limits from Specific Targets

7.4.4.1 Limitation by Bearing Forces

If the objective of balancing is primarily to limit the unbalance-related bearing forces, the determination of the permissible unbalance can be based on this objective. Once the

[6] Other questions—e.g. how the service life depends on the unbalance tolerance—usually cannot be answered on the short term, but require long-term tests.

permissible bearing forces have been determined, they must be converted into unbalances. In the simplest case—with a sufficiently rigidly mounted rotor—the permissible unbalance is determined via the centrifugal force:

$$U_{per\,A} = F_A/\Omega^2; \quad U_{per\,B} = F_B/\Omega^2 \qquad g \cdot mm \tag{7.6}$$

With:

$U_{per\,A}$ Permissible residual unbalance at bearing A
$U_{per\,B}$ Permissible residual unbalance at bearing B
F_A Permissible bearing force at bearing A caused by the unbalance
F_B Permissible bearing force caused by the unbalance at bearing B
Ω Angular velocity at (highest) service speed

Assumption

The permissible unbalance-related bearing forces for the rotor in Fig. 7.4 are:

Permissible force on bearing A: $F_A = 2000$ N; permissible force on bearing B: $F_B = 1200$ N.

Question

What are the permissible unbalances for service speed $n = 3000\ \text{min}^{-1}$?

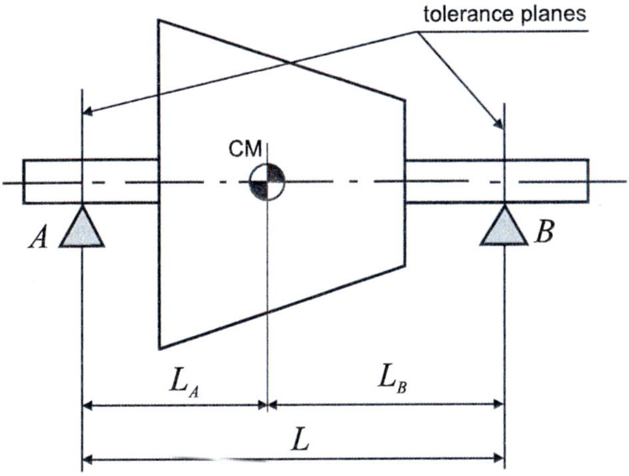

Fig. 7.4 Inboard rotor: the mass centre lies (asymmetrically) between the bearings

Solution

With Eq. 7.6 the permissible residual unbalances in the bearing planes are:

$$U_{per\,A} = \frac{F_A}{\Omega^2} = \frac{2000}{314.2^2} = 20.3 \cdot 10^{-3}\ \text{kg} \cdot \text{m} = 20.3 \cdot 10^3\ \text{g} \cdot \text{mm}$$

$$U_{per\,B} = \frac{F_B}{\Omega^2} = \frac{1200}{314.2^2} = 12.2 \cdot 10^{-3}\ \text{kg} \cdot \text{m} = 12.2 \cdot 10^3\ \text{g} \cdot \text{mm}$$

In more complicated cases, the dynamic behaviour of the system must be sufficiently taken into account.

7.4.4.2 Limitation Through Vibrations

Sometimes the aim of balancing is mainly to limit the unbalance-induced vibrations in certain planes, e.g. in hand-held machines.

ISO 21940-11 only provides general information on this, as there is no reliable material for simple considerations, and more complex approaches require a modelling effort that cannot be presented within the framework of a standard—and also not in this book.

7.4.5 Proven Experience

If a company has sufficient experience in determining the balancing quality of its products, it can build on this experience. If the results are well documented and the balancing objective is still the same, the method can be applied to other rotors. For a few cases, suggestions are made for handling.

7.4.5.1 Almost Identical Rotor Size

The *same* limit values can be set for similarly situated correction planes as for almost identical rotors that have already been successfully balanced.

7.4.5.2 Similar Rotor Size

ISO 21940-11 mentions two different ways of deriving permissible unbalance values: Extrapolation and calculation.

Extrapolation The dependence of the unbalance tolerances on data of known rotors (e.g. diameter, package length, mass, power) can be displayed graphically. If there is a good correlation to one of the data, the required unbalance tolerance for a new rotor size can be derived from such a diagram.

Different diagrams are usually required for different rotor types.

Calculation If new rotors are similar to those already balanced, the similarity laws regarding mass and speed (Sect. 7.3) can be used, according to Eq. 7.7:

$$U_{per\ new} = U_{per\ known} \frac{m_{new}}{m_{known}} \cdot \frac{n_{known}}{n_{new}} \qquad g \cdot mm \tag{7.7}$$

7.5 Allocation to Tolerance Planes

With the various possibilities to define the permissible unbalance, in many cases the tolerance planes can be used directly, so that no further action is necessary.

But especially with quality levels G, a total permissible unbalance is first determined, which then has to be allocated if necessary.

7.5.1 Rotors with a Single Tolerance Plane

For some rotors with rigid behaviour, only the ***resulting unbalance*** is out of tolerance, not the couple unbalance.

This case typically occurs with disc-shaped rotors when the following conditions are met:

- The bearing distance is sufficiently large.
- The disc is sufficiently perpendicular to the shaft axis (i.e. has a sufficiently small axial run-out).
- The correction plane for the resulting unbalance can be suitably selected so that the couple unbalance remains small enough.

7.5.1.1 Practical Review

Whether these conditions are fulfilled can be examined in the practical case:

After a larger number of rotors of the type of interest have been balanced in one (well-chosen) plane, the largest remaining moment unbalance is determined and divided by the bearing distance. If the unbalance of this couple unbalance C is not greater than half the permissible residual unbalance U_{per}, even in the worst case, then a single-plane balancing is normally sufficient. The complete permissible value of U_{per} may be present in this plane.

7.5.1.2 Acceptance

A fan type of 20 kg mass is to be balanced to $e_{per} = 40$ g · mm/kg. The bearing distance is L = 800 mm. After balancing in one plane, the moment unbalance is checked on a larger number and a maximum value $P = 240{,}000$ g · mm^2 is determined.

Question

Is balancing in *one* plane sufficient?

Solution

The permissible unbalance is: $U_{per} = e_{per}\, m = 40 \cdot 20 = 800\ \text{g} \cdot \text{mm}$.

The magnitude of the unbalances in the bearing planes (the moment unbalance, related to the bearing distance) is:

$$U_{A,B} = \frac{U_m}{L} = \frac{240{,}000}{800} = 300\ \text{g} \cdot \text{mm}, \quad \text{this means:}\ U_{A,B} \le \frac{U_{per}}{2},$$

i.e. balancing in *one* plane is probably sufficient, the 800 g · mm determined can be allowed in this single plane.

The magnitude of the moment unbalance depends on the position of the *single* correction plane (see Sect. 5.2). If there are several planes to choose from, it must be determined experimentally for which planes the remaining moment unbalance is typically sufficiently small.

7.5.2 Rotors with Two Tolerance Planes

If a rotor with rigid behaviour does not fulfil the conditions of Sect. 7.5.1 the moment unbalance must also be corrected. For this purpose, the resulting unbalance and the moment unbalance are usually combined to form a dynamic unbalance, i.e. two unbalances in two planes (called complementary unbalances). The tolerance planes are special reference planes, or simplified the bearing planes (see Sect. 7.2.1).

However, the permissible unbalance is usually determined for another plane first: From the quality grade G, one obtains a permissible unbalance for the entire rotor in the plane of the mass centre.

The permissible unbalance in the mass centre plane must therefore be allocated to the tolerance planes (bearing planes *A* and *B*). This is done according to the mass distribution with the help of the leverage laws, where *L* is the distance of the bearing planes, L_A and L_B are the distances of the bearings from the mass centre:

$$U_{per\,A} = U_{per}\,\frac{L_B}{L} \qquad \text{g} \cdot \text{mm} \tag{7.8}$$

$$U_{per\,B} = U_{per}\,\frac{L_A}{L} \qquad \text{g} \cdot \text{mm} \tag{7.9}$$

For a rotor with mass centre ***between*** the bearings (inboard rotor), see Fig. 7.4, for a rotor with an ***overhung*** mass centre (outboard rotor), see Fig. 7.5.

7.5.2.1 Restrictions on Inboard Rotors

If in Fig. 7.4 the mass centre is near a tolerance plane (a bearing plane), the tolerance calculated for this plane is very large, but the value for the other tolerance plane is very small, it can approach zero (Eqs. 7.8 and 7.9). However, this would only be correct with regard to the static unbalance—a moment unbalance would result in equal tolerance values in both bearings.

In order to avoid such an extreme distribution—one would not be able to reliably achieve the smaller value and would also disregard the influence of a moment unbalance—it is stipulated that the:

- Larger value shall not be larger than 0.7 U_{per}.
- Smaller value shall not be smaller than 0.3 U_{per}.

Question

How is the permissible residual unbalance of an inboard rotor with asymmetrical position of the mass centre to the bearing planes with $L=750$ mm, $L_A=150$ mm and $L_B=600$ mm to be allocated to the bearings A and B?

Solution

$$U_{per\,A} = U_{per}\frac{600}{750} = 0.8\,U_{per}; \qquad U_{per\,B} = U_{per}\frac{150}{750} = 0.2\,U_{per}$$

Since these values are outside the limits considered reasonable, revisions are made, i.e. the above mentioned limit values apply in both cases:

$$U_{per\,A} = 0.7\,U_{per}; \qquad U_{zul\,B} = 0.3\,U_{zul}$$

7.5.2.2 Restrictions on Outboard Rotors

If in Fig. 7.5 the mass centre is near the right tolerance plane (bearing plane), the value for the left tolerance plane is very small, it can approach zero (Eqs. 7.8 and 7.9). However, this would only be correct in relation to the static unbalance—a moment unbalance would result in equal values in both bearings.

In order to avoid such an extreme distribution—one would not be able to reliably achieve the smaller value and would also disregard the influence of a moment unbalance—the:

- Larger value shall not be larger than 1.3 U_{per}.
- Smaller value shall not be smaller than 0.3 U_{per}.

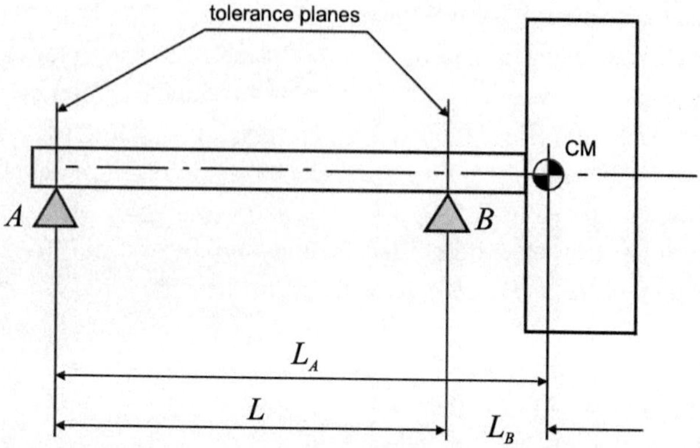

Fig. 7.5 Outboard rotor: the mass centre is in an overhung position

The upper limit value is defined differently here than for the inboard rotor for the following reason: It is assumed that bearing *B* and its components are designed to absorb the higher static load due to the overhung mass. Then they can presumably also bear correspondingly higher dynamic loads due to unbalances.

Question

How is the permissible residual unbalance of an outboard rotor with a position of the mass centre to the bearing planes with $L = 700$ mm, $L_A = 900$ mm and $L_B = 200$ mm to be distributed to the bearings *A* and *B*?

Solution

$$U_{per\,A} = U_{per}\frac{200}{700} = 0.28\ U_{per}; \qquad U_{per\,B} = U_{per}\frac{900}{700} = 1.28\ U_{per}$$

Bearing A is revised to:$U_{per\,A} = 0.3\ U_{per}$, the value for B remains.

Note

If the above assumption about the load capacity of the bearings is not correct, the limit values for inboard rotors should also be used for outboard rotors.

7.6 Assignment of the Unbalance Tolerance to the Correction Planes

ISO 21940-11 (2016) strongly recommends using special reference planes for the *determination of unbalance tolerances* (see Sects. 7.2.1 and 7.5) and no longer the previously used correction planes.

If, nevertheless, correction planes are to be used to state the tolerances, the single-plane case and the two-plane case are to be considered separately.

7.6.1 Single-Plane Case

When correcting in a ***single*** plane, the total permissible unbalance U_{per} can be allowed for this plane.

Note

For the verification of the moment unbalance, according to Sect. 7.5.1 both tolerance planes must still be used as a basis.

7.6.2 Two-Plane Case

When balancing in ***two*** planes, ISO 21940-11 today only considers cases where the balancing planes are close to the bearing planes.

In these relatively simple cases, it is recommended for the transmission of the permissible unbalances:

- In case of correction planes between the bearing planes, allow the same value as in the respective adjacent bearing, Fig. 7.6.
- For correction planes (with distance b) outside the bearing planes (distance L) use the value of the respective adjacent bearing multiplied by the factor L/b, see Fig. 7.7.

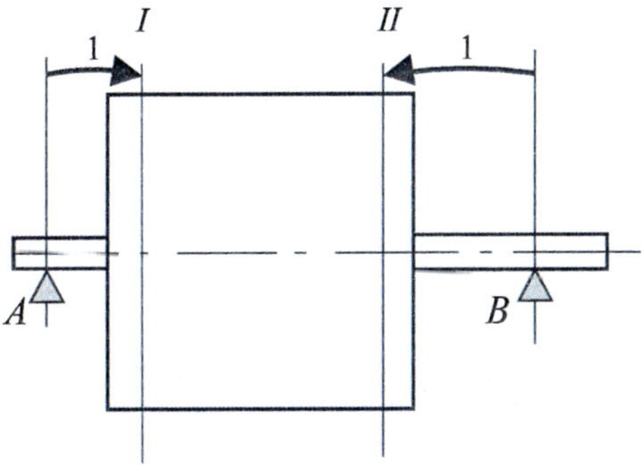

Fig. 7.6 Permissible unbalances from tolerance planes *A* and *B* are converted to the inboard and nearby correction planes *I* and *II* with the factor 1

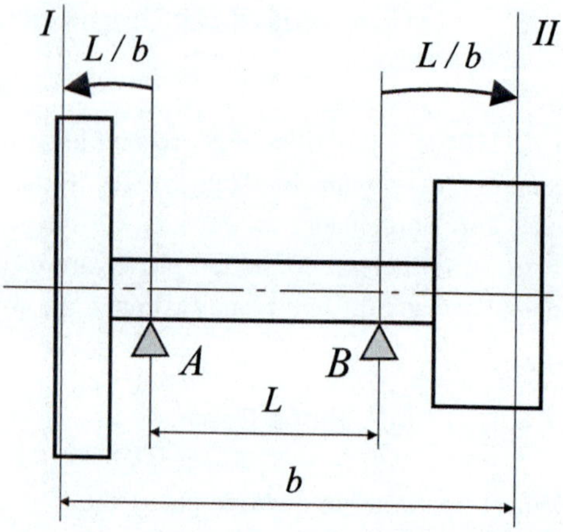

Fig. 7.7 Permissible unbalances from tolerance planes *A* and *B* are converted to the outboard and nearby correction planes *I* and *II* with the factor *L*/*b*

7.7 Assembled Rotors

Rotors assembled from individual parts and/or subassemblies can be balanced as a whole, or by balancing individual parts and subassemblies individually. During assembly, the unbalances of the individual parts are superimposed and additional unbalances occur due to assembly errors—e.g. due to radial and axial run-out errors and play (see Sect. 9.3).

If components are balanced individually, ISO 21940-11 points out:

1. Normally, the same tolerance value (permissible related residual unbalance e_{per}) is used for all components. However, if assembly errors have a visible negative effect, the permissible residual specific unbalance of the individual parts must be correspondingly smaller than that of the rotor.
2. If this does not make sense for all parts—e.g. a light fan on a heavy electric motor—then the unbalance tolerance may be distributed as desired as long as the total unbalance of the assembly remains in tolerance (see example in Sect. 9.3).
3. The manufacturer and the customer must agree on the condition of the rotor to which the tolerance specification refers, e.g. which connecting elements are included in the assembly (for keys see Sect. 9.4).

If the unbalance tolerance cannot be ensured by balancing the individual parts and subassemblies, it is necessary to correct the completed assembly. If it still makes sense to also balance the individual parts first, there are unbalance tolerances for the assembled rotor *and* (possibly others) for the individual parts.

7.8 Unbalance Readings for the Balancing Process

As explained in Sect. 4.8 and mentioned in Sect. 7.2.3, the tolerances described so far (e.g. U_{per}) are intended for ***delivery*** of the rotor (the machine). For the ***balancing process,*** the permissible unbalance reading must be reduced by the errors ΔU (which occurs during balancing compared to delivery): errors when measuring the unbalance and due to a different constellation of the rotor.

$$U_{reading} = U_{per} - \Delta U \qquad g \cdot mm \tag{7.10}$$

ISO 21940-14:2012 describes in general terms the errors that can occur during balancing and—for a number of common problems—their systematic reduction to acceptable, or unavoidable, residuals (see Sect. 16.4).

Furthermore, a method is explained to establish error limits of the balancing machine. The remaining errors determine which measured values are permissible at the end of the balancing process, but also whether the process—as intended—can function at all.

7.8.1 Example

For the outboard rotor in Sect. 7.5.2.2 (Fig. 7.5), a total permissible unbalance U_{per} was derived from the quality grade and a permissible unbalance—from the mass centre position—in bearing A: $U_{per\ A}=0.3U_{per}$, in bearing B: $U_{per\ B}=1.3U_{per}$. These tolerances apply to the delivery.

Assumption

The errors caused by the balancing machine amount to $0.1U_{per}$ per tolerance plane. In addition, there is an error due to the cardan shaft drive of $0.2U_{per}$ in plane A.

Question

What are the permissible unbalance readings for the balancing process?

Solution

The combined error in plane A is $\Delta U_A=0.3U_{per}$. This value is exactly as large as $U_{per\ A}$, i.e. in plane A $U_{reading\ A}=0U_{per}$ remains for the reading of the residual unbalance.

Conclusion
It doesn't work like that.

Assumption

However, if the rotor is turned (around the vertical axis), the permissible unbalance readings in the bearings are reversed (bearing A is still on the left hand).

Solution

This results in the following requirement for the delivery: permissible unbalances in bearing *A*: $U_{\text{per A}} = 1.3U_{\text{per}}$, in bearing *B*: $U_{\text{per B}} = 0.3U_{\text{per}}$.

With the same errors ($\Delta U_A = 0.3U_{\text{per}}$, $\Delta U_B = 0.1U_{\text{per}}$), this then results in: in plane *A* $U_{\text{reading A}} = 1.0U_{\text{per}}$, in plane *B*, $U_{\text{reading B}} = 0.2U_{\text{per}}$ remain for the indicated residual unbalances.

Conclusion
Feasible.

The permissible unbalance readings for the balancing process are thus derived from the tolerances specified for the delivery, but beyond that they depend on the balancing process itself. As a result, they can only be planned in conjunction with a balancing machine, its respective use and all the boundary conditions associated with the process.

At the same time, this planning process can uncover mistakes using the balancing machine and its accessories, making the balancing process unnecessarily difficult or unreliable.

Conclusion
Good planning of the ***balancing process*** and its permissible unbalance readings guarantees the achievement of the balancing target for the ***delivery.***

7.9 Checking the Residual Unbalance

After balancing, the residual unbalance reading is checked for quality assurance purposes to ensure that the permissible unbalances for *delivery* are complied with. Since every measurement contains errors, these errors must also be taken into account appropriately when checking the residual unbalance.

According to ISO 21940-14 the *amount of uncorrected combined error* for the rotor ΔU is formed[7] (see Sect. 16.4).

ISO 21940-11:2016 adopts this approach—but emphasises more clearly the ***intentional*** errors of the rotor configuration during balancing—and formulates the corresponding requirements, this time, however, distributed over the two tolerance planes (bearing planes) A and B. They are more far-reaching in application than the example in Sect. 7.8, see Sect. 4.8.

7.9.1 Acceptance Criteria

When determining the *permissible indications* when checking the residual unbalance, the following quantities are used:

$U_{\text{per A}}$	Amount of the permissible residual unbalance in plane *A*
$U_{\text{per B}}$	Amount of the permissible residual unbalance in plane *B*
$U_{\text{reading A}}$	Amount of the unbalance reading in plane *A*
$U_{\text{reading B}}$	Amount of the unbalance reading in plane *B*
ΔU_{A}	Amount of the combined error in plane *A*
ΔU_{B}	Amount of the combined error in plane *B*

The amounts of the combined errors ΔU_{A} and ΔU_{B} can differ—for ***one and the same*** rotor—even for identical machines. Considerable differences can occur with different machine designs and deviating boundary conditions.

If ΔU_{A} or ΔU_{B} is less than 10% of $U_{\text{per A}}$ or $U_{\text{per B}}$, the respective error may be disregarded.

Previously, different rules applied to the check following balancing and a separate recheck: this has been superseded by the new version of Part 11, see Sects. 7.9.2 to 7.9.4.[8]

7.9.2 Unbalance Readings in Tolerance

The unbalance is clearly ***within the tolerance,*** i.e. does not exceed the specified tolerance U_{per}, if for the unbalance readings $U_{\text{reading A}}$ and $U_{\text{reading B}}$ ***both*** Eqs. 7.11 and 7.12 hold true:

$$U_{reading\,\text{A}} \leq U_{per\,\text{A}} - \Delta U_{\text{A}} \quad \text{g} \cdot \text{mm} \tag{7.11}$$

[7] ISO 21940-14:2014 describes in detail different possibilities to estimate the combined error. The term *measuring planes* used in this context is to be understood as the tolerance planes (usually the bearing planes).

[8] A revision is in preparation for Part 14.

$$U_{reading\,\mathrm{B}} \leq U_{per\,\mathrm{B}} - \Delta U_{\mathrm{B}} \quad \mathrm{g} \cdot \mathrm{mm} \tag{7.12}$$

The size of both ΔU must therefore be limited to ensure a reliable balancing process.

7.9.3 Unbalance Readings Outside Tolerance

The unbalance is clearly ***out of tolerance,*** i.e. exceeds the specified tolerance U_{per}, if for the unbalance readings $U_{\mathrm{reading\ A}}$ and $U_{\mathrm{reading\ B}}$ at least ***one*** of the two Eqs. 7.13 and 7.14 holds true:

$$U_{reading\,\mathrm{A}} > U_{per\,\mathrm{A}} + \Delta U_{\mathrm{A}} \quad \mathrm{g} \cdot \mathrm{mm} \tag{7.13}$$

$$U_{reading\,\mathrm{B}} > U_{per\,\mathrm{B}} + \Delta U_{\mathrm{B}} \quad \mathrm{g} \cdot \mathrm{mm} \tag{7.14}$$

7.9.4 Region of Uncertainty

The field between *within the tolerance* and *outside the tolerance* is the ***region of uncertainty.***

The region of uncertainty is given by ***both*** Eqs. 7.15 and 7.16:

$$U_{per\,\mathrm{A}} - \Delta U_{\mathrm{A}} < U_{reading\,\mathrm{A}} \leq U_{per\,\mathrm{A}} + \Delta U_{\mathrm{A}} \quad \mathrm{g} \cdot \mathrm{mm} \tag{7.15}$$

$$U_{per\,\mathrm{B}} - \Delta U_{\mathrm{B}} < U_{reading\,\mathrm{B}} \leq U_{per\,\mathrm{B}} + \Delta U_{\mathrm{B}} \quad \mathrm{g} \cdot \mathrm{mm} \tag{7.16}$$

Here, too, it can be seen that the combined errors ΔU_{A} and ΔU_{B} must be kept within narrow limits, in order to minimise this region of uncertainty.

7.9.5 Particularities when Measuring Unbalances

Is there a special situation when balancing is performed?

Unfortunately, the various standards do not yet provide any information on this. But at least some particularities can be mentioned.

Particularities when measuring unbalances:

1. There is no generally stated and accepted standard for unbalances.
2. Most permissible residual unbalances *could be* measured on a balancing machine with errors of 5 to 10%.[9] For economic reasons, however, this capability is often not used, e.g. by applying lower balancing speeds than would be optimal.

[9] These percentages are apparently generous. One must bear in mind that only the tolerance field is to be measured here, and that the next quality grade is far above that (at 250%), or with the finest gradation still at 160%.

3. Added to this are the errors caused by the boundary conditions during balancing (see Sect. 4.8), which usually are much greater.
4. In almost all balancing machines, measuring and processing the unbalance coincide, i.e. the unbalance is corrected on the basis of the measured values.
5. Measuring the residual unbalance is usually done after balancing in the same balancing machine, sometimes later—for checking—in another one.
6. There are neither standardised regulations nor quality levels for checking unbalances.

7.9.6 Checking on a Balancing Machine

ISO 21940-21 specifies two characteristic values of the balancing machine; the minimum achievable residual unbalance U_{mar} and the unbalance reduction ratio R_{UR} (see Sects. 13.4 and 13.5).

If these two characteristic values correspond to the task, the residual unbalance can be measured directly on the balancing machine. Of course, the influences of the boundary conditions during balancing are added for the combined error.

The conditions in Sects. 7.9.2 to 7.9.4 still apply.

7.9.7 Checking Without a Balancing Machine

The residual unbalance can also be determined without a balancing machine, e.g. at the installation site with means and procedures used for in-situ balancing (see Chap. 18).

This eliminates the errors that occur during the balancing process due to other boundary conditions. However, the service condition can also deviate from the delivery condition: these influences must be thoroughly considered.

The errors caused by the measuring process are also not easy to assess in this case. In addition to the errors of the measuring device in the specific task, they include the individual errors when setting the test unbalances and the influence of the scatter from run to run.

8 Tolerances for Rotors with Flexible Behaviour

Contents

H. Schneider, *Rotor Balancing*, https://doi.org/10.1007/978-3-662-66049-2_8

8.1 Preliminary Note

The term *flexible behaviour* is now again understood as an umbrella term by the relevant ISO committee and was also used in this sense when restructuring all balancing standards as parts of ISO 21940.

8.2 General

Today, the following rotor behaviours are summarized by the term ***flexible behaviour***, see Table 4.3:

1. ***Shaft-elastic behaviour,*** see Sect. 6.4,
2. ***Component-elastic behaviour,*** see Sect. 6.3
3. ***Settling behaviour,*** see Sect. 6.2.

8.2.1 Balancing Target

When balancing rotors with flexible behaviour, the aim is to achieve acceptable values for *all* effects generated by the unbalances of the rotor, the vibrations, the forces, the deflections.

There is a whole series of standards for the measurement and evaluation of vibrations in the *service state,* see Sect. 18.1. These are mainly measurements on previously balanced rotors.

Rotors can also be balanced in service condition, but for rotors with flexible behaviour this can be very expensive, sometimes it is even impossible to realise. That is why it is only very rarely used in production. The following is therefore about tolerances during balancing in a *balancing machine* or a *balancing system.*

8.2.2 Balancing Procedures

ISO 21940—Part 12 deals with all three types of behaviour (Sect. 8.2) and assigns different procedures to them, see Sect. 10.4:

- Rotors with ***shaft-elastic*** behaviour the balancing procedures A to H.
- Rotors with ***component-elastic*** behaviour the balancing procedure I.
- Rotors with ***settling behaviour*** a settling procedure (without a code letter).

The ***shaft-elastic*** behaviour is by far the most important for practice, obvious from the number of possible procedures. Procedures A to F use low speeds, procedures G and H use high speeds, mostly up to the service speed.

The ***component-elastic*** behaviour is basically not a balancing problem, but a problem of design, manufacturing and assembly. For this reason, "correct" balancing is generally not possible; it is usually an attempt to find a compromise over the speed range. Procedure I therefore is a kind of high-speed balancing.

The rotor with ***settling behaviour*** loses this behaviour by performing a settling procedure and then shows one of the other behaviours—from rigid behaviour to component-elastic behaviour. No *balancing procedure* is used here, and no identification letter exists.

Each of these nine procedures has different conditions, so that different rules apply when setting the tolerances.

In unbalance correction, an attempt must *always be made*—even with procedure G, which looks relatively permissive—to correct the unbalances as precisely as possible in *those* planes in which they occur, so that the modal unbalances become small. Whether this endeavour has succeeded can only be determined if the tolerance planes are assumed in these unbalance planes. This situation is thus fundamentally different from rotors with rigid behaviour (see Chap. 5), where modal considerations play no role.

8.3 Tolerance Criteria

As explained in Sect. 8.2.1 this book is limited to tolerances during balancing in a *balancing machine* or a *balancing system.*

Often balancing machines are used for a whole range of rotors—e.g. for rotor weights in the range of 1:20 to 1:200. The data of the bearing support therefore rarely or never match the respective service data of the rotors and lead—for a given unbalance condition—to a different vibration pattern than in the service condition (see Sect. 8.3.1.2).

ISO 21940—Part 12 uses two different tolerance criteria for rotors with shaft-elastic behaviour: vibrations and unbalances.

D. Wiese (see literature, special cases) has pointed out that unbalances (low-speed as well as modal unbalances) are a much more reliable criterion than vibrations—when it comes to assessing a shaft-elastic rotor under different bearing conditions than it runs in service. Nevertheless, for the assessment of the balancing result—to my experience—other criteria are often used by the ***manufacturer*** of the rotors.

These are, in order of incidence:

1. Predominantly: vibrations at service speed.
2. Sometimes: residual unbalances in the correction planes.
3. Rarely: low-speed and equivalent modal residual unbalances.

Up to now, only vibrations have been assumed and accepted as a criterion by most ***customers,*** supported by some ideas from ISO to API.

8.3.1 Vibrations

For many decades, the vibrational state[1] at service speed has been used to assess the smoothness of machine operation—and for its *monitoring*—measurement is taken either:

- On the shaft (ISO 7919 series of standards), e.g. as peak-to-peak value of the vibration displacement *in the bearing.*
- On non-rotating parts (ISO 10816 series of standards), generally broadband as an effective value.

It was therefore obvious to use such vibration limits also for the assessment of rotors in the balancing system.

---

[1]In 2017, the introductory parts of ISO 7919-1 and ISO 10816-1 were revised and published together as ISO 20816-1.

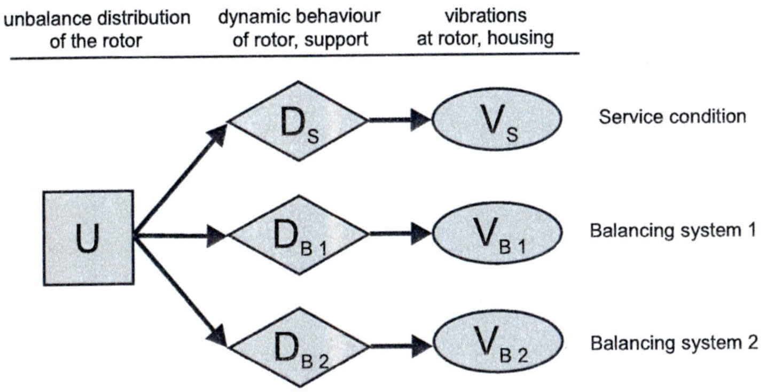

Fig. 8.1 One and the same unbalance condition U results in the non-comparable vibration responses in service V_S, in balancing system V_{B1} and in balancing system V_{B2}

However, the limit values for the service condition cannot be used without further ado for the balancing system, especially because here:

- Necessary measurement is frequency-selective: looking for the unbalance-related part of the vibration.
- Other bearing and/or coupling conditions prevail, so that resonances can be displaced and differently pronounced.
- Other measuring planes need to be used.

However, the response behaviour of a rotor with flexible behaviour is not only different from the service condition to one balancing system, but also from one balancing system to another. Even if permissible vibrations on *one balancing system* are established to define the tolerances, they cannot be transferred to *another balancing system*, see Fig. 8.1.

8.3.1.1 Vibrations According to ISO 21940-12

ISO 21940—Part 12 attempts to provide assistance for the transfer from vibrations in service condition X to vibrations in a balancing system Y with a chain of factors:

$$Y = X \cdot K_0 \cdot K_1 \cdot K_2 \quad \text{mm/s} \tag{8.1}$$

The factors K_0 and K_1 vary depending on the machine class, whereby the division into four classes was taken from ISO 10816-1.

However, these factors are not sufficient as long as there is insufficient knowledge about the behaviour of the rotors. Therefore this approach is not explained here.

Furthermore, there is a fundamental problem at all with the attempt to use vibrations as a tolerance criterion when balancing rotors with shaft-elastic behaviour, see Sect. 8.3.1.2.

8.3.1.2 Problems with Vibrations

Although vibrations are predominantly used for assessment, the question must be asked: Are vibrations suitable as criterion in a balancing system at all?

In practice, for shaft-elastic rotors, measurements are usually taken at service speed on two bearings; ISO 21940-12 does not make any suggestions beyond this. In my view, however, such a measurement cannot provide a clear statement on the balancing quality, because for each ***measured vibration state,*** an infinite number of ***unbalance states*** are conceivable.

The following consideration may help to clarify: The unbalance-related vibrations of a rotor—i.e. those that can be changed by the balancing process—are caused by various unbalances in a shaft-elastic rotor.

Example

- Resulting unbalance.
- Moment unbalance.
- 1st modal unbalance.
- 2nd modal unbalance.
- Possibly further modal unbalances. ◀

Each of these vibration vectors has a different characteristic curve (amplitude, phase) over the rotational speed of the rotor. Fig. 8.2 shows typical amplitude curves of five unbalances.

In Fig. 8.2 a cumulative curve is also drawn as an example. However, it is not *generally* valid, because the vibration components add up vectorially for each speed. Since each of the unbalances can have an infinite number of different angles (not possible to show this here because of the flat representation), this cumulative curve is only one of an *infinite number of* possibilities.

At any speed (i.e. also at the most commonly used service speed), a sum value between zero and the maximum sum of all vibration amplitudes can occur. This also applies if measurements are taken at *two* measuring points—e.g. at both bearings. However, generally neither zero nor the maximum will occur at both bearings at the same time.

Figure 8.3 shows examples of unbalances 1 to 5 for both bearings of a rotor with centre-symmetrical flexural modes (e.g. same mass and stiffness distribution, same support properties). As a result of this assumption, the vibrations of both bearings behave as follows: The *amplitudes* are the same in each case, the *angular positions* of the vibration vectors 1, 3 and 5 are the same, and those of 2 and 4 are offset from each other by 180° for both bearings.

As can be seen, only a moderate vector sums S occurs at both bearings, hardly larger than the individual vectors. However, the maxima of the vector sums S can also be smaller, or up to 4 times as large—in this example, with other angular positions of the individual vectors.

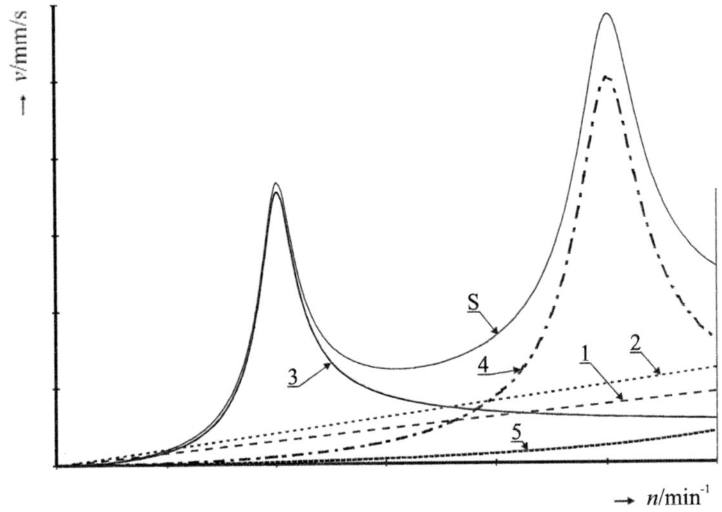

Fig. 8.2 Course of the vibration responses (amplitudes) of five different unbalance types of the example: resultant unbalance **(1)**; moment unbalance **(2)**; 1st modal unbalance **(3)**; 2nd modal unbalance **(4)**; 3rd modal unbalance **(5)**. ***One*** of infinitely many ***possible*** sum curves **(S)**

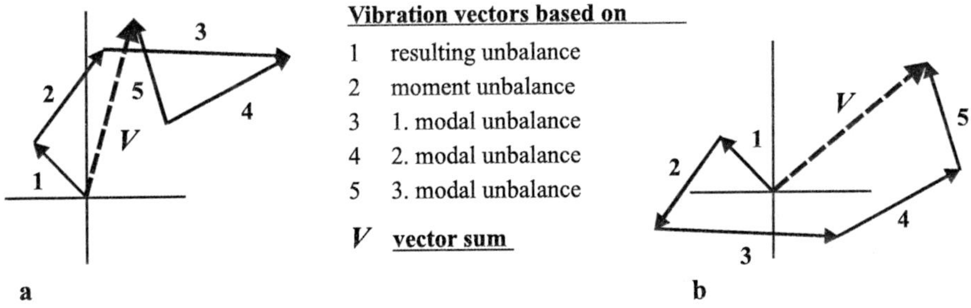

Fig. 8.3 Example of the vector sums of the vibrations of both bearings at service speed: *left* bearing **(a)**; *right* bearing **(b)**. Assumption for better clarity: rotor with equal distribution of mass and stiffness

8.3.1.3 Conclusion

A low vibration level at service speed *can* therefore mean that *all vibration components* are low, but also that only the *vector sums* give small values.

How should it possible to draw conclusions about the balancing quality from this vibration measurement? This deduction is definitely not possible.

That this conclusion is nevertheless is attempted in many cases of balancing shaft-elastic rotors, in my opinion, due to the following reasons:

- This measurement is proven in machine monitoring—but disregarded: this it is a different task and has different boundary conditions.
- Alternatives are not sufficiently known.
- It is assumed that the balancing process just completed was "well executed".

Notes

- If, after the balancing process has been completed, its result is to be ***checked,*** then it must not be assumed that it has been *carried out well.*
- There are more possibilities in the dynamic system than only at service speed because practically all speeds can be used for measurement: One could select other speeds in addition to the service speed, e.g. those used for correction, and include the vibrations measured there in the evaluation.
- There is a reference in ISO 21940-12 that attention should also be paid to the run-up and resonance passages. However, no procedure is mentioned on how to use this information—and the author has never heard of any procedures. ◀

8.3.2 Unbalances

ISO 21940—Part 12 is based on unbalances for all low-speed balancing procedures (A to F). The same applies to the high-speed balancing procedures; for procedures G it is one of two criteria.

However, for procedure G only unbalance tolerances for the various unbalances are dealt with in detail in ISO 21940-12, see Sect. 8.5.1.

8.3.2.1 Total Permissible Unbalance

The quality grades for rotors with rigid behaviour (Sect. 7.4.2) are also used for rotors with flexible behaviour according to ISO 21940—Part 12. The total permissible unbalance U_{per} of a rotor is derived from the quality grade.

It should be noted that:

- For rotors with shaft-elastic behaviour, tolerance specifications are required for one, two or more tolerance planes, sometimes—procedure G—not only for tolerance planes but also for different types of unbalance.
- For rotors with component-elastic behaviour, tolerance specifications for two tolerance planes are required.

8.3.2.2 Tolerance Planes

As explained in Sect. 8.2.2 the tolerance planes and the correction planes coincide in case of flexible rotors. This situation is therefore fundamentally different from rotors with rigid behaviour (see Chap. 5), where the modal unbalances *do not* play a role.

8.3.2.3 Distribution of the Total Permissible Unbalance

For rotors with *flexible* behaviour, the distribution rules depend on the respective task and the balancing procedure used, i.e. they are very different, see Sects. 8.4 and 8.5.

8.3.2.4 Modal Influence on the Permissible Unbalances

In addition to the distribution principles for rotors with *rigid* behaviour, there is also the modal aspect for rotors with *shaft-elastic* behaviour, see Fig. 8.4.

Two boundary conditions must be observed, as detailed in balancing procedure G:

- The distance of the service speed from the flexural speed is to be considered, expressed as a modal weighting factor G_n, see Eq. (4.1), *and*
- The effectiveness of the respective tolerance plane k in relation to the most sensitive plane, expressed as a quotient of the bending arrows $\phi_{n\,max}/\phi_{n\,k}$, taken from Eq. (4.23).

$$U_{per\,k} = U_{per}\,G_{n\,k}\frac{\phi_{n\,max}}{\phi_{n\,k}} \qquad \mathrm{g\cdot mm} \tag{8.2}$$

With:

$U_{per\,k}$	Permissible unbalance in the tolerance plane k
U_{per}	Total permissible unbalance, see Sect. 8.3.2.1
$\phi_{n\,max}$	Max. bending arrow of the nth flexural mode
$\phi_{n\,k}$	Bending arrow of the nth flexural mode in the tolerance plane k

Note

The quotient $\phi_{n\,max}/\phi_{n\,k}$ is the reciprocal of the quotient in Eq. 4.23: There, the equivalent modal unbalance is calculated, here the permissible tolerance—and this is the greater the smaller the effect of the plane is, i.e. the value $\phi_{n\,k}$. ◀

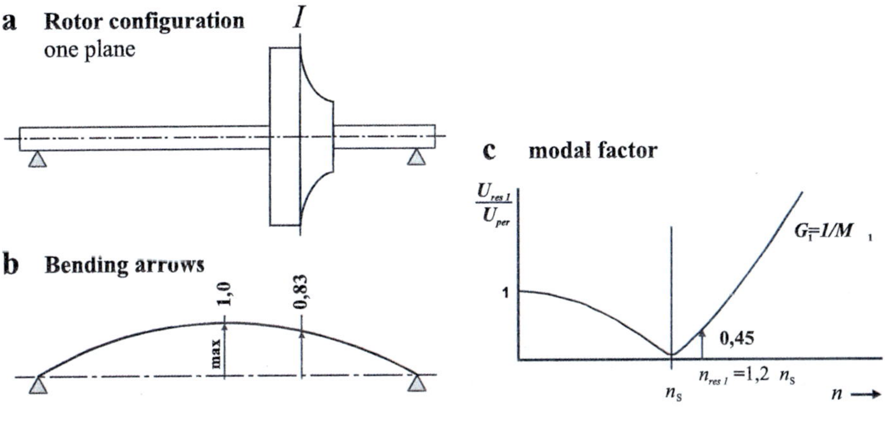

Fig. 8.4 Example of a disc-shaped rotor with *one* correction plane: rotor configuration (**a**); bending arrows (**b**); modal factor (**c**)

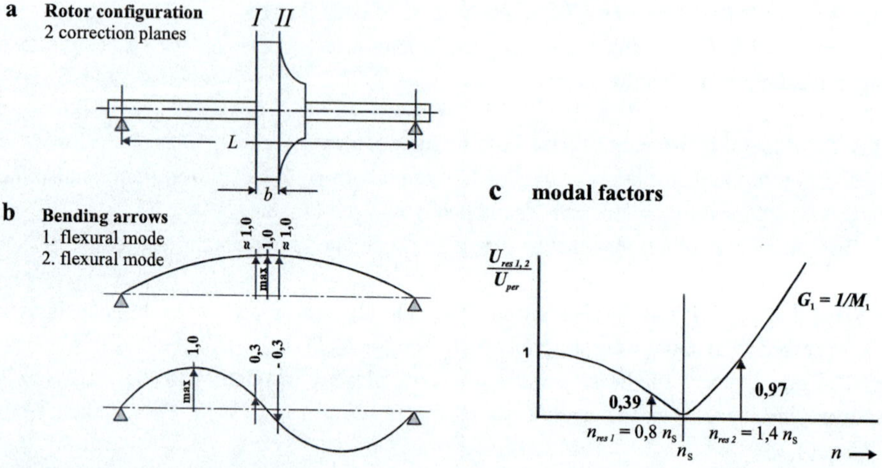

Fig. 8.5 Example of a disc-shaped rotor with *two* balancing planes: Rotor configuration (**a**); bending arrows (**b**); modal factors (**c**)

If not only a single tolerance plane is considered, but two planes together, (Eq. 8.2) is not sufficient:

In the case of a disc, or a rigid section (Sect. 10.3), in addition to the resulting unbalance, the moment unbalance (or the couple unbalance) also have to be taken into account Fig. 8.5.

If more types of unbalance have to be taken into account, the total tolerance is distributed between the types—either equally (e.g. according to Table 8.1), or weighted.

Two-plane case: If we describe both planes individually with (Eq. 8.2), we do not always get the right impression. So we summarise the effect of the *couple unbalance* in the two tolerance planes I and *II*:

$$U_{per\ C} = U_{per}\ G_{n\,k} \frac{\phi_{n\ max}}{\phi_{n\ I} - \phi_{n\ II}} \qquad \text{g} \cdot \text{mm} \tag{8.3}$$

With:

$U_{per\ C}$ Permissible unbalance in the planes of the couple unbalance
U_{per} Total permissible unbalance, see Sect. 8.3.2.1
$\phi_{n\ max}$ Max. Bending arrow of the nth flexural mode
$\phi_{n\ I,II}$ Bending arrows of the nth flexural mode in tolerance planes *I* and *II of* the couple unbalance

Table 8.1 Factor per tolerance plane, depending on their number, equally distributed

Number of tolerance planes	2	3	4	5	6	7	8	9	10
Factor per plane	0,71	0,58	0,50	0,45	0,41	0,38	0,35	0,33	0,32

Here, the difference of the bending arrows is used, which means that for a couple unbalances, the *gradient of the flexure line* is important, in contrast to the resulting unbalance, where the *size of the single bending arrow* counts.

8.4 Unbalance Tolerances for Procedures A to I

ISO 21940—Part 12 distinguishes between low-speed and high-speed procedures. The low-speed procedures are primarily used to treat *some* rotors with *shaft-elastic* behaviour, while the high-speed procedures are used to treat *all flexible* behaviours.

The classification—low-speed/high-speed—is also advantageous for the discussion of unbalance tolerances; unfortunately, the standard does not provide any further design information here.

Therefore, here the **suggestions of the author.**

8.4.1 Tolerances of Low-Speed Balancing Procedures

During low-speed balancing, only resulting unbalances and moment unbalances (or couple unbalances) can be *measured.* All tolerances must therefore be expressed in these unbalances.

Principle

The tolerances is first determined the same way as for the rotor with *rigid* behaviour, see Sect. 7.4, but must also be *modally* weighted, see Sect. 8.3.2.4. The smaller tolerance value is then to be used.

Moreover, such tolerances can occur *several times* on different elements of the rotor, depending on the balancing procedure.

8.4.1.1 Procedure A: Single-Plane Balancing

Procedure A (see Sect. 10.4.1) uses *a single* tolerance plane, the total permissible unbalance U_{per} applies to this plane.

For a rotor with *rigid* behaviour, this statement would be sufficient. For the *shaft-elastic* rotor, modal aspects also play a role.

Example

A rotor with an impeller and *a single* tolerance plane *I* (Fig. 8.4) runs subcritically in service. The bending arrow in the tolerance plane *I* of the impeller is 0.83 of the maximum deflection of the bending line (Fig. 8.4), so the quotient $\phi_{n\,max}/\phi_{n\,k}$ is 1.2. The 1st flexural resonance speed is $n_{1,\,crit}=1.2\;n_S$, the modal damping is $D_1=0.05$ (Fig. 8.4). ◄

Task

What is the permissible unbalance $U_{per\,I}$ in the impeller plane?

Solution

Rigid rotor behaviour: $U_{per\,I}=U_{per}$
Shaft-elastic rotor behaviour:
Using Eq. 4.1: $G_1=0.45$; Eq. 8.2: $U_{per\,I}=U_{per}\cdot 1.2\cdot 0.45=0.54U_{per}$

Conclusion
The modal consideration results in a significantly smaller tolerance: thus the modal tolerance is to be used.

8.4.1.2 Procedure B: Two-Plane Balancing

Procedure B (see Sect. 10.4.2) requires two tolerance planes.

Here, too, the rigid properties *and* the modal properties of the rotor must be taken into account. In contrast to the first example (Sect. 8.4.1.1), *two* flexural modes have to be considered here. Since the resulting unbalance and moment unbalance (couple unbalance) often each have *a* dominant influence on *one* bending mode, it is recommended in this case to divide the total permissible unbalance into these two types of unbalance.

Example

A rotor with *a single* impeller and tolerance planes *I* and *II* (Fig. 8.5) runs above the 1st critical and below the 2nd critical in service. The plane spacing is $b=L/15$.

The 1st flexural speed is at $n_{1,\,crit}=0.8\;n_S$, the modal damping is $D_1=0.1$; and the 2nd flexural speed is at $n_{2,\,crit}=1.4\;n_S$, the modal damping is $D_2=0.05$:

- For the 1st flexural mode, the bending arrows in planes *I* and *II are:* ≈ 1 (Fig. 8.5).
- For the 2nd flexural mode, the bending arrows in planes *I* and *II are:* 0.3 and −0.3; the quotients $\phi_{n\,max}/\phi_{n\,k}$ are therefore 3.3 and −3.3. ◄

Task

How large are the permissible unbalances in the impeller planes?

Solution

Due to the small plane distance, resulting unbalance and moment unbalance are taken as a basis and provided with their own (here equal) tolerances.

Rigid rotor behaviour:

1. Resulting unbalance
   $U_{per\,R} = ½\, U_{per} = 0.5U_{per}$; and distributed to both planes: $U_{per\,I,\,II} = 0.25U_{per}$
2. Couple unbalance
   $U_{per\,C} = ½\, U_{per} \cdot L/b$; in this case $7.5U_{per}$. For each plane, therefore, approximately: $U_{per\,I,\,II} = 3.8U_{per}$

Shaft-elastic rotor behaviour:

1. Resulting unbalance: 1st flexural mode
   Using Eq. (4.1), $G_1 = 0.39$; Eq. (Eq. 8.2) gives: $U_{per\,R} = ½\, U_{per} \cdot 0.39 \cdot 1.0 = U_{per} \cdot 0.20$. Divided between the two planes: $U_{per\,I,\,II} = U_{per} \cdot 0.2 \cdot 0.5 = 0.10U_{per}$.
2. Couple unbalance: 2nd flexural mode
   Using Eq. (4.1), $G_2 = 0.97$; Eq. (8.3) yields:
   $U_{per\,C} = ½\, U_{per} \cdot 0.97\,[3.3 - (-3.3)] = 3.2U_{per}$. This figure is already related to the planes.

Conclusion

The modal considerations result in a significantly smaller tolerance for the resulting unbalance, for the couple unbalance it is only slightly smaller: “modal” is to be taken as a basis for both types of unbalance.

Note

If both flexural modes, excited by the resulting unbalance and the couple unbalance, are more strongly mixed than assumed here (e.g. in the case of clear asymmetry), the permissible unbalances of *both* types of unbalance must be calculated and compared for *both* flexural modes. ◀

8.4.1.3 Procedure C: Balancing Individual Components Before Assembly

Procedure C (see Sect. 10.4.3) requires as many tolerance planes as the individual parts of the rotor have. If the impellers of a 4-stage pump rotor are balanced in two planes each, the rotor will have 8 tolerance planes.

With an even distribution, the total permissible unbalance U_{per} would have to be divided by 8, i.e. per plane $U_{\text{per E}} = U_{\text{per}}/8$. However, such small values can usually no longer be handled reliably—also because of the errors occurring when balancing individual parts.

In this case, one can make use of a rule suggested by DIN ISO 21940—Teil 12, Beiblatt 1[2] for handling a large number of unbalances (Sect. 4.1.4): Assuming that the unbalances in the different planes are not dependent on each other in size and angle, factors for the permissible residual unbalances per plane can be determined with the help of the *sum of the error squares.* Analogously, factors can be calculated here for the individual tolerance planes, see Table 8.1.

These factors result from an equal distribution. If the bending mode shapes indicate that the tolerances should be adjusted in some planes, a modal consideration can be used in addition, according to Sect. 8.3.2.4.

Since the planes of each individual part are usually close together, the moment unbalance would be tolerated much more narrowly than the resultant unbalance if distributed to two planes. It is therefore recommended to specify the permissible residual unbalance per part as resultant unbalance and moment unbalance (couple unbalance).

8.4.1.4 Procedure D: Balancing After Limiting the Starting Unbalance

Procedure D (see Sect. 10.4.4) requires two tolerance planes to measure the ***starting unbalance***[3] and to distributed the total permissible unbalance U_{per} (often evenly).

An additional modal consideration as in Sect. 8.3.2.4 is necessary unless the tolerance planes are close to the nodes of the relevant flexural mode shapes so that they excite the flexural mode shapes only negligibly.

[2] DIN ISO 21940: Mechanische Schwingungen—Auswuchten von Rotoren—Teil 12: Verfahren und Toleranzen für Rotoren mit nachgiebigem Verhalten, Beiblatt 1 (2015): Verfahren zum Auswuchten bei mehreren Drehzahlen. (English: DIN ISO 21940—Part 12: Procedures and tolerances for rotors with flexible behaviour, Supplement 1 (2015): Methods for balancing at multiple speeds).

[3] ISO 21940-12 uses the term "initial unbalance" here. However, this term refers to the condition after the production and assembly of a rotor, *before* balancing. Here, however, the condition is meant *after the* individual parts have already been balanced. I therefore use the term *starting state* for the beginning of this *new* balancing step.

8.4.1.5 Procedure E: Sequential Balancing During Assembly

Procedure E (see Sect. 10.4.5) requires *at least* as many tolerance planes as the individual parts of the rotor would need if they were balanced individually, see example in Sect. 8.4.1.3.

Additional plane requirements can arise due to the assembly. If, for example, two narrow panes would have required only one correction plane, but the assembly results in an inclined position and thus in an additional moment unbalance: two tolerance planes are required for proper correction.

For the distribution to all tolerance planes, the same rules apply as in Sect. 8.4.1.3, also the modal properties shall be taken into account, see Sect. 8.3.2.4.

8.4.1.6 Procedure F: Balancing in Optimal Planes

ISO 21940—Part 12 distinguishes between a basic procedure with two planes and an extended procedure with three planes, see Sect. 10.4.6.

The basic procedure requires two tolerance planes to distribute the total permissible unbalance U_{per} evenly.

The extended procedure requires three tolerance planes. The distribution should be done according to the weighting of the third plane.

In both cases, the modal properties (see Sect. 8.3.2.4) should be taken into account and included in the tolerances.

8.4.2 Tolerances of High-Speed Balancing Procedures

ISO 21940—Part 12 combines three very differently significant procedures here, and the tolerance specifications are accordingly different.

8.4.2.1 Procedure G: Multiple-Speed Balancing

The tolerances for this extremely important procedure (see Sect. 10.4.7) are discussed in Sect. 8.5 in detail.

8.4.2.2 Procedure H: Balancing at Service Speed

Procedure H (see Sect. 10.4.8) requires two tolerance planes and evenly distributes the total permissible unbalance U_{per}.

Sometimes aspects of 8.4.1.6 are taken into account and adapted to three tolerance planes; always the modal properties—resonance distance and effectiveness of the tolerance planes—have to be considered (see Sect. 8.3.2.4).

8.4.2.3 Procedure I: Balancing at a Fixed Speed

Procedure I (see Sect. 10.4.9) requires two tolerance planes and distributes evenly the total permissible unbalance U_{per}.

Here, the modal properties—resonance distance and effectiveness of the tolerance planes—must also be taken into account (see Sect. 8.3.2.4).

8.5 Unbalance Tolerances for Procedure G

8.5.1 Unbalance Tolerances According to ISO 21940-12

ISO 21940-12 recommends to first calculate a total permissible residual unbalance U_{per} for the rotor with shaft-elastic behaviour, just as for a rotor with rigid behaviour.

Starting from this value, the following steps are defined:

- Permissible unbalances for low-speed condition (according to standard: not mandatory).
- Equivalent modal residual unbalances for the different flexural mode shapes.

An overview of the graduations, depending on the number of flexural modes, is shown in Table 8.2.

Annex D of ISO 21940-12 describes the data of a gas turbine and shows a run-up curve, see Fig. 8.6.

Example

Mass 1625 kg; service speed 10,125 min^{-1}; two resonances below service speed; quality grade G 2,5; four correction planes, one in overhung position (IV). ◀

Table 8.2 Overview of recommended unbalance tolerances according to ISO 21940—Part 12

Rotor is influenced by low-speed unbalances and unbalance in the	Permissible residual unbalance [in % U_{per}]			
	Low-speed	1st flexural mode	2nd flexural mode	3rd flexural mode
Single flexural mode	100	60	Not needed	Not needed
Two flexural modes	100	60	60	Not applicable
Three flexural modes	———no details———			

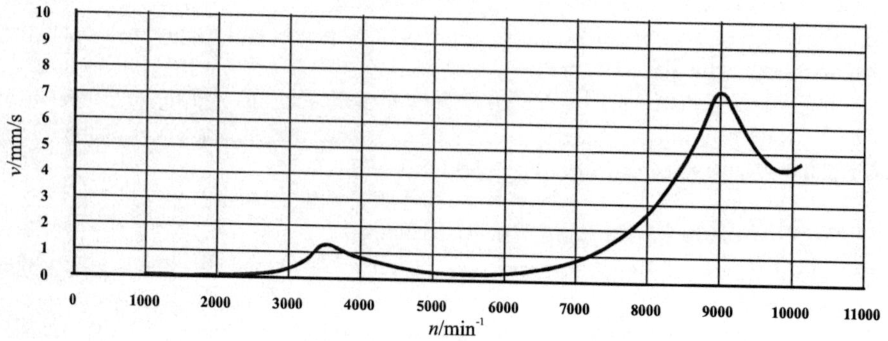

Fig. 8.6 Run-up curve of a gas turbine, taken from ISO 21940-12, Annex D

Task

What are the unbalance tolerances according to ISO 21940—Part 12?

Solution

Total permissible unbalance $U_{per} = 3850\ \mathrm{g \cdot mm}$;

- Permissible low-speed residual unbalance: 3850 g · mm, i.e. per plane I and III: 1925 g · mm.
- Permissible equivalent unbalance in the 1st flexural mode: 60%, i.e. 2311 g · mm.
- Permissible equivalent unbalance in the 2nd flexural mode: 60%, i.e. 2311 g · mm.

8.5.2 Unbalance Tolerances According to DIN ISO 21940-12, Beiblatt 1

ISO 21940—Part 12 does not sufficiently take into account the very different resonance distances in practice. A proposal made by the author at the VDI Vibration Conference year 2000 was included in VDI Guideline 3835, which was published in draft form in 2009 and is now available—revised—as DIN ISO 21940-12, Beiblatt 1.[4]

The following objective was pursued: The range of rotor behaviour—from rigid to shaft-elastic, and correction from one plane to many planes—is presented in a compact manner, as are the tolerances. The systematics of this representation is explained in Sect. 4.2, here it is applied to the tolerances of procedure G.

8.5.2.1 Distribution to Several Unbalances

The requirement is formulated as follows: If several unbalances simultaneously determine the unbalance condition of the rotor, the sum of their *effects at service speed* should not exceed the effect of the permissible unbalance U_{per} (on a rotor with rigid behaviour).

However, the superposition of the effects is quite complicated due to the phase shifts crossing resonance, so that the necessary calculation processes can certainly only be carried out in few cases. Therefore, a simple rule is given for the general case.

Assuming that the different unbalances are not dependent on each other in size and angle, one can determine the permissible residual unbalances according to the mathematical rule *sum of the error squares:*

[4] DIN ISO 21940: Mechanische Schwingungen—Auswuchten von Rotoren—Teil 12: Verfahren und Toleranzen für Rotoren mit nachgiebigem Verhalten, Beiblatt 1 (2015): Verfahren zum Auswuchten bei mehreren Drehzahlen. (English: DIN ISO 21940—Part 12: Procedures and tolerances for rotors with flexible behaviour, Supplement 1 (2015): Methods for balancing at multiple speeds).

$$\sqrt{U_{r,per}^2 + \left(\frac{P_{r.per}}{L}\right)^2 + \sum_{n=1}^{N}(U_{ne,per} \cdot M_n)^2} = U_{per} \quad \text{g} \cdot \text{mm} \tag{8.4}$$

Instead of the permissible resulting unbalance $U_{\text{r, per}}$ and the permissible resulting couple unbalance $C_{\text{r per}} = P_{\text{r, per}}/L$, the permissible dynamic unbalance $U_{\text{dyn, per}}$ can also be used as a basis, distributed to two tolerance planes, e.g. bearing planes 1 and 2:

$$\sqrt{U_{1,per}^2 + U_{2,per}^2 + \sum_{1}^{n} U^2{}_{\text{mod}\ ,e,n,per}} = U_{per} \quad \text{g} \cdot \text{mm} \tag{8.5}$$

8.5.2.1.1 Even Distribution

If *two* low-speed unbalances and additionally N modal unbalances are to be considered for balancing, the calculated factor per unbalance is: $1/\sqrt{2+N}$ (see Eq. 4.3).

Depending on the total number of unbalances to be taken into account, the factors are stated in Table 8.3). For each unbalance the tolerance is determined by multiplying U_{per} by this factor.

If one wants to consider the example from ISO 21940-12, Annex D (Fig. 8.6), with the systematics of DIN ISO 21940-12, Beiblatt 1, the flexural resonance speeds must also be named in addition to the rotor data.[5] The modal dampings are not known, they are estimated (they are particularly significant in the case of a resonance near the service speed).

Example

Rotor from Sect. 8.5.1: mass 1625 kg, service speed 10,125 min^{-1}; two (known) resonances: approx.. 3500 min^{-1} and 9100 min^{-1}; assumed modal dampings 0.05; quality grade G 2,5. ◀

Task

What would be the tolerances according to DIN ISO 21940-12, Supplement 1?

Solution

Total permissible unbalance is $U_{\text{per}} = 3850$ g · mm; four tolerances, i.e. factor 0.5 corresponding to Table 8.3):

- Permissible low-speed residual unbalance: per plane *I* and *III*: 1925 g · mm

[5]According to DIN ISO 21940-12, Supplement 1, at least *one* flexural resonance speed above the service speed would have to be evaluated here. This would replace the evaluation at service speed originally intended in the example in ISO.

Table 8.3 Factors per unbalance as a function of total number of unbalances

Number of unbalances	1	2	3	4	5	6
Factor calculated	1	0,71	0,58	0,50	0,45	0,41
Recommended factor (rounded)	1	0,70	0,60	0,50	0,45	0,40

- Permissible equivalent modal unbalance in the 1st flexural mode:
  $n_{\text{crit, 1}}/n_{\text{B}} = 3500/10{,}125 = 0.35$; with Eq. (4.1) $G_1 = 0.88$.
  $U_{\text{1e, per}} = 0.5 \cdot G_1 \cdot U_{\text{per}} = 0.5 \cdot 0.88 \cdot 3850\ \text{g} \cdot \text{mm} = 1694\ \text{g} \cdot \text{mm}$.
- Permissible equivalent modal unbalance in the 2nd flexural mode:
  $n_{\text{crit, 2}}/n_{\text{B}} = 9100/10{,}125 = 0.90$; with Eq. (4.1) $G_2 = 0.21$.
  $U_{\text{2e, per}} = 0.5 \cdot G_2\ U_{\text{per}} = 0.5 \cdot 0.21 \cdot 3850\ \text{g} \cdot \text{mm} = 404\ \text{g} \cdot \text{mm}$.

In comparison to ISO 21940-12, it becomes clear that DIN ISO 21940-12, Supplement 1 specifies in a more differentiated manner:

- The low-speed unbalances are 50% in this case (with a total number of 4).
- The permissible equivalent modal unbalance in the 1st flexural mode is somewhat reduced in the Supplement 1.
- The permissible equivalent modal unbalance in the 2nd flexural mode is largely reduced in the Supplement 1: only approx. 1/6—because of the proximity to the service speed!

If one follows DIN ISO 21940-12, Supplement 1 and takes into account a 3rd modal unbalance (above the service speed), this results—for five unbalances—in a general factor of 0.45.

These tolerances would follow:

- Permissible low-speed residual unbalance: per plane I and III: 867 g · mm
- Permissible equivalent modal unbalance for the 1st flexural mode:

$$U_{1e,per} = 0.45\ G_1\ U_{per} = 0.45 \cdot 0.88 \cdot 3850\ \text{g} \cdot \text{mm} = 1525\ \text{g} \cdot \text{mm};$$

- Permissible equivalent modal unbalance for the 2nd flexural mode:

$$U_{2e,per} = 0.45\ G_2\ U_{per} = 0.45 \cdot 0.21 \cdot 3850\ \text{g} \cdot \text{mm} = 364\ \text{g} \cdot \text{mm};$$

- permissible equivalent modal unbalance for the 3rd flexural mode:

  $U_{3e,per}$ *cannot be calculated* because the resonance speed is unknown.

In the representation of DIN ISO 21940-12, Supplement 1, the information then looks as follows, Fig. 8.7.

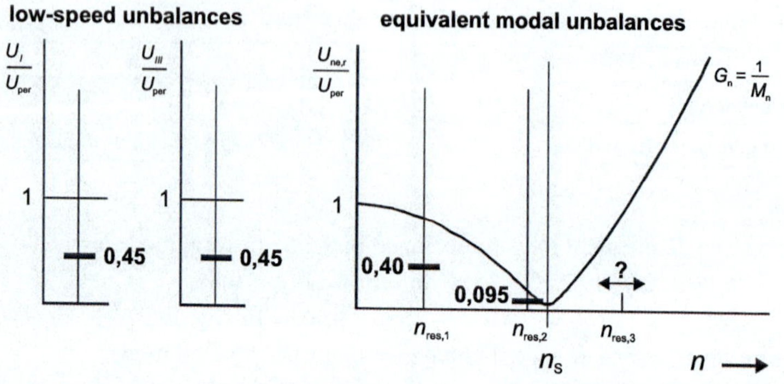

Fig. 8.7 Gas turbine from ISO 21940-12, Annex D. Representation of *three* critical speeds (position of $n_{crit,3}$ unclear) and *all* unbalance tolerances according to DIN ISO 21940-12, Supplement 1

8.5.2.1.2 Weighted Distribution

For individual cases, the unbalances can be weighted differently, e.g.:

- To particularly calm down a flexural resonances, that is passed running up to service speed.
- If unbalances that change to a greater extent than others during a service period of the machine (see Sect. 8.8), they should be provided with a larger tolerance value.

DIN ISO 21940-12, Supplement 1 provides information on such a weighting.

8.6 Tolerances for the Balancing Process

As explained in Sect. 4.7, the tolerances described so far in this chapter—U_{per} and all values derived from it—are intended for the *delivery of* the machine.

For the *balancing process,* the permissible unbalance must be reduced by the error ΔU that occurs during balancing compared to *delivery:* Errors when measuring the unbalance and due to a different constellation of the rotor. The indicated unbalance U after the balancing process may then be max. (see Sect. 7.9.1):

$$U_{\text{indication}} = U_{\text{per}} - \Delta U \qquad \text{g} \cdot \text{mm} \tag{8.6}$$

ISO 21940-14[6] describes in general terms the errors that can occur during balancing and—for a number of common problems—their systematic reduction to unavoidable

[6]The predecessor, ISO 1940-2, was assigned to ISO 1940-1, so it was at that time only intended for rotors with rigid behaviour. However, most of the indications can also be used for rotors with flexible behaviour. When this standard was revised and published as Part 14 of ISO 21940, it was made applicable to all rotor behaviours.

residues (see Sect. 16.5). In addition, a procedure of determining the error limits of the balancing machine is explained. The remaining errors determine which indications are permissible at the end of the balancing process.

The permissible readings for the balancing process are therefore dependent on the process itself. They can only be planned in conjunction with a given balancing machine, its respective use and all the boundary conditions associated with the process.

At the same time, this planning process can prevent mistakes from being made when using the balancing machine and accessories, making the balancing process unnecessarily difficult or unreliable.

Good planning of the balancing process and its unbalance tolerances guarantees the achievement of the balancing target for the *delivery*.

For rotors with flexible behaviour, the examples described in the fundamentals (Sect. 4.8) and those for rotors with rigid behaviour (Sect. 7.7) can be used to a large extent.

However, attention must be paid to the sometimes very small plane distances. For these cases, the splitting into resulting unbalance and moment unbalance can be advantageous (see example under Sect. 8.4.1).

The situation with *balancing procedure G is* more complex and will be discussed below, e.g. the errors due to measurement and evaluation:

- For the *low speed* unbalances of procedure G the procedures of Sect. 4.8 and the examples of Sect. 7.8 are applicable, if the low speed unbalances are measured and corrected directly (separately).
- For the *modal* unbalances—and even for *all* unbalances if evaluated *simultaneously* (Fig. 10.12)—a calculation on *error propagation* must show how the errors of the individual measurements affect the results for the different unbalances.

In addition to the errors during measurement and evaluation, there are of course the deviations that occur due to a different state of the rotor in the balancing machine compared to service.

Typical examples are, in addition to those mentioned in Sect. 4.8:

- Instead of sleeve bearings used in service, the rotor is balanced in open rollers, or in a different sleeve bearing system, or with different sleeve bearing operational data. The differences in the contact pattern may lead to other journal centres. These errors can influence all unbalances.
- The rotor is not completely assembled for balancing; e.g. on a turbo generator the fans are disassembled as they interfere at high speed and may suck oil out of the bearings.

Again, one should not be satisfied with a *percentage deduction* from the tolerance, but carefully analyse the situation. If afterwards the usual balancing routine can be expressed by an error percentage, this is permissible.

When outsourcing the balancing process, the targets for the *delivery* must of course also be specified. The subcontractor must then reduce according to *his* process and define *his* permissible unbalance indications.

8.7 Assessment of the Unbalance State Achieved

According to ISO 21940-12, the assessment of the achieved unbalance state (as well as the balancing process) can be carried out according to two different criteria, by:

- Vibrations.
- Residual unbalances.

DIN ISO 21940-12, Supplement 1, does not consider vibration to be suitable, but is based on residual unbalances only. In Sect. 8.3.1 the problem is explained in more detail.

The rules of ISO 1940-2 were originally developed for rotors with rigid behaviour; they are explained in detail in Sect. 7.9.

However, most of the comments are also applicable to rotors with flexible behaviour. When this standard was revised and published as Part 14 of ISO 21940, it was made applicable to all rotors.

Rotors with flexible behaviour often have more than two tolerance planes or more than two types of unbalance, so the task is more complicated, but can be solved with the same tools.

Again, for the statement ***unbalance indication in tolerance*** the following applies:

- When checking directly *after the balancing* process, the errors due to the balancing process must be *deducted* from the unbalance tolerances.
- When ***checking***—on the same balancing machine, or on another one—the errors due to the checking process must *also* be *deducted* from the unbalance tolerances.

Note

Before this revision, the deviations were accounted for as minus during balancing and as plus during a check.

The principle is currently modified (see Sect. 7.9). ◄

8.7.1 Assessment by Vibrations

According to ISO 21940-12, measurements of the vibrations under one of the following conditions are possible for an assessment of the achieved unbalance condition, depending on the type and use of the respective rotor:

1. In a high-speed balancing machine or system.
2. In a test field on an assembled machine.
3. On site in final assembly condition.

8.7.1.1 Assessment in a High-Speed Balancing Machine

As explained in Sect. 8.3.1.2 vibrations in high-speed balancing machines—in contrast to unbalances—are not a reliable criterion. This approach of ISO 21940-12 is therefore not recommended and not explained here.

8.7.1.2 Assessment in the Test Field

ISO 21940-12 gives various instructions on installation, handling, measuring equipment and the test procedure. The boundary conditions are similar to those for an assessment in service condition, Sect. 8.7.1.3.

8.7.1.3 Assessment in Service Condition

There are two series of standards for the measurement and evaluation of vibrations in the service condition, e.g.: ISO 7919, various parts; and ISO 10816, various parts.

Most standards specify limit values for overall vibrations (effective value of the vibration velocity) of a rotating machine. They are primarily used for acceptance of the machine after installation and for checks during service.

They are only suitable to a limited extent for assessing the *balancing quality* of a rotor with *flexible behaviour:*

- They usually only allow a general statement and do not provide a sufficiently detailed picture.
- In order to assess the unbalance state, the level of the rotational frequency component alone must be known—and this component can also be influenced by other phenomena in addition to the unbalance.
- It can be generally stated: the simpler the rotor behaviour, the rotor configuration is, and the balancing procedure that can be used to correctly balance a rotor with flexible behaviour, the more likely it is that its unbalance state due to vibrations can be adequately assessed in service. This is due to the fact that in service only very limited speeds and correction planes are available to check the rotor behaviour and the distribution of unbalances in detail.

Author's Estimation

For assessing the unbalance state by vibration measurement in service condition:

- Procedure A: Single-plane balancing → pretty good.
- Procedure B: Two-plane balancing → pretty good.

- Procedure C: Balancing individual components before assembly → poor:
  Neither the rotor behaviour can be assessed nor whether the balancing procedure has been carried out correctly.
- Procedure D: Balancing after limiting the starting unbalance → moderate:
  It is not possible to assess whether the two-plane balancing procedure was carried out correctly.
- Procedure E: Sequential balancing during assembly → bad:
  Neither the rotor behaviour can be assessed nor whether the sequential balancing procedure has been carried out correctly.
- Procedure F: Balancing in optimal planes → moderate:
  It cannot be assessed whether the planes are optimal for the balancing procedure.
- Procedure G: Balancing at multiple speeds → extremely poor:
  Neither the rotor behaviour nor the low-speed and modal unbalances can be assessed.
- Procedure H: Balancing at service speed → moderate:
  It is not possible to assess whether a good compromise has been found over the speed range.
- Procedure I: Balancing at a fixed speed → moderate:
  It is not possible to assess whether a good compromise for component-elasticity has been found.

8.7.2 Assessment by Unbalances

One of the following environments and conditions can be considered for an assessment of the unbalance state achieved:

1. Low-speed balancing machine and if possible the most complete rotor.
2. High-speed balancing system and if possible the most complete rotor.
3. Test field: the complete machine.
4. Final installation: the final assembly state.

8.7.2.1 Assessment in a Low-Speed Balancing Machine

The balancing machine should meet the requirements of ISO 21940-21, especially with regard to the minimum achievable residual unbalance U_{mar} (Sect. 13.4) and the unbalance reduction ratio R_{UR} (Sect. 13.5).

Only shaft-elastic rotors that are low-speed balanced with procedures A to F can be assessed on a low-speed balancing machine. For the fully assembled rotor, however, only the final condition can be checked.

The usual individual steps carried out on individual parts and assemblies cannot be checked without dismantling the rotor. This makes careful quality assurance of these steps all the more important.

The assessment of the initial condition—the initial unbalance or starting unbalance—is also limited in some cases, i.e. it must be checked and recorded.

One can therefore conclude:

For all rotors, the residual unbalance after the balancing process can be measured directly in up to two planes.

- Procedures A and B can clearly be assessed.
- Procedures C and D: it is not possible to check whether the prerequisites—limited unbalance—have been fulfilled.
- Procedure E: there is no information whether the sequential process has been carried out correctly.
- Procedure F: it is not possible to check whether the planes have been chosen optimally or whether the distribution to 3 planes is correctly done.

8.7.2.2 Assessment in a High-Speed Balancing System

In a high-speed balancing system, *all* rotors can basically be assessed, regardless of whether they have been balanced at low speed or high speed, and by which procedure. One can determine:

Low Speed Procedures

- Procedures A and B can clearly be assessed.
- Procedures C and D: it is also not possible to check whether the prerequisites—limited unbalance—have been kept.
- Procedure E: it can be checked up to service speed whether the equivalent modal tolerances are kept, i.e. whether the sequential process has been planned and completed correctly.
- Procedure F: a high-speed check can be made to see whether the planes have been optimally selected or whether the distribution between the three planes has been correct.

High Speed Procedures

- Procedure G: it can be checked whether the low-speed tolerances and the equivalent modal tolerances are observed. The determined residual unbalances in the individual planes are converted into the various unbalance components, see Sect. 6.4.6.
- Procedures H and I: the tolerance planes used and the relevant speeds must be known. Then high-speed testing can be carried out in two planes.

8.7.2.3 Assessment in the Test Field

An assessment of the unbalance condition in the test field can be carried out with in-situ balancing means. In most cases, the possibilities are severely limited, either because many tolerance planes are not accessible or because the speed cannot be varied to the required extent.

8.7.2.4 Assessment in Service

Here, additional restrictions usually apply compared to the test field, e.g. due to the drive, other coupled machines and the service conditions.

8.8 Susceptibility and Sensitivity of Machines to Unbalance

Small unbalance or vibration values during acceptance of a machine do not guarantee undisturbed running over a longer period of time, because all rotors change their unbalance state to a lesser or greater extent—depending on the rotor type, the design and the service conditions:

- Under load.
- With time.

Machines also react very differently to unbalances, with resonance proximity and damping ratio being the most important factors.

In this very difficult field, ISO 21940-31[7] attempts to describe the behaviour of machines with a systematic approach, to define sensible limit values and to check them. That is why the standard is assigned to tolerances in this book.

The range around the service speed is primarily considered and the influences of nearby flexural speeds and the occurring damping are determined. The derivation and representation of the tolerances of rotors with shaft-elastic behaviour according to DIN ISO 21940-12, Supplement 1 (Sect. 8.3.2) also shows how sensitive this range can be.[8]

This standard should already help during the design of a machine to plan for sufficient resonance distances and damping.

If unfavourable data later arise during service, the possibilities are limited: sometimes resonance shifts are possible, rather a change in damping (bearing).

[7] According to the standard, these rules should only be used for simple systems that have *one* resonance in the speed range. However, the author is of the opinion that this approach can also be used for multiple resonances if these resonances are far enough away from each other, e.g. more than 20% to 60% depending on the susceptibility.

[8] In the new edition of the standard, the range below resonance is supplemented by permissible vibration curves from other standards. The author does not think this is a good idea: the resonance range is clearly in the foreground in this standard.

8.8.1 Classification of the Susceptibility of Machines

ISO 21940-31 forms three different classes of machines which differ in the likelihood of changing the unbalance condition, i.e. which are differently susceptible to unbalance. The classes with characteristics and examples:

I. **Low susceptibility**
   Characteristics: These machines have typically large rotor masses compared to the bearing housings, operate in a clean environment, have negligible wear and show minimal deformation due to temperature changes.
   Examples: Paper machine rolls, printing machine rolls, high-speed vacuum pumps.
II. **Medium susceptibility**
   Characteristics: These machines operate in an environment with large temperature differences and/or with medium wear.
   Examples: Pumps in clean medium, electric armatures, gas and steam turbines, small turbo generators for industrial application, turbo compressors.
III. **High susceptibility**
   Characteristics: These machines operate in environments where deposits occur, or high wear, or severe corrosion.
   Examples: Centrifuges, fans, decanters, hammer mills.

8.8.2 Modal Sensitivity Ranges

Another criterion in ISO 21940-31 is the sensitivity to unbalance, i.e. the change in the vibration state when the unbalance changes. The modal sensitivity (the modal magnification factor) is:

$$M_n = \frac{\left(\frac{n_S}{n}\right)^2}{\sqrt{\left[1-\left(\frac{n_S}{n}\right)^2\right]^2 + 4D^2\left(\frac{n_S}{n}\right)^2}} \quad \text{dimensionless} \tag{8.7}$$

The modal sensitivity in the resonance Q_n is only dependent on the damping ratio D, it results in:

$$Q_n = \frac{1}{2D} \quad \text{dimensionless} \tag{8.8}$$

Five ranges are defined for the modal sensitivity M_n to cover all practically occurring cases:

A. **Very low sensitivity**
   Expected running behaviour: very smooth resonance speed difficult to detect.
B. **Low sensitivity**
   Expected running behaviour: smooth, low and stable vibrations.
C. **Medium sensitivity**
   Expected running behaviour: acceptable, moderate and slightly unsteady vibrations.
D. **High sensitivity**
   Expected running behaviour: sensitive to unbalance; regular in-situ balancing may be required.
E. **Very high sensitivity**
   Expected running behaviour: too sensitive to unbalances; this area must be avoided.

ISO 21940-31 gives some more practical advice on these ranges of modal sensitivity:

- Although theoretically range A always seems desirable, consideration of cost and feasibility often make it necessary to work with a higher sensitivity.
- For high performance machines—e.g. those that have a short time period between planned maintenance cycles—higher values of modal sensitivity may be permissible.
- For machines where in-situ balancing is not feasible or economical, smaller values of modal sensitivity may need to be applied.
- Considerations of modal sensitivity do not always provide sufficient certainty that the vibration limits are observed on all parts of the machine. Supplementary local sensitivities are therefore defined and used, i.e. the ratio between the change in vibration due to a change in unbalance in certain planes.

8.8.3 Limit Curves

For each class of machines—with different susceptibility to unbalance, Sect. 8.8.1—limit curves are formed in ISO 21940-31 which limit the different ranges of modal sensitivity from each other (Figs. 8.8, 8.9 and 8.10).

These limit curves represent different modal sensitivities (modal magnifications), i.e. all points on a curve have the same modal sensitivity. With increasing resonance distance, the degree of damping may therefore decrease.

Class I allows higher modal magnifications (for the modal sensitivity ranges A to E) than Class II because of its low susceptibility to unbalances, and Class II in turn allows higher modal magnifications than Class III.

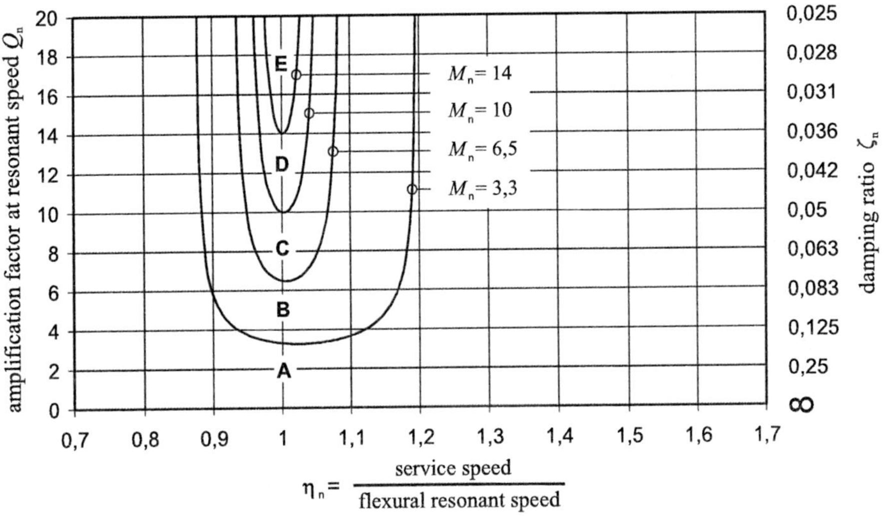

Fig. 8.8 Susceptibility Class I with ranges of modal sensitivity A to E

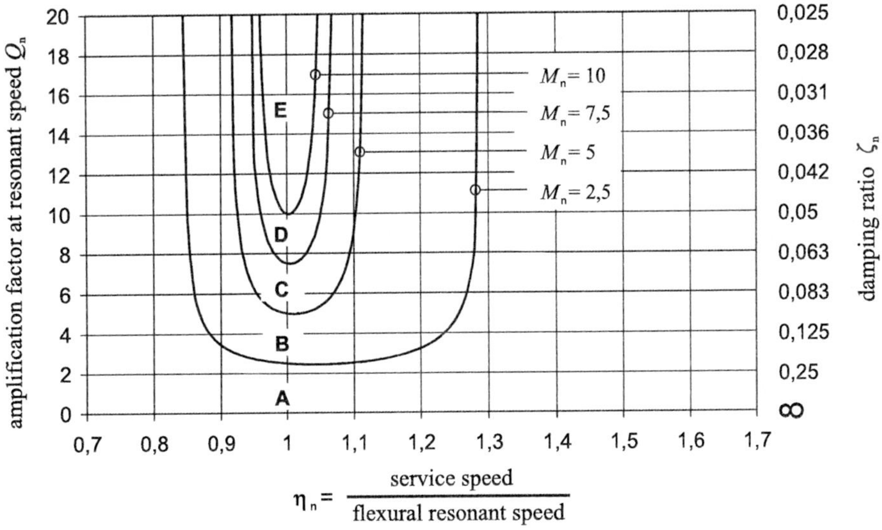

Fig. 8.9 Susceptibility Class II with ranges of modal sensitivity A to E

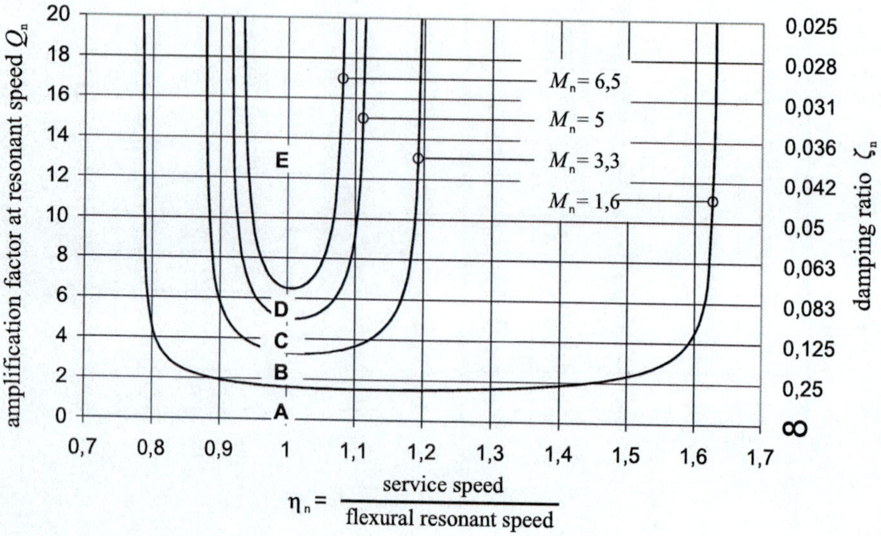

Fig. 8.10 Susceptibility Class III with ranges of modal sensitivity A to E

8.8.3.1 Example 1

Steam turbine, service speed 3000 min^{-1}, 1st resonance speed 2730 min^{-1}, damping ratio D=0.04, measured magnification in resonance, $Q_n = 12.5$ (see Sect. 8.8.4).

Task

Is the design acceptable?

Solution

The machine belongs to Class II with regard to susceptibility to unbalance, Fig. 8.9 applies. The resonance ratio $\eta_1 = 3000/2730 = 1.1$. From Fig. 8.9 results range C, i.e. a moderate modal sensitivity.

Conclusion

The system is acceptable.

8.8.3.2 Example 2

However, if there is no constant service speed, but a speed range which includes the resonance, the most unfavourable situation must be evaluated.

Steam turbine, variable service speed up to 3000 min^{-1}, the 1st resonance speed is at 2730 min^{-1}, damping ratio D=0.04.

Task

Is the design acceptable?

Solution

The machine belongs to Class II with regard to susceptibility to unbalance, Fig. 8.9 is valid. The range of service speeds includes the resonance: the resonance ratio therefore is $\eta_1 = 1$. From Fig. 8.9 results range E; i.e. an extremely high modal sensitivity.

Conclusion

The system is *not* acceptable.

8.8.3.3 Special Case Acceleration

If resonances are passed so quickly that the vibration amplitudes do not fully develop, this effect can also be taken into account, see Fig. 8.11. The curves represent angular accelerations related to the square of the angular frequency (in resonance).

ISO 21940-31 evaluates acceleration and deceleration differently, as can be seen from the curves.

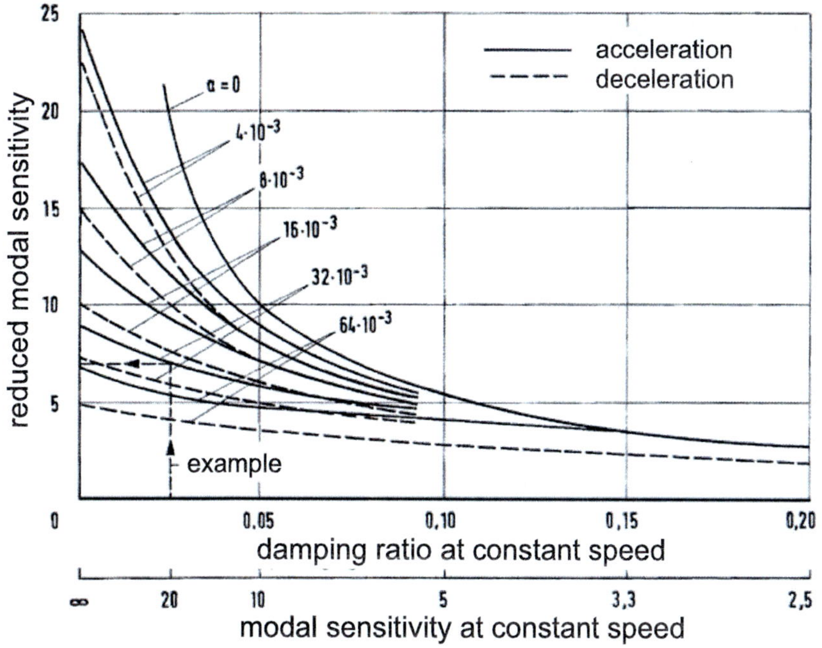

Fig. 8.11 Reduction of modal sensitivity as a function of the damping ratio and the specific angular acceleration and deceleration *a*

8.8.3.4 Example

Gas turbine with service speed 4000 min^{-1}, 1st resonance speed 2730 min^{-1}, damping ratio D=0.025, acceleration from 1200 min^{-1} to 3600 min^{-1} within t=0.1 s, deceleration same magnitude (somewhat unrealistic, but chosen this way to show the calculation).

Task

How large is the modal sensitivity?

Solution

The machine belongs to Class II with regard to susceptibility to unbalance, see Fig. 8.9.

Angular acceleration $\alpha = \frac{\Delta\omega}{t} \approx \frac{\Delta n}{10\,t} = \frac{2400}{10 \cdot 0.1} = 2400 \quad \text{rad/s}^2$.

Ratio a $\quad a = \frac{\alpha}{\omega_n^2} \approx \frac{2400}{273^2} = 32 \cdot 10^{-3}$.

Figure 8.11: Starting from D=0.025 on the x-axis, the curves for $32 \cdot 10^{-3}$ yield values on the y-axis of about 6.9 (acceleration) and 5.8 (deceleration),. The worse (larger) value is used as a basis.

This consideration includes resonance, thus the resonance ratio η_1=1. From Fig. 8.9 results range C, i.e. a moderate modal sensitivity.

Conclusion

The system is acceptable.

8.8.4 Experimental Determination of the Modal Sensitivity

ISO 21940-31 formulates: If the rotor is driven slowly through the resonance and measured values can be recorded—represented e.g. as a polar diagram (Nyquist diagram)—the modal sensitivity in the resonance can be calculated from certain speeds:

$$Q_n = \frac{\omega_n \cdot \Omega_{45}}{\omega_n^2 - \Omega_{45}^2} \qquad \text{dimensionless} \tag{8.9}$$

where Q_n is the maximum modal sensitivity (at resonance), ω_n is the angular frequency of resonance speed and Ω_{45} is the angular frequency of that speed at which the phase position has changed by 45° compared to resonance, see Fig. 8.12.

8.8.4.1 Example 1

From Fig. 8.12 we take: resonance speed n_n=3000 min^{-1}; speed with 45° phase shift n_{45}=2710 min^{-1}.

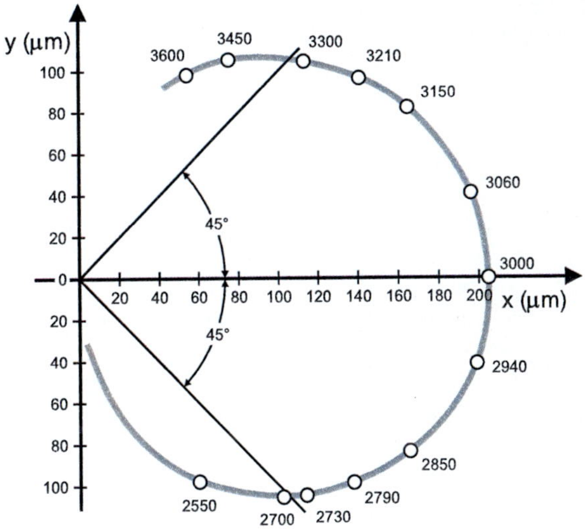

Fig. 8.12 Polar diagram when passing resonance: straight lines for 45° deviation from the resonance angle (X-axis) and the speeds found there serve to determine the magnification in the resonance and thus the degree of damping

Task

How large is Q_n ?

Solution

Instead of the angular frequencies, the speeds can be used directly:

$$Q_n = \frac{n_n \cdot n_{45}}{n_n^2 - n_{45}^2} = \frac{3000 \cdot 2710}{3000^2 - 2710^2} = 4.91$$

Conclusion
In resonance, a magnification of approx. 4.9 occurs.

Alternatively, according to ISO 21940-31, it is possible to calculate with the magnitudes of the displacements versus speed alone, but the errors to be expected are somewhat larger. According to Fig. 8.13 those speeds are searched for at which the amplitudes have decreased to 0.707 S_{res} compared to the amplitude at resonance S_{res}.

With the corresponding angular frequencies (or speeds), the magnification factor in the resonance results in:

$$Q_n = \frac{n_{res}}{n_2 - n_1} \quad \text{dimensionless} \tag{8.10}$$

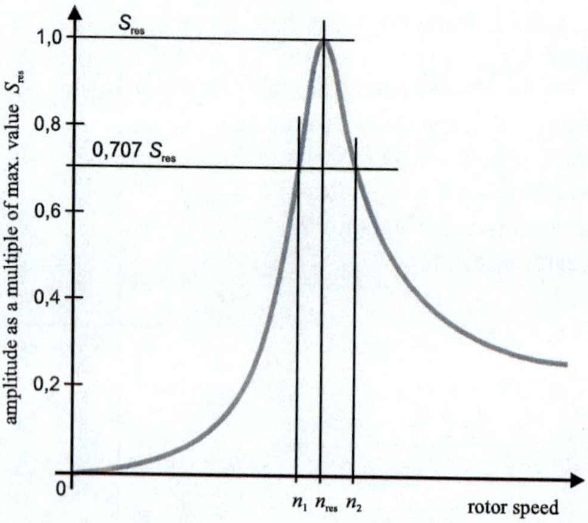

Fig. 8.13 Amplitude curve when passing the resonance. The speed for the resonance and the speeds at which the amplitude is 0.707 of the resonance amplitude are searched for

8.8.4.2 Example 2

An amplitude curve (Fig. 8.13) was recorded.

Task

How large is Q_n ?

Solution

The run-up shows: resonance speed $n_r = 3000\ \text{min}^{-1}$, $n_1 = 2800\ \text{min}^{-1}$, $n_2 = 3400\ \text{min}^{-1}$:

$$Q_n = \frac{n_{res}}{n_2 - n_1} = \frac{3000}{3400 - 2800} = 5$$

Conclusion

This results in a magnification of 5: sufficiently accurate.

9 Procedures for Balancing Rotors with Rigid Behaviour

Contents

H. Schneider, *Rotor Balancing*, https://doi.org/10.1007/978-3-662-66049-2_9

9.1 Preliminary Note

With the basics (see Chap. 2), the theory of balancing (see Chap. 4), the theory of the rotor with rigid behaviour (see Chap. 5) and the notes on its unbalance tolerances (see Chap. 7), the basic requirements for balancing rotors with rigid behaviour are described.

In practice, however, problems can still occur that require special procedures. The most important problems are presented below and suitable procedures are explained.

9.2 Bodies Without Own Bearing Journals

9.2.1 General

Many bodies to be balanced, e.g. pulleys, fan impellers, flywheels etc. do not have their own bearing journals. To be able to balance such a part, a *shaft axis* must be defined.

There are two different ways to do this, which also determine the type of balancing machine:

- The body is mounted to an auxiliary shaft and placed into a horizontal balancing machine, e.g. on twin-roller.
- The body is mounted on the spindle of a horizontal or vertical balancing machine using an adapter.

The following sections describe basics when using auxiliary shafts and adapters.

The methods and possibilities explained here can also be used in other contexts.

9.2.2 Unbalances Due to Assembly

When mounting a body on an auxiliary shaft or an adapter, unavoidable errors occur due to radial play as well as radial and axial run-out errors. The resulting unbalances can be calculated from the displacements and the mass data of the body (Eqs. 5.14 and 5.16).

When measuring the unbalance state, these unbalances add up to the unbalances of the body itself. If the total unbalance is now corrected, the rotor appears balanced, but this is only true together with the current adapter. Without this adapter, i.e. related to the body itself (its bore axis), the unbalance condition is not so good, the assembly error has been balanced into the body by the correction Fig. 9.1.

If the body is subsequently mounted for service, e.g. a fan on its service shaft, new displacements occur which cause a new state of unbalance. The unbalance *acting* in service is the *vectorial* sum of the unbalances in both cases.

If, on the other hand, one wants to know the ***maximum*** fit-related unbalance, the angular position is irrelevant, as it is generally not controlled. The maximum fit-related

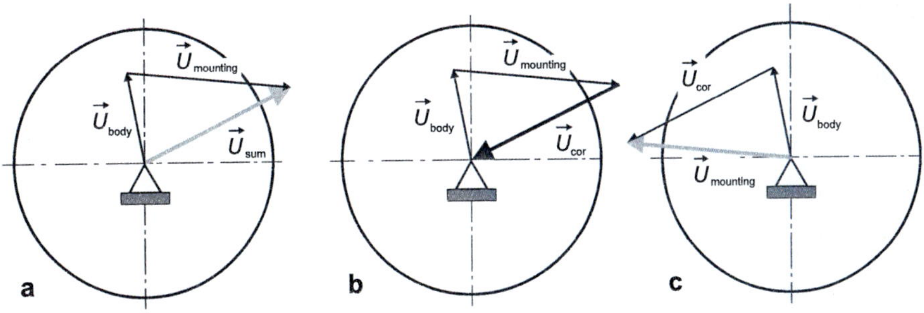

Fig. 9.1 Unbalances due to assembly, shown for a single plane: **a** The unbalances of the body and due to adapter add up to a total unbalance, **b** This unbalance is corrected, **c** The body alone (related to its bore) has an unbalance in the size of the adapter unbalance

unbalance results from the sum of the maximum possible unbalance amounts in both cases—in the balancing machine and in the operating state—or from the amounts of the misalignments.

When balancing the body on the auxiliary shaft (or a support) in the balancing machine, the unbalances caused by the fit can be detected and eliminated by an appropriate procedure (index balancing, see Sect. 9.2.3).But it can occur in full size in the service state without being able to be taken into account during balancing.

The permissible unbalance for this condition must therefore be divided between the unbalance of the individual part and the unbalance due to the fit. Care must be taken to ensure a sensible ratio between the two.

If the permissible residual unbalance is e.g. 30 g · mm, and the unbalance due to the fit is max. 28 g · mm, then only 2 g · mm would remain for the permissible residual unbalance of the body itself.

Since it is certainly not sensible and perhaps not possible to balance the individual part so precisely (e.g. changes from run to run), either:

- The fit tolerances are checked and tightened.
- The rotor to which this body is to be assembled is balanced with a compensation for this mounting error (see Sect. 9.3.4).
- The rotor can be balanced with the individual body already mounted (see Sect. 9.3.4).

Example

A fan has an operating speed $n = 650\ \text{min}^{-1}$ and is to be balanced to quality class G 6.3; see Fig. 9.2.

Further data: Mass $m = 100$ kg, moment of inertia about the shaft axis $J_z = 15\ \text{kg} \cdot \text{m}^2$, mass moment of inertia about the transverse axis $J_x = J_y = 10\ \text{kg} \cdot \text{m}^2$. The fan sits on a shaft of 100 mm Ø, fit H7/h6.

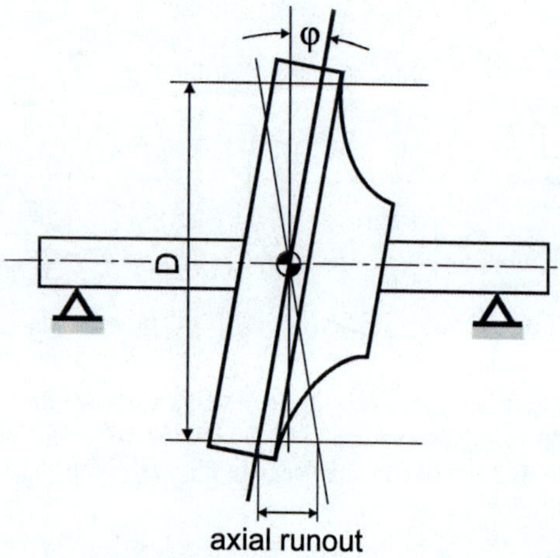

Fig. 9.2 Rotor with axial runout *pl* on diameter *D* leads to angle φ

Concentricity deviation to the shaft axis (bearing points) 0.06 mm. Axial run-out of the flange on 200 mm Ø: 0.04 mm. The distance between the tolerance planes (bearing planes) is $L = 800$ mm, the two bearing planes are approximately equidistant from the mass centre. ◀

Task

How big is the unbalance of this fan due to the fit?

Solution

From the service speed and quality class follows a permissible eccentricity of the mass centre $e_{per} = 92$ µm. Approach: permissible residual unbalance evenly distributed, see Sect. 7.4.2.

1. Fit clearance at 100 mm and H7/h6 is max. 35 µm + 22 µm = 57 µm; the eccentricity due to the clearance $e_{cl} = 28.5$ µm. The unbalance due to the maximum possible displacement within the clearance:

$$U_{cl} = e_{cl}\, m = 28.5 \cdot 100 = 2850\ \text{g} \cdot \text{mm}$$

2. Concentricity: The eccentricity is max. $e_{ecc} = 30$ µm at the permissible runout of 60 µm. The unbalance due to misalignment as a result of the eccentricity is therefore:

$$U_{ecc} = e_{ecc}\, m = 30 \cdot 100 = 3000\ \text{g} \cdot \text{mm}$$

3. Axial run-out: The angle φ by which the body is tilted:

$$\varphi = \frac{pl/2}{D/2} = \frac{pl}{D} = \frac{0.04}{200} = 2 \cdot 10^{-4} \text{ rad}$$

The *moment unbalance* due to the axial runout is: $P_{aro} = \varphi\,(J_x - J_z)$ and for the *couple unbalance* in the two correction planes, results in:

$$C_{aro} / - C_{aro} = \frac{\varphi\,(J_x - J_z)}{L} = \frac{2 \times 10^{-4}\,(10 - 15)10^9}{800} = 1250 \text{ g} \cdot \text{mm}$$

Note

The moments of inertia are conveniently entered in $\text{g} \cdot \text{mm}^2$: $1 \text{ kg} \cdot \text{m}^2 = 10^9 \text{ g} \cdot \text{mm}^2$.

These individual errors add up vectorially. If they are independent of each other, one could use statistics to calculate the probable size and dispersion for larger numbers of bodies. For single rotors, however, one must start from the worst possible value, i.e. the sum of the amounts.

The total fit-related unbalances per bearing planes A, B can therefore be at maximum:

$$U_{fit\,A,B} = U_{ce}/2 + U_{ecc}/2 + C_{aro} = 4180 \text{ g} \cdot \text{mm}$$

For the fan, this means, expressed as centre-of-gravity eccentricity (Fig. 9.2):

$$e_{fit} = \frac{2\,U_{fit\,A,B}}{m} = \frac{2 \cdot 4180}{100} \approx 84\,\mu\text{m} \blacktriangleleft$$

Conclusion

This leaves only 8 μm for the fan itself, a much too small value. In order to create reasonable conditions for this case, the measures mentioned (before the example) must be taken, or the fan must be in-situ balanced (see Chap. 18).

9.2.3 Index Balancing

9.2.3.1 Single Plane with Unbalances

This procedure, which separates the errors of a mandrel from the unbalance of a body, is best made clear first for *one* plane and only for *unbalances* on both parts (Fig. 9.3).[1]

[1] In this case the reference for the angular position (for the coordinate system) is connected to the mandrel. If it is connected to the body—e.g. the scanning mark for photo scanning—the conditions are reversed: The unbalance of the body remains in the angle, the unbalance of the mandrel is measured offset by 180° for the second reading. The principle of evaluation is identical (Fig. 9.4), but now the vector (from the origin) describes the unbalance of the body, the other vectors the unbalance of the mandrel.

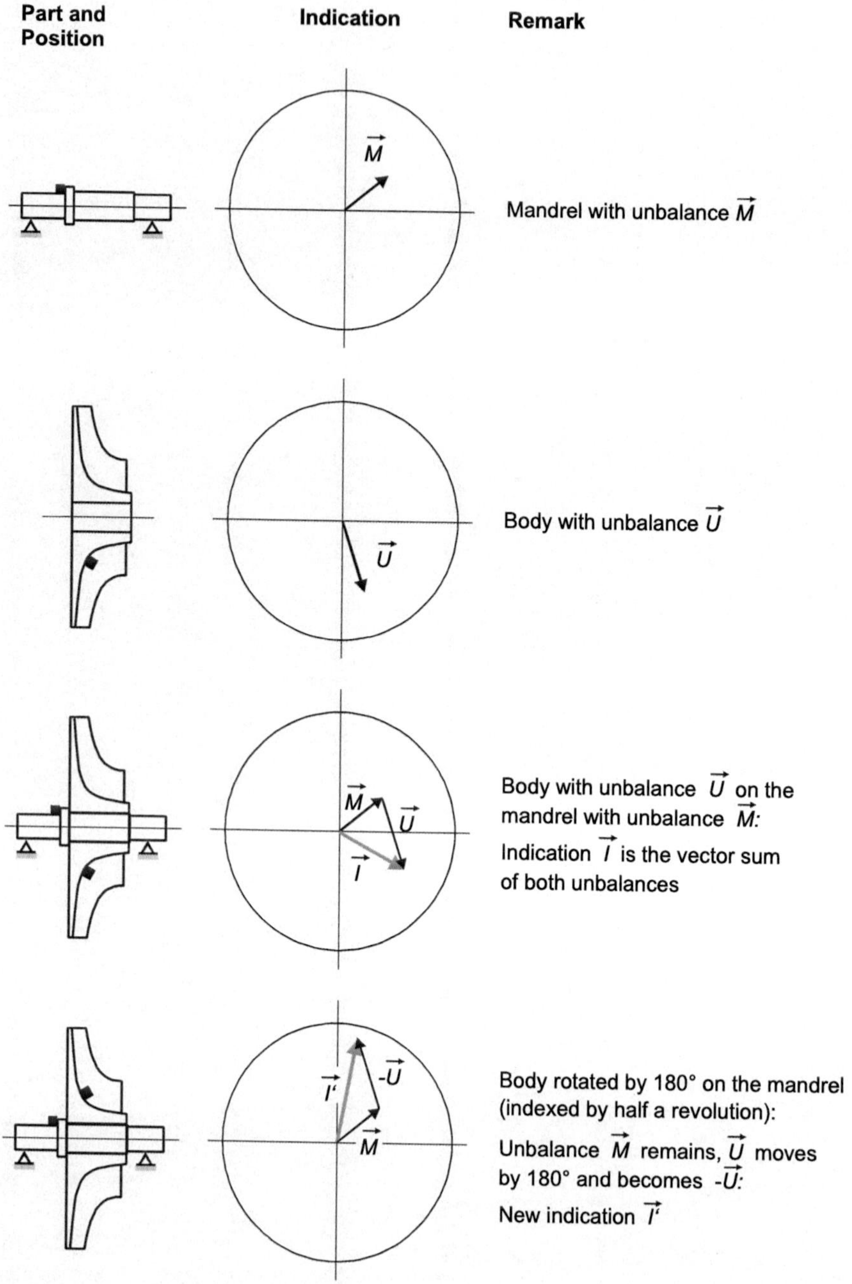

Fig. 9.3 Separating the unbalances of an assembly by index balancing: The procedure is broken down into four mental steps to make it easier to imagine (index balancing really only includes the last two steps). The coordinate system for the angle display is firmly connected to the mandrel

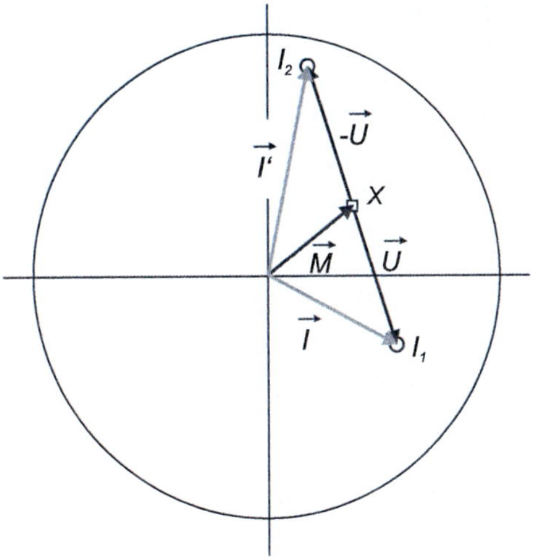

Fig. 9.4 Evaluation of the measurements of Fig. 9.3: The vectors $\vec{M}, \vec{U}$ and $-\vec{U}$ are not directly visible, but are constructed

The result of the two measurements and the evaluation is taken from Fig. 9.4. The difference between the measuring points I_1 and I_2 (distance between the ends of the arrows) corresponds to $2U$ ($\vec{U}$ and $-\vec{U}$ mean the same unbalance of the body, only measured in 180° different angular positions). When correcting towards point X (centre of the connecting line of measuring points I_1 and I_2), the unbalance in the body will be eliminated.

Check

If the balanced body is rotated again by 180°, the display remains at X.

The unbalance $\vec{M}$ of the mandrel can be corrected on the mandrel itself. The next bodies (of the same type) can then be balanced on this mandrel without having to make an index turn.

In this case (there is only one unbalance on the mandrel, Fig. 9.3) it would have been easier to balance the mandrel empty first and only then to fix the body.

9.2.3.2 Single Plane with Fit-Related Errors

Other errors, fit-related errors—play, axial and radial runout errors—only become visible with the body in place. The procedure is basically the same in this case, but considerably more influences have to be considered, see Fig. 9.5.

The evaluation of the measurement results of Fig. 9.5 can be seen in Fig. 9.6: The distance between I_1 and I_2 again corresponds to $2U$. It is balanced towards the point X, whereby the correction is made on the body. The point X *is caused* by the unbalance of the mandrel $\vec{M}$, the eccentric seat of the body, which generates the unbalance $\vec{E}$ and by the play caused by the unbalance $\vec{P}$. $\vec{E}$ and $\vec{P}$ are dependent on the body (as a product with the body mass).

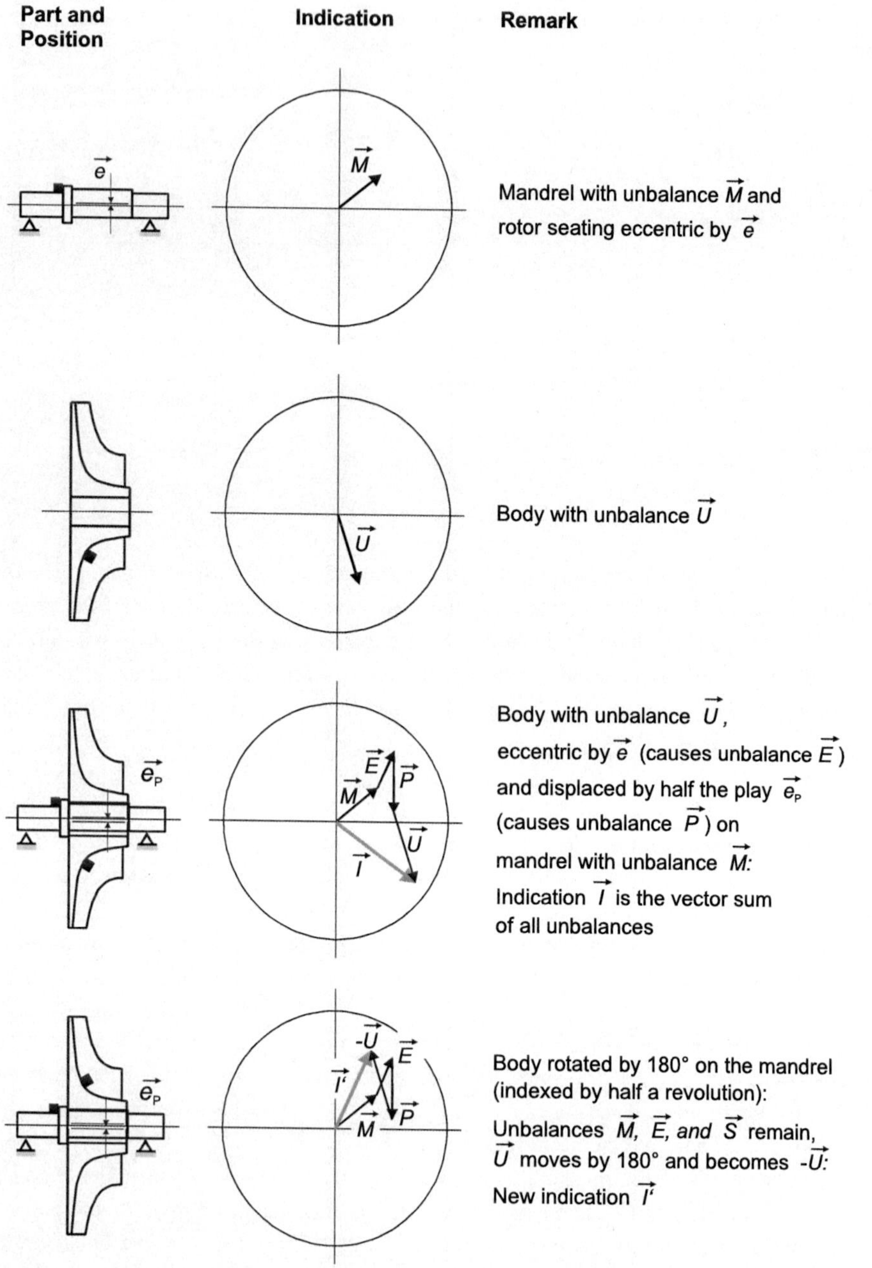

Fig. 9.5 Index balancing with unbalances, eccentricity and play: The procedure is broken down into four mental steps to make it easier to imagine (index balancing really only includes the last two steps). The coordinate system for the angle display is firmly connected to the mandrel

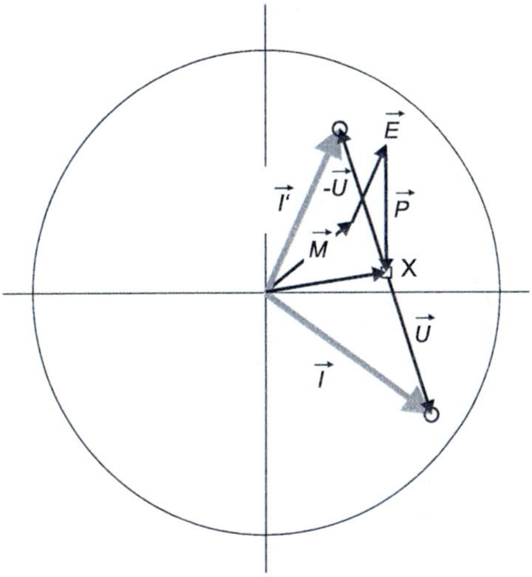

Fig. 9.6 Evaluation of the measurement results of Fig. 9.5

If only ***one*** body type is balanced on this mandrel, the unbalance $\vec{E}$ can be corrected together with the body-independent unbalance $\vec{M}$ by a correction on the mandrel. This makes balancing the body easier.

The unbalance $\vec{P}$ cannot be exactly corrected for because it depends on the size of the clearance and therefore varies greatly in amount. But one can correct an average value and thus reduce the error.

The important point here is that the play is "pushed out" each time in the same direction (in relation to the mandrel) before the body is clamped tightly, so that the remaining unbalance appears to be connected to the mandrel.

9.2.3.3 Generalisation

A body with ***two*** correction planes can also be index balanced, whereby the evaluation is carried out separately for each correction plane, following the same principle as described above.

Angles deviating from 180° can also be used. The correct assignment of the rotational direction of the body is decisive for the evaluation of the measured values.

9.2.4 Further Use of the Index Balancing Method

This method—index balancing—is also used to:

- Eliminate the influence of the drive of a balancing machine (cardan shaft, driver) on the rotor (see Chap. 16),
- Separate the unbalance components of different rotor components of assemblies (see Sect. 9.3).

Modern balancing machines also support this method even for single bodies.

For series, the index value can be saved so that not every part has to be indexed for balancing.

9.2.5 Auxiliary Shafts, Adapters

All parts to accommodate bodies without their own journals, i.e. auxiliary shafts, mandrels and adapters, must:

- Either be manufactured so precisely that the permissible residual unbalance can be achieved.
- Or allow an index balancing so that this error can be detected and eliminated.

It is important not to forget to provide a ***correction*** for the unbalance, which can compensate not only for the unbalance of the auxiliary part, but also for the displacement of the body to be balanced. The unbalances due to displacement of the body are often a decade greater than those of the auxiliary part.

9.3 Assemblies

9.3.1 General

If a rotor consists of several individual parts, all parts can naturally be balanced individually. During assembly, all unbalances of the individual parts add up vectorially. However, since the residual unbalances of the individual parts can have any position, in the worst case they add up fully with their respective amounts.

In addition, there are the unbalances caused by the fit (see Sect. 9.2). If the balancing quality required for the assembly cannot be achieved by balancing the individual parts, the assembly must be balanced as a whole or at least the main components together.

9.3.2 Procedure

It is very important that the assembly is not disassembled after balancing.

If disassembly cannot be avoided, carefully marking the position of the individual parts in relation to each other and ensuring that they are in exactly the same position when reassembling may help.

A high-speed electric motor mounted in antifriction bearings can already be regarded as an assembly in this sense.

It is also necessary to check which errors are caused by play.

9.3.2.1 Example 1

An electric armature with an operating speed $n = 15{,}000\ \text{min}^{-1}$ is to be balanced to quality class G 2.5. The permissible eccentricity error of the antifriction bearings (of the inner rings) is 3 µm.

Task

Does the armature need to be balanced with its service bearings?

Solution

The permissible eccentricity of the mass centre is $e_{\text{per}} = 1.6\ \mu\text{m}$ (see Sect. 7.3.2.3). Since the eccentricity of the antifriction bearings is greater than the permissible eccentricity of the mass centre, the answer here is clear.

Conclusion

The armature must be balanced with its service bearings.

While the permissible eccentricity of the mass centre for the assembly is usually also used for the individual parts (reduced by the fitting errors, see Sect. 9.2), a different distribution may be preferable if the individual parts have very different masses. If, e.g. to the armature from the last example a light pulley is mounted, the larger mass, the armature, can be balanced somewhat more accurately, so that for the lighter part, the pulley (which may have to be replaced more often), a normal balancing quality remains, which can be achieved easily on a mandrel.

9.3.2.2 Example 2

The mass of the armature is $m_1 = 5$ kg, that of the pulley $m_2 = 0.1$ kg, the eccentricity due to the fit is $e_{\text{fit}} = 10\ \mu\text{m}$, no play because of a conical connection. The unbalance of the pulley is fully attributable to *a single* tolerance plane of the armature, as the pulley is arranged overhung. The armature and the tolerance planes (bearing planes) are almost symmetrical.

Task

How should the permissible unbalance of the assembly be distributed so that the pulley can be balanced separately?

Solution

1. The permissible unbalance per tolerance plane (bearing planes A, B) is:

$$U_{per\ A,B} = \frac{1}{2}\, e_{per}\, (m_1 + m_2) = \frac{1}{2}\, 1.6\, (5 + 0.1) \approx 4.1\ \text{g} \cdot \text{mm}$$

2. The pulley can be balanced as an individual part (index balancing) to about 5 μm. In addition, there is the fit-related eccentricity of 10 μm. In the worst case, both values add up, so that a total eccentricity of the pulley $e_{pu} = 15$ μm must be expected.
   The unbalance of the pulley has therefore a maximum:

$$U_{pu} = e_{pu}\, m_2 = 15 \cdot 0.1 = 1.5 \text{ g} \cdot \text{mm}$$

3. The armature must be better balanced by this value:

$$U_{cor\ A,B} = U_{per\ A,B} - U_{pu} = 4.1 - 1.5 = 2.6 \text{ g} \cdot \text{mm}$$

It may be useful to allow the full value of 4.1 g · mm in the plane opposite to the pulley.

9.3.3 Interchangeability of Parts

In the above example, the armature and pulley can be balanced separately. Since the assembly error is taken into account when determining the individual tolerances, this pulley can—if necessary—be exchanged for another one treated in the same way. It is not necessary to know the angles at which the residual unbalances of the individual parts lie, nor is it necessary to pay attention to a specific angular position during assembly.

However, if the unbalance caused by the fit is of the same order of magnitude as the unbalance tolerances (or even greater), such a simple procedure is no longer sufficient; the assembly error must also be corrected.

9.3.4 Correction of the Assembly Error

If the pulley is mounted when balancing the armature (in the above example), the different unbalances are separated by index balancing (see Sect. 9.2.2).

Since the unbalances caused by the fit—due to its incorrect seating for the pulley—are caused by the armature, these unbalances change the angular position just as little as the unbalance of the armature when the pulley is indexed: The balancing machine measures the vectorial sum of both unbalances.[2]

If the armature is balanced this way, each *individually* balanced pulley will then fit on the armature without exceeding the permissible residual unbalance.

Note that the armature alone does not have to be in tolerance; it is only balanced for later assembly.

---

[2]Here it is assumed that the reference for the unbalance measurement (e.g. the scanning mark for photo scanning) is firmly connected to the armature.

9.3.5 Dummies (Substitute Masses)

Strictly speaking, the correction of the assembly error only applies to identical parts. Each part with a different mass, mass centre position or moments of inertia would cause different unbalances due to the fit and therefore require different correction.

In the case of very expensive parts (or non-existent originals) it may make sense to work with dummies (substitute masses) that correspond to the original. The deviations of the physical data must be all the smaller, the greater the fit-related unbalances U_{fit} are in relation to the tolerance U_{per}.

Example

Assumption: $U_{fit}/U_{per}=5$. ◀

Recommendation

Recommended permissible deviations of the physical data approx. 5%. ◀

9.4 Rotors with Parallel Keys

9.4.1 General

Torques between shafts and mounted parts—e.g. electric motors + pulleys—are often transmitted by keys. If one wants to balance the armature and the pulley each as an individual part, an agreement must be made about the connecting piece (i.e. the parallel key).

Three possibilities are conceivable in principle; they have different practical meanings:

1. The key is completely allocated to the shaft.
2. Half of the key is allocated to the shaft, the other half to the pulley.
3. The key is allocated entirely to the pulley.

In practice, only both variants 1) and 2) are applied, each with a number of advantages and disadvantages, as to be seen below.

9.4.2 Shaft with Complete Key

This handling used to be widespread in Europe and prescribed in standards (e.g. ISO 2373 for electrical machines).

Advantages

- The shaft can be balanced and supplied with the original key.
- A test run (without pulley) shows whether balancing has been done correctly (without changing the key).
- A mounted part—balanced separately, without key—can have different hub widths; the balance state is always OK.

Disadvantages

- For the complete key, unbalance correction must be carried out on the rotor, which can cause problems with large shaft overhang and/or narrow correction planes.
- In the case of shaft-elastic rotors (see Chap. 6), inadmissibly large excitations of the flexural modes may occur.

9.4.3 Shaft with Half Key

This method, which used to be applied mainly in the USA, is now standardised internationally (ISO 21940-32). The advantages and disadvantages compared to Sect. 9.4.1 are just reversed.

Advantages

- No unnecessary balancing on the parts.
- No unnecessary excitation of the flexural modes.

Disadvantages

- Special keys required for balancing.
- Special keys during test run.
- Specially contoured keys required if hub length deviates from length of key.

ISO 21940-32 requires parts to be marked according to the method used so that there are no unpleasant surprises when parts are paired.

9.4.4 Influence on the Unbalance Condition

In a typical electric armature, the full key means an unbalance corresponding to a centre of gravity eccentricity of about 10 μm (depending on masses and geometry, it may be less or more in individual cases).

This means that the key should be taken into account during balancing if the permissible specific residual unbalance is less than 30 μm, e.g. with quality class G 6.3 and an service speed of more than 2000 min^{-1}.

Table 9.1 Estimated permissible deviation of the key mass, depending on the permissible mass centre eccentricity of the rotor, example: electric armature

Permissible mass centre eccentricity of the rotor e_{per} (μm)	30	10	3	1
Permissible deviation of the mass of the key (%)	<100	<30	<10	<3

However, it is also possible to estimate from this how *exactly* the key should correspond to the ideal state (see Table 9.1).

The key should be considered (whichever of the methods is used) as a source of error that should only consume an appropriate percentage—e.g. less than 30%—of the unbalance tolerance.

9.4.5 Bias for a Parallel Key

Modern balancing machines support this procedure, either by entering the unbalance values or by measuring the unbalance with and without a key.

Storing the unbalance value of the parallel key (bias) in the measuring device, following motors can then be balanced without having to fit a parallel key each time.

The prerequisite for this is that the parallel key has a fixed position to the angular reference of the balancing machine.

9.4.6 Constructive Measures

The problems (of both methods) can be avoided, e.g. if the parts are equipped with symmetrical (180° offset) keys.

This method is recommended especially for shaft-elastic rotors or if the part to be mounted later is not sufficiently known. Alternatively, a different type of torque transmission can be used that causes less unbalance.

Procedures for Balancing Rotors with Flexible Behaviour

10

Contents

H. Schneider, *Rotor Balancing*, https://doi.org/10.1007/978-3-662-66049-2_10

10.1 Preliminary Note

For ***flexible rotors,*** the *shaft-elastic rotor behaviour* (see Chap. 6) is by far the most common, so that most of the procedures refer to this behaviour, they are explained first.

The procedures for the *body-elastic behaviour* (see Sect. 10.4.9) and the *settling behaviour* (see Sect. 10.4.10) are dealt with at the end of this chapter.

10.2 General

Balancing rotors with shaft-elastic behaviour has been practised for many decades. Consequently, typical characteristics of the respective rotors as well as the balancing equipment used have led to different procedures.

After a long struggle for a generally understandable classification, the current ISO 21940-12[1] describes six low-speed procedures and three high-speed procedures.

But when is a low-speed procedure sufficient, when is high-speed balancing needed? The time required and the equipment costs for high-speed balancing are significantly higher than for low-speed balancing. Sometimes this ratio is even one to two powers of ten, so that it is economically very important to find a procedure that is optimally adapted to the respective task.

Particularly difficult is the search for a sensible solution when there is no experience available with shaft-elastic rotors.

The new description of the balancing theory (see Chap. 4) opens up additional insights that go beyond the current ISO 21940-12; they are essentially based on ***Beiblatt 1 of*** DIN ISO 21940-12.[2]

---

[1] ISO 21940: Mechanical vibration - Balancing of rotors, part 12 (2015): Procedures and tolerances for rotors with flexible behaviour.

[2] DIN ISO 21940: Mechanische Schwingungen—Auswuchten von Rotoren—Teil 12: Verfahren und Toleranzen für Rotoren mit nachgiebigem Verhalten, Beiblatt 1 (2015): Verfahren zum Auswuchten bei mehreren Drehzahlen. (English: DIN ISO 21940—Part 12: Procedures and tolerances for rotors with flexible behaviour, Supplement 1 (2015): Methods for balancing at multiple speeds).

10.3 Rotor Configurations

To facilitate the selection of the appropriate procedures in practical cases, ISO 21940-12 shows typical rotor configurations, describes characteristics and then lists possible balancing procedures in order of increasing effort.

10.3.1 Basic Elements Of Shaft-Elastic Rotors

Five basic elements with specific characteristics represent all important behaviours of shaft-elastic rotors in terms of *balancing technology*, Fig. 10.1.

By combining these elements, any conceivable rotor configuration can be represented.

The elements rigid disc, Fig. 10.1b and rigid section, Fig. 10.1c, must always be joined by an elastic shaft, Fig. 10.1a; otherwise no shaft-elastic rotor can be created.

The shaft-elastic roll, Fig. 10.1d and the integral rotor, Fig. 10.1e are shaft-elastic by themselves, but can also be equipped with additional rigid elements (discs and sections).

10.3.2 Balancing Principles

The aim is always to correct the unbalances wherever possible in *the* planes in which they occur in order to also keep the correction masses sufficiently small.

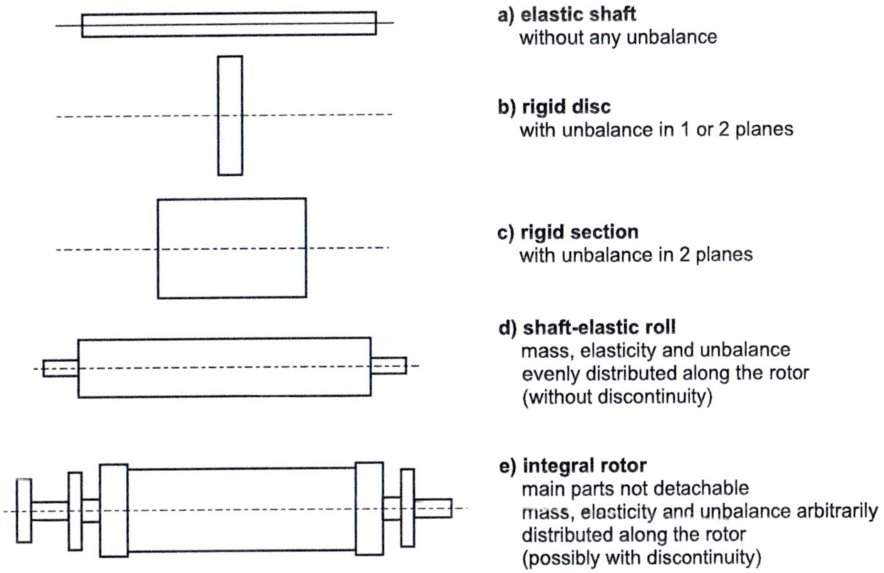

Fig. 10.1 Basic elements of shaft-elastic rotors

If this is not successful, care must at least be taken that there is as little elasticity (elastic element) as possible between the unbalance plane and the (next) correction plane.

In principle, any shaft-elastic rotor can be balanced at high speed using the procedure described in Sect. 10.4.7. However, whether certain shaft-elastic rotors can also be balanced at low speeds depends on the number of planes in which unbalances may occur and whether they can be measured and corrected for in sufficient detail.

A low-speed balancing machine can only measure in one, or two planes. With several elements and a correspondingly large number of unbalance planes, ways must therefore be found to differentiate between these unbalances and to treat them in a targeted manner, otherwise satisfactory high-speed running cannot be expected.

Depending on the configuration and characteristics of the rotor, different balancing principles can be formulated:

1. Rotor of rigid elements and elastic shafts, Fig. 10.1a to 10.1c:
   - Balancing principle: low-speed correction of the unbalances of each rigid rotor element *in this* element.
2. Rotor with predictable type of unbalance distribution, Fig. 10.1d
   - Balancing principle: correction of the unbalance in 2 (or 3) planes at low speed in such a way that the modal unbalances are also reduced (which cannot be measured at low speed).
3. Rotor with arbitrary (unpredictable) type of unbalance distribution, Fig. 10.1e
   - Balancing principle: correct the rigid unbalance components at low speed and the modal unbalance at high speed (at several speeds), possibly all together.

With these tools, let us now go through the different configurations of rotors.

10.3.3 Rotor with Discs

Assumption: the discs are so narrow that their moment unbalances are negligible. The decisive factor is then how the disc is mounted in the service state, i.e. how it is positioned in relation to the shaft axis (see Sect. 9.2.1).

Only deviations that lead to unbalance values outside the unbalance tolerances are considered:

- If the disc is perpendicular, only a resulting unbalance can occur.
- If it shows a run-out error, a moment unbalance also occurs.

In case of a rotor with discs (Fig. 10.2), these disks can be balanced as individual parts if the unbalances caused by the subsequent assembly are sufficiently small for the service condition (see Sect. 9.2.1).

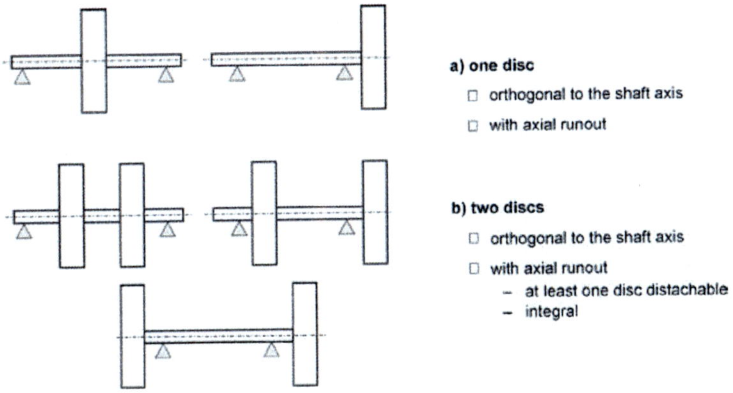

Fig. 10.2 Rotors with one (**a**) or two discs (**b**)

If the unbalances cannot be neglected due to assembly, the rotor must be designed in such a way that it can be assembled and balanced in such small steps that never more than two unbalance planes are added (for problems see Sect. 10.4.5).

10.3.3.1 A Single Disc

A rotor with ***a single*** disc on an elastic shaft (Fig. 10.2a) can form *two* flexure resonances. If they cannot be neglected—depending on the service speed, the unbalance excitation and the unbalance tolerances—they must be reduced by balancing:

Case 1

If the disc is perpendicular to the shaft, only a resulting unbalance occurs. If this unbalance is outside the tolerance, a single-plane balancing is sufficient.

Case 2

If the disc has a runout, then a resultant unbalance and a moment unbalance will occur. If both are out of tolerance, a two-plane balancing is required.

> **Conclusion**
> For these two cases, the rotor can be readily mounted for balancing.

10.3.3.2 Two Discs

A rotor with ***two*** discs on an elastic shaft (Fig. 10.2b) can develop *four* bending resonances. If they cannot be neglected—depending on the service speed, the unbalance excitation and the unbalance tolerances—they must be reduced by balancing:

Case 1

If both discs are perpendicular to the shaft axis, then only one resulting unbalance occurs in each disc. If both unbalances are out of tolerance, two-plane balancing is required.

> **Conclusion**
> The rotor can be fully assembled for this balancing task.

Case 2

If both discs have axial run-out errors, then one resulting unbalance and one moment unbalance occur per disc. If at least one resulting unbalance and one moment unbalance are out of tolerance, the measured unbalances can no longer be reliably assigned to the planes from which they originate.

> **Conclusion**
> There are two alternatives for this balancing task:
>
> 1. Assumption: at least one disc is removable.
>   Sequential balancing during assembly is possible: The rotor with one disc is treated according to Sect. 10.3.3.1. Then the second disc is mounted and in the next step the unbalance is corrected only on this disc in two planes (for problems see Sect. 10.4.5).
> 2. Assumption: none of the discs can be removed (i.e. an integral rotor).
>   Only high-speed multi-plane balancing can be used.

10.3.3.3 More than Two Discs

A rotor with ***many*** discs on an elastic shaft, e.g. a multi-stage pump or compressor rotor (Fig. 10.3), can develop *many* bending resonances. If they disturb—depending on the service speed, the unbalance excitation and the unbalance tolerances—then they must be reduced by balancing.

Even though the task already looks quite complicated, it can be solved with the same logic as before:

Case 1

If all discs are perpendicular to the shaft axis, then only one resulting unbalance occurs in each disc. If more than two unbalances are out of tolerance, multi-plane balancing is required.

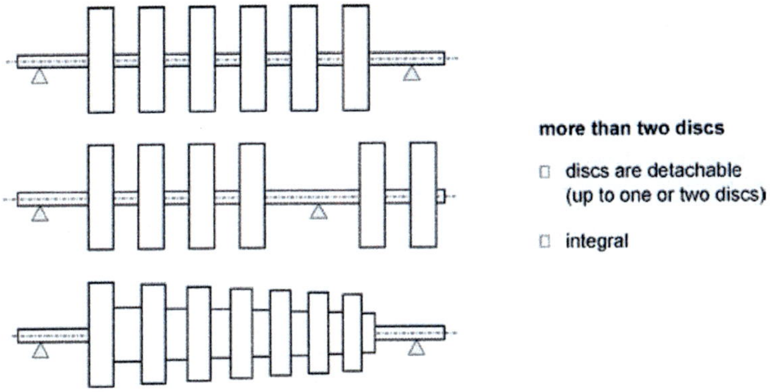

Fig. 10.3 Rotors with more than two discs

Conclusion
There are two alternatives for this balancing task:

1. Assumption: all discs (except one) can be removed. Sequential balancing balancing during assembly is possible: The rotor with one disc is treated according to Sect. 10.3.3.1 (case 1), then the second disc is mounted and in the next step the unbalance is corrected only on this disc in one plane (for problems see Sect. 10.4.5).[3]
2. Assumption: not enough discs can be removed (i.e. integral rotor).
   Only high-speed multi-plane balancing can be used.

Case 2
If all discs have axial run-out errors, then one resulting unbalance and one moment unbalance occur per disc. If at least one resulting unbalance and one moment unbalance are out of tolerance, the measured unbalances can no longer be reliably assigned to the planes from which they originate.

Conclusion
There are two alternatives for this balancing task:

[3] Two discs can always be mounted also and corrected in these two planes by two-plane balancing.

1. Assumption: all discs (except one) can be removed. Sequential balancing during assembly is possible: The rotor with one disc is treated according to Sect. 10.3.3.1 (Case 2), then the second disc is mounted and in the next step the unbalance is corrected only on this disc in two planes. This is continued until the rotor is complete (for problems see Sect. 10.4.2).
2. Assumption: not enough discs can be removed (integral rotor). Only high-speed multi-plane balancing can be applied.

10.3.4 Rigid Sections

A rigid section, see Fig. 10.1c, basically contains—because of its length—a moment unbalance in addition to a resultant unbalance. Furthermore, it is assumed that the modal unbalances *within the* rigid section are negligible, as they do not influence the flexural modes to be balanced. Various configurations are shown in Fig. 10.4.

Thus, a single rigid section behaves like a disc with axial runout in terms of balancing and is to be treated in exactly the same way (see Sect. 10.3.3).

For rotors with multiple rigid sections, the same rules apply as for rotors with multiple rigid discs that have axial run-out errors.

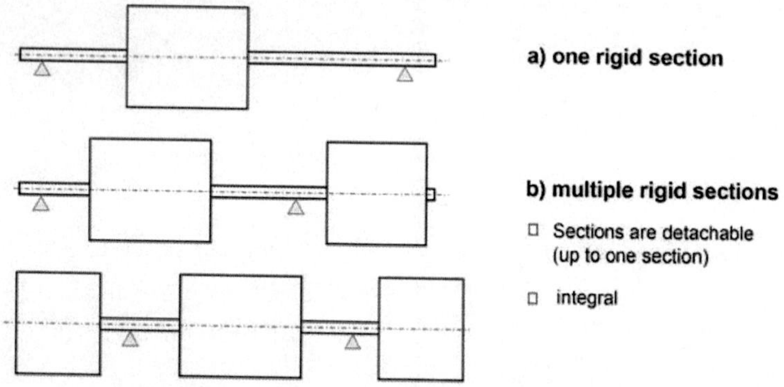

Fig. 10.4 Rotors with one (**a**) or with several rigid sections (**b**)

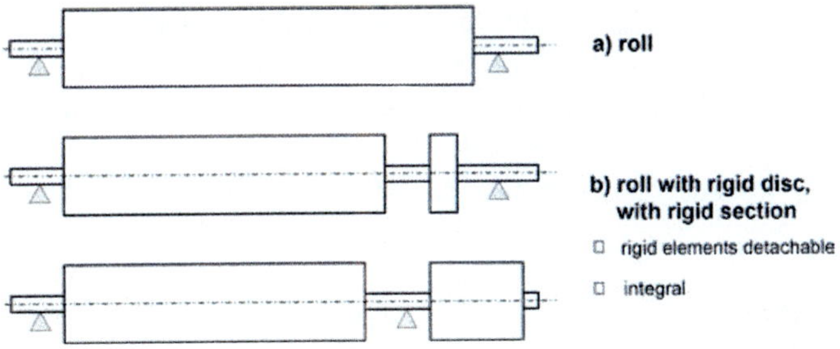

Fig. 10.5 Roll (**a**); with rigid disc, or rigid section (**b**)

10.3.5 Rolls

In the case of rolls, in contrast to rigid sections, it is assumed that the modal unbalances cannot be neglected, see Fig. 10.1d.

In practice, cylindrical rotors are often designed as solid rolls or as tubes. In this case, they have a uniformly distributed mass and stiffness due to their design. The unbalance is also usually evenly distributed, e.g. due to an offset, oblique or curved bore.

In such a case, the low-speed unbalances (resultant unbalance and moment unbalance) and the modal unbalances can be dependent on each other, i.e. have a fixed relationship (in size and angle).

Then it may be possible to find two planes (optimal planes) in which all types of unbalance are reduced together in a low-speed correction to such an extent that the rotor also runs well at high speed.

The appropriate *low-speed* balancing procedure can often only be determined after one or more rotors have been balanced at *high speed* and the unbalances has been analysed (Fig. 10.5).

10.3.6 Integral Rotor

An integral rotor is either one-piece, or composed of several parts, but which cannot be disassembled to a required extent for the balancing process in order to perform a procedure according to Sect. 10.3.3 or Sect. 10.3.4, e.g. a turbogenerator, Fig. 10.6.

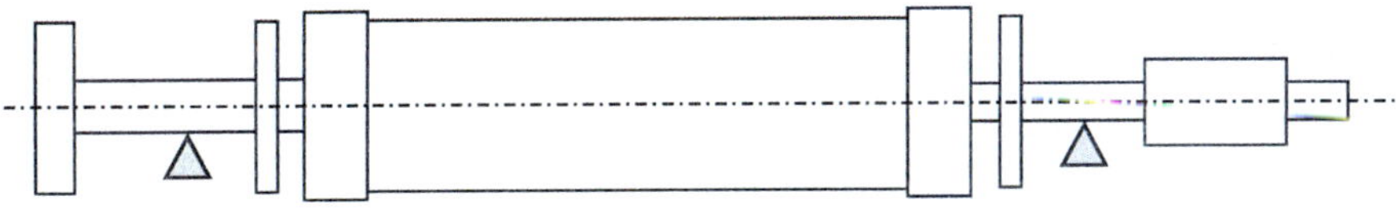

Fig. 10.6 Integral rotor: a turbogenerator

Mass and stiffness are distributed; the unbalance distribution can make jumps and is not predictable. Consequently, there is no typical correlation between the low-speed unbalances (resultant unbalance and moment unbalance) and the modal unbalances, so that a procedure as with rolls is ruled out.

Integral rotors can only be balanced at high speed.

10.3.7 Combinations

The difficulty of the balancing task for the various basic elements increases significantly from the disc towards the integral rotor.

It should be noted here: Combinations of the different basic elements always require the procedure of the most difficult element.

10.3.8 Repairs

Even if a shaft-elastic rotor is balanced after production—according to its configuration—following a very specific procedure, e.g. high-speed balancing, a simpler procedure may be possible in case of repair.

Example

A dual-flow LP steam turbine is balanced in production as an integral rotor at high speed in 4 correction planes.

During a repair, both last stages are fitted with new blades, an inspection of the geometry—bearing points, runout deviations on the rotor—shows no faults. ◀

Task

How should the situation be assessed?

Solution

In this case, it can be expected that the new unbalances will only occur in the planes of both last stages, they can then also be corrected in these planes at low-speed.

Of course, this rotor can be balanced even better at low speed if the correction data and residual unbalances of the initial balancing are available and taken into account in this repair balancing.

10.4 Balancing Procedures

ISO 21940-12 lists and explains a number of balancing procedures, split into low-speed and high-speed procedures. In addition to ISO 21940-12, the background to the procedures will be explained and processes specified in the following—the quoted sections therefore refer to this book.

Low Speed Procedures

A Single-plane balancing: Sect. 10.4.1
B Two-plane balancing: Sect. 10.4.2
C Balancing individual components before assembly: Sect. 10.4.3
D Balancing after limiting the initial unbalance: Sect. 10.4.4
E Sequential balancing during assembly: Sect. 10.4.5
F Balancing in the optimum planes: Sect. 10.4.6

High Speed Procedures

G Balancing at several speeds: Sect. 10.4.7
H Balancing at the service speed: Sect. 10.4.8
I Balancing at a fixed speed: Sect. 10.4.9

10.4.1 Procedure A: Single-Plane Balancing

Before balancing, a high-speed measurement of the initial unbalance shows a resulting unbalance and modal unbalances (see Chapter 4), which can be out of tolerance.

However, if the resulting unbalance is corrected in the best plane, i.e. in the plane in which it occurs (see Sect. 10.3.3.1), the modal unbalances are automatically reduced as well: a low-speed balancing is therefore sufficient.

10.4.2 Procedure B: Two-Plane Balancing

Before balancing, a high-speed measurement of the initial unbalance reveals a resultant unbalance, a moment unbalance and modal unbalances (see Chap. 4) out of tolerance.

There are two possibilities:

1. A disc or rigid section has a resultant unbalance and a moment unbalance.
2. Two discs each have a resulting unbalance. Since these unbalances are independent of each other, in the general case a moment unbalance also occurs on this rotor.

However, if these unbalances are corrected in the correct planes (see Sect. 10.3.3.1 and 10.3.3.2), the modal unbalances are automatically reduced as well: a low-speed balancing is sufficient.

10.4.3 Procedure C: Balancing Individual Components Before Assembly

The essential principles and boundary conditions for this procedure are described in Sect. 9.3.

10.4.4 Procedure D: Balancing After Limiting The Initial Unbalance

Limiting the start unbalance[4] can be useful for various reasons:

1. This is an assembled rotor whose individual parts have been balanced before assembly. The start unbalance of the assembly should therefore be within certain limits. If it is outside, it can be assumed that:
   - The individual parts have not been properly balanced, or
   - the assembly process is faulty.

   Then the assembly usually is checked, or the rotor is disassembled again and the unbalances of the individual parts are checked.
2. It is a rotor that should actually be balanced with a higher-quality procedure. The procedure used is therefore not sufficiently precise for the balancing task. Only if the start unbalance is small enough can the desired tolerance still be achieved and low-speed balancing can succeed.

10.4.5 Procedure E: Sequential Balancing During Assembly

With this procedure, attention must be paid to the problem with *transfer unbalances,* which can easily lead to incorrect balancing results, see Sect. 10.4.5.1.

Unfortunately, such a fault cannot be detected in a low-speed balancing procedure. Sadly, if an unsatisfactory running condition occurs in service after correct application of this procedure E, the unbalance tolerances are often drastically reduced without looking for the real reasons.

Particularly in case of rotors, where the E procedure is widely used, tolerances are therefore sometimes stated that are beyond any realisation limit.

[4] ISO 21940-12 uses the term "initial unbalance" here. However, this term refers to the condition after production, *before* balancing starts. Here it refers to the condition after the individual parts have already been balanced. I therefore use the term "start unbalance" here for the beginning of a new balancing step.

10.4.5.1 Problem with Transfer Unbalances

Each balancing step of procedure E is terminated when the achieved unbalances are in tolerance, either as dynamic unbalance, or as resulting unbalance and moment unbalance (often couple unbalance).

However, these measured unbalances only apply to the planes used in the process. In other planes—e.g. in which the next correction step is to take place—completely different unbalances are measured for the same unbalance state, they are called *transfer unbalances*, see Fig. 10.7.

How to understand this?

- When shifting the planes, a moment unbalance does not change in magnitude (it can be shifted to any position on the rotor axis), the couple unbalance derived from it are only dependent on the distances of the respective correction planes (old or new) Fig. 10.7.
- On the other hand, when a resulting unbalance is displaced, a moment unbalance also arises; it is proportional to the displacement length l, Fig. 10.7. With large displacement lengths and small distances between the correction planes, very large unbalances (dominating as a couple unbalance) can occur, often far outside the desired tolerance.

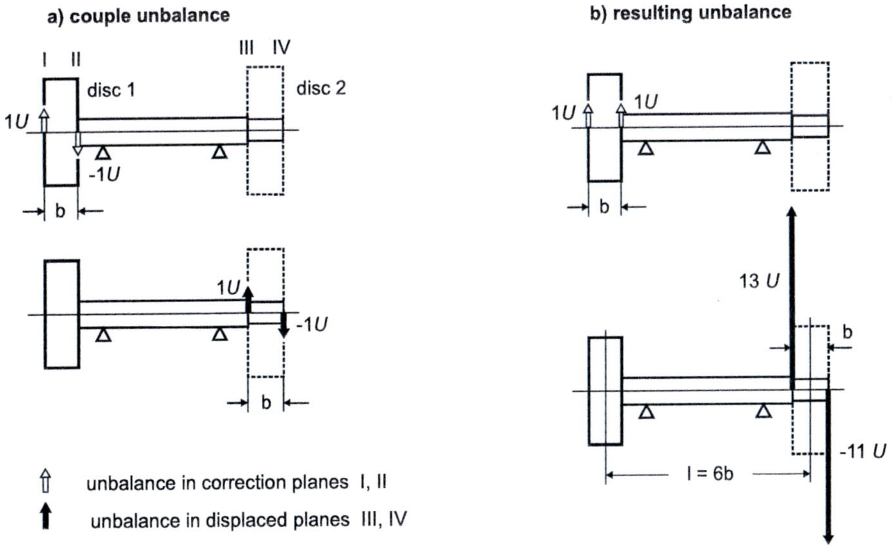

Fig. 10.7 Transfer unbalances: Change in unbalances when measured in displaced planes. A moment unbalance (**a**) changes little, but a resulting unbalance (**b**) can appear in multiple magnitudes, predominantly as a couple unbalance

If the unbalance is measured in these planes only after the next element has been mounted, these transfer unbalances will overlap with the unbalances that have been added by this new element and the assembly.

Correcting this total unbalance also corrects the unbalance caused by the new element, but leaves additional modal unbalances in the rotor.

If there is an elastic rotor element between both pairs of planes, these modal unbalance may not be acceptable.

10.4.5.2 Solution

So what is to be done to avoid this error? Fig. 10.7 shows a rotor with two discs, each with two correction planes. Disc 1 (left) is fixed, disc 2 (right) can be removed.

A special procedure is used, shown for *a single* plane in Fig. 10.8:

Procedure for detecting and eliminating transfer unbalances:

- The rotor is balanced (without disc 2) in two planes (*I*, *II*) of the left disc: Residual unbalance measurement 1.
- This unbalance condition is then measured in planes (*III*, *IV*) of disc 2 before this disc is assenbled: Measurement 2.
- Disc 2 is mounted and the unbalance is measured again in these planes (*III*, *IV*): Measurement 3.
- The vector difference between measurements 2 and 3 is the unbalance added by disc 2 and its assembling.
- Only *this* unbalance is corrected at disc 2 in planes *III* and *IV*.

In practice, it is not always easy to perform this balancing: the vector difference must be formed correctly and if the tolerance cannot be achieved in one step, a corresponding evaluation must be repeated for the next balancing.

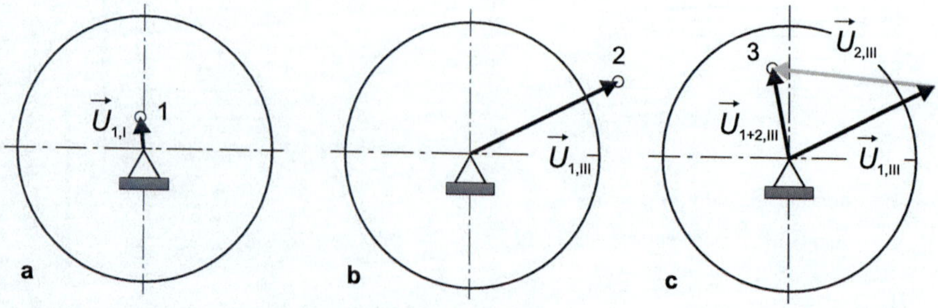

Fig. 10.8 Step-by-step elimination of the transfer unbalances, shown for one plane: (**a**) Residual unbalance of part 1 in plane *I*; (**b**) display of this residual unbalance in plane *III*; (**c**) unbalance display in plane *III* after assembly of part 2. The measurement of the total unbalance 3 is not corrected, but only the vector difference between measurements 2 and 3

However, modern balancing machines can support this process by compensating measurement 2 (in planes *III* and *IV*) in the measuring device so that it subsequently displays zero. After putting on disc 2, the display shows exactly *the* unbalances that have to be corrected on disc 2—even if several correction steps are needed.

If the transfer unbalances are switched off in this way, only the respective residual unbalances remain with each balancing step, in this case the residual unbalances of planes I, II and those of planes III, IV.

10.4.5.3 Problem Assembly

When sequential balancing is applied, problems associated with assembly often occur.

A distinction must be made between two types of design:

1. Parts are placed one after the other on a base body with two *initial* bearing planes, for example as in the case of a multi-stage pump rotor.
2. Balancing starts with a part *without original* bearing planes (or only one). Further parts are connected to corresponding fitting planes of the previous assembly, for example as in the case of jet engines.

In the case of the *pump* impeller, there is a risk of distortion of the base body due to the assembly of a part (impeller).

The concentricity must then be checked at each assembly step. The required accuracy results from the unbalance tolerance and should be better by a factor of 3–5 than the permissible specific residual unbalance (in μm).

If the basic body distorts to an unacceptable extent, unbalances will occur which are not only related to the newly mounted part and therefore must *not* be corrected. One can calculate the new balancing target based on the runout measurement and knowledge of the masses and lever arms and take it into account when correcting. However, the safer way is to use one of the high-speed balancing procedures.

In the case of *jet engine rotor*, auxiliary parts are used on horizontal balancing machines to create the missing bearing positions. Errors of these parts (unbalances, runout errors) are eliminated by index balancing.

Sometimes the rotor parts are assembled on rotary tables to detect misalignments and take them into account applying different procedures.

10.4.6 Procedure F: Balancing in Optimal Planes

The typical characteristics of the rotors are described in Sect. 10.3.5. Only if these characteristics occur permanently in a series (for this, high-speed balancing is sometimes required on a trial basis), the optimal planes for a low-speed procedure can be determined with sufficient experience.

10.4.7 Procedure G: Balancing at Multiple Speeds

The typical characteristics of the rotors are described in Sect. 10.3.6. The unbalance situation is described in Sect. 4.4.

Balancing shaft-elastic rotors at multiple speeds is based on a modal concept: the flexural modes are specifically used for measuring the unbalance distribution and correction (see Sect. 10.4.7.3).

Two different approaches have been used for decades:

1. First the low-speed unbalance of the rotor is corrected, then (one after the other) N modal unbalances. This is also called 2+N procedure.
2. Balancing starts directly with the N modal unbalances, which are treated one after the other. At service speed, another correction is applied. This is also called N+2 procedure.

ISO 21940-12 unfortunately describes these two procedures together, blurring the serious differences.

Therefore, here some explanations, but first a general criticism of a practice that is mostly practised today:

1. If a rotor is operated *subcritically*, but flexural resonances are expected above the service speed, it is quite natural to check whether these flexural resonances need to be treated in terms of balancing.
2. If a rotor is operated supercritically, i.e. has already undergone flexural resonances up to the service speed, but the flexural resonances above the service speed are ignored (e.g. ISO 21940-12, Appendix D).

This is illogical and can lead to considerable errors, which then have to be ironed out later on the test floor or at the final service place by in-situ balancing.

10.4.7.1 2 + N procedure and N+2 Procedure

In Sect. 4.1.5 a special representation of the unbalance state of a shaft-elastic rotor and its different unbalance tolerances is described, an example shows Fig. 10.9.

Up to the service speed n_S, two flexural speeds are passed, three more are above n_S. In this example, both low-speed unbalance types of the rotor are out of tolerance and the first three modal unbalances, the 3rd being above the service speed. The other modal unbalances need not be considered.

If the 2+N procedure is used, the low-speed unbalances of the rotor are corrected first, then the modal unbalances of the three flexural resonance speeds one after the other. For the 3rd resonance speed, the service speed is used (see Sect. 4.6).

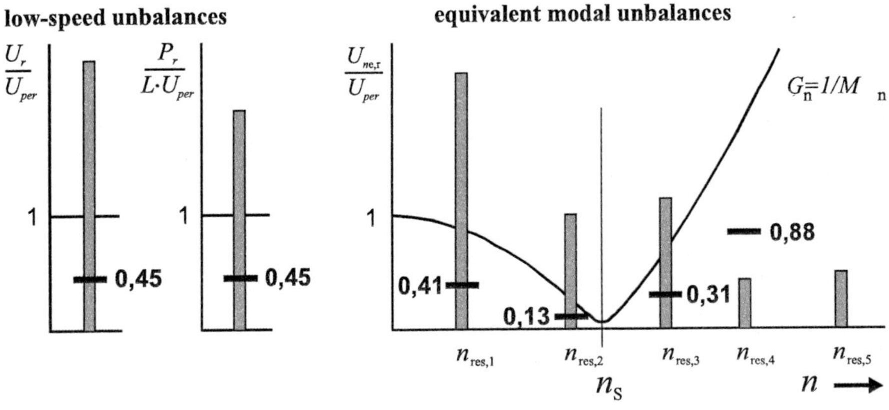

Fig. 10.9 Example of the unbalance state of a shaft-elastic rotor. The low-speed unbalances are out of tolerance, as are two modal unbalances below the service speed n_S and one above

The correction always aims for zero, or for corresponding small residual values of the unbalances. Each step is clearly defined and can be applied to any rotor—even an unknown one.

The N+2 procedure starts directly with the modal unbalances, not working towards display zero, but only towards the disappearance (or sufficient reduction) of the respective resonance loop.

At service speed, the unresolved low-speed unbalances of the rotor *and* the modal unbalance of the 3rd critical (above service speed) must then be dealt with. There is no *general* technical approach for this task (3 different unbalances), a compromise must be found in each case, depending on the circumstances.

The N+2 procedure can therefore not be regarded as a generally applicable procedure.

10.4.7.2 Correction Ratio

Rotors with shaft-elastic behaviour can have very different initial unbalances and unbalance tolerances. The correction ratio is the ratio of initial unbalance to permissible residual unbalance.

If the correction ratio is small, compromises can be made; if the correction ratio is large, very precise work is required, i.e.:

- If this ratio is small (e.g. 2–3), a satisfactory result can sometimes be achieved *without* correcting the low-speed unbalance.
- If this ratio is large (e.g. 5–20), a systematic treatment of the different unbalances is only possible *with* the correction of the low-speed unbalance.

10.4.7.3 Recommendation

One can therefore conclude:

- Low-speed balancing of the rotor may increase the excitation of the flexural modes, but allows a better view to the flexural modes during the balancing process.
- Balancing near each flexural resonance gives the clearest information about the excitations of the different flexural modes.
- The correction of one to two modal unbalances above the service speed succeeds, even if these resonances cannot be approached: The tolerances required there are comparatively rough. However, the other unbalances may have to be corrected first so that the modal unbalances above can be sufficiently revealed and measured (see also the different tolerances in Fig. 10.9).

For a systematic procedure with an *unknown* rotor and with rotors with a *large correction ratio,* the 2+N procedure with correction of the low-speed unbalance of the rotor should therefore always be used. Only this procedure is explained in detail in this book.

The individual steps are:

1. Correction of the low-speed unbalance of the rotor.
2. Measurements are obtained near the 1st flexural speed, this flexural mode is calibrated with a set of test weights and corrected by a specific correction (without worsening the low-speed condition).
3. The other flexural modes are treated the same way, whereby correspondingly more correction planes are required and care must be taken that the achieved state—of the low-speed state and the lower flexural modes—is not deteriorated.

After having corrected the low-speed unbalances and also the flexural modes with appropriate weight sets, the rotor is completely balanced. The result is a low vibration level over the entire speed range, when passing through flexural resonances as well as at service speed.

Figure 10.10 shows the measured values at the beginning and in the course of the different process steps. Here it was assumed that the low-speed balancing of the rotor takes place in planes which almost do not influence the 1st and 2nd eigenmode, i.e. the vectorial run-up curve is only shifted in parallel as a result.

In Fig. 10.9 some details become obvious, that must be taken into account when balancing these rotors:

- The flexural mode very near the service speed is heavily weighted, so this modal unbalance must be reduced to a small tolerance value, (see Sect. 4.1.4).
- The importance of flexural modes above the service speed is rapidly decreasing. If these modal unbalances need correction, they only need to be reduced to correspondingly larger tolerance values (see Sect. 4.1.4).

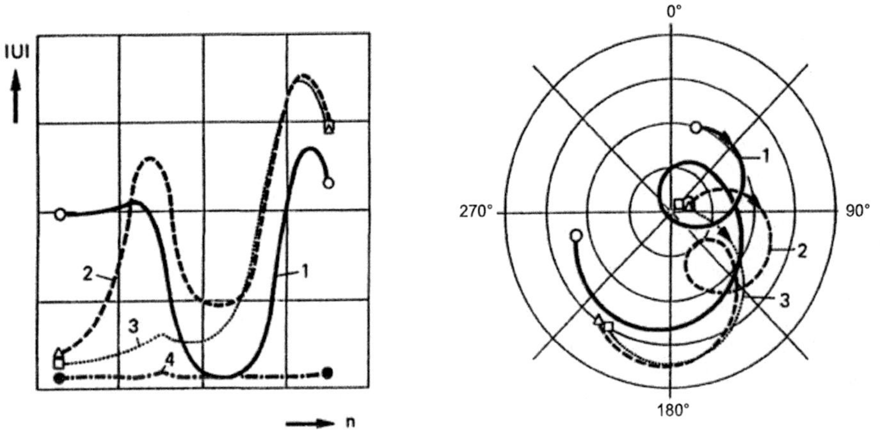

Fig. 10.10 Balancing at multiple speeds. *Left:* Amount of the balancing machine display, *right:* polar diagram. 1 Initial state, **2** after correction of the low-speed unbalances, **3** after correction of the 1st flexural mode, **4** after correction of the 2nd flexural mode (omitted in the vector diagram, as almost only a point in this scale)

10.4.7.4 Computer Support

When balancing shaft-elastic rotors, a large amount of data has to be collected and processed, a typical task for computers.

Consequently, computers of different sizes and capabilities with special balancing programmes have been used for this purpose for several decades, with some typical advantages:

- With the effect of the individual masses (or mass sets) on the correction planes on all modes, influence coefficients can be systematically established and the correction masses calculated from the measured values of the initial unbalance.
- In addition, rotor-typical data can be stored in the process, which can be reused for balancing the same type of rotor, thus eliminating some of the test runs.
- If it is possible to run through the entire speed range with all individual masses (or mass sets), the computer can give correction instructions for all correction planes together at the end of the measurement runs.
- Here, the instruction for *each* correction plane contains all the proportions calculated for the correction of the low-speed unbalances of the rotor and for the modal unbalances considered, combined into one vector.
- Correction of the individual flexural modes one after the other—as is necessary in a manual process—is no longer necessary here, which makes work much easier.
- It also eliminates the need to combine the correction masses in the different planes, which is sometimes necessary at the end.

When using a computer, it can be problematic if the individual flexural modes have very different tolerances. Due to the automatic summation, it is always necessary to work with the precision of the most sensitive part, which is not always possible.

In such a case, it may make sense to split the correction again and correct special modes separately.

10.4.7.5 Beiblatt 1 to DIN ISO 21940-12[5]

A German working group had set itself the goal (with the original VDI Guideline 3835) of detailing procedure G to such an extent that all the basic principles and notes required for practice are included. The VDI draft was published in 2009. After a thorough revision, the final publication took place in 2015 as Beiblatt 1 to DIN ISO 21940-12.

In the very comprehensive paper, a flow chart contrasts a ***sequential*** approach (step-by-step processing of the relevant unbalances) with a ***simultaneous*** approach (simultaneous processing), Fig. 10.11. In practice, there are also mixed forms, e.g. if balancing starts with simultaneous processing, but the fine balancing of certain flexural modes—for example with particularly narrow tolerances—is then carried out sequentially.

However, it is also made clear that jumps between the procedures are conceivable, which cannot all be marked.

10.4.8 Procedure H: Balancing at Service Speed

In principle, a shaft-elastic rotor can be balanced at any chosen speed in two correction planes in such a way that its bearing forces and bearing vibrations are sufficiently small. However, this condition only applies to *the bearing* support in which the rotor actually is running (see Sect. 6.3.2) and at the same time means that the flexure of the rotor has *not* been treated.

Procedure H is therefore not a generally applicable procedure, but is limited to very specific cases where:

- The passage of flexural resonances is fast enough.
- Flexural resonances are without significance, e.g. they are well damped.
- There is no flexural resonance near the service speed.
- The remaining flexures of the rotor do not bother.

---

[5] DIN ISO 21940: Mechanische Schwingungen—Auswuchten von Rotoren—Teil 12: Verfahren und Toleranzen für Rotoren mit nachgiebigem Verhalten, Beiblatt 1 (2015): Verfahren zum Auswuchten bei mehreren Drehzahlen. (English: DIN ISO 21940—Part 12: Procedures and tolerances for rotors with flexible behaviour, Supplement 1 (2015): Methods for balancing at multiple speeds).

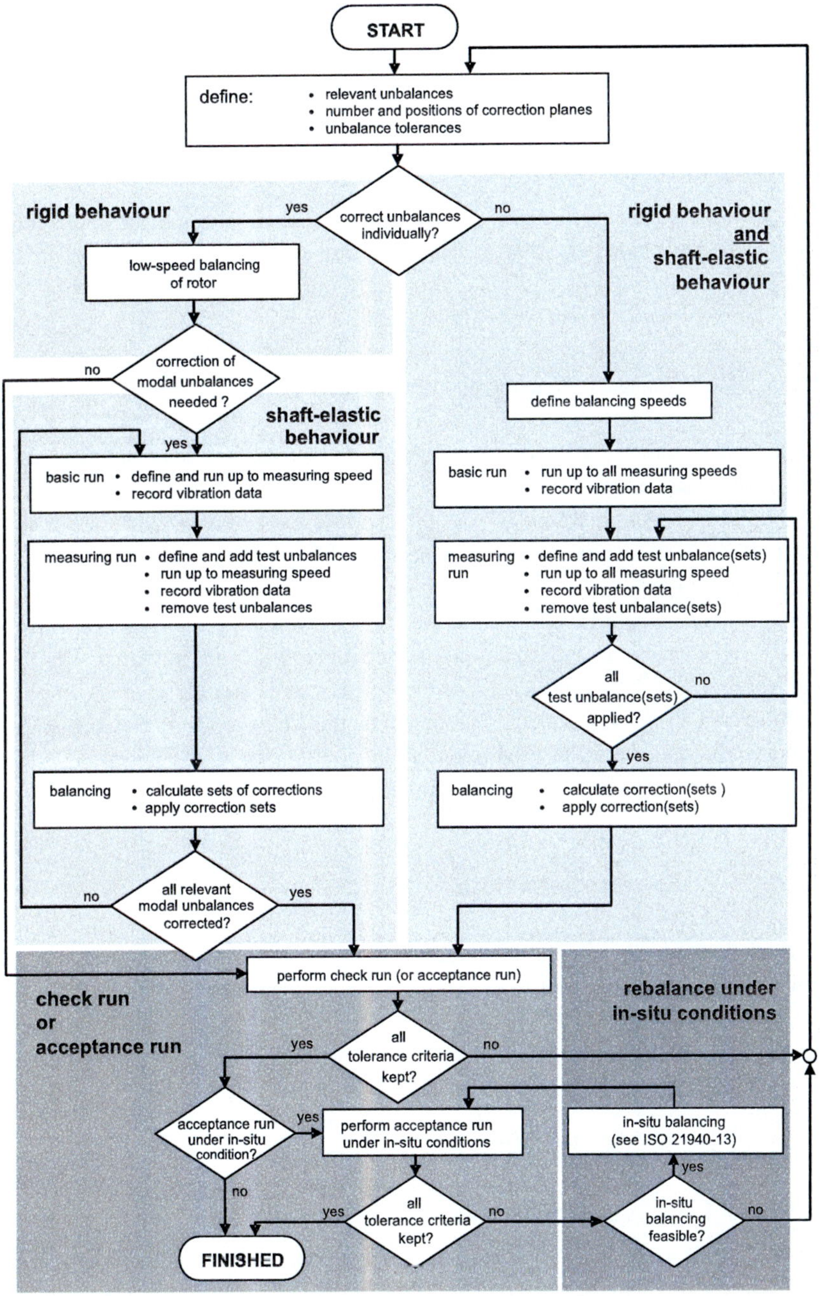

Fig. 10.11 Flow chart for sequential processing *(left)* and simultaneous processing *(right)*. Extract from DIN ISO 21940-12, Beiblatt 1

In order to minimise influences of modal unbalances, also some principles of low-speed balancing procedures can be applied, e.g. the choice of optimal planes.

10.4.9 Procedure I: Balancing at a Fixed Speed

The body-elastic behaviour of these rotors (see Sect. 6.2) results from a single component (or several components), that is elastic in itself or is elastically connected to the main rotor.

There are two very different types of behaviour:

1. Rotors whose unbalance changes continuously with speed, e.g. fans with rubber blades,
2. Rotors whose unbalance changes at a certain speed and then remains constant, e.g. rotors with a centrifugal switch.

Group 1 rotors should be balanced at the speed at which smooth running is required, i.e. mostly at service speed.

Group 2 rotors should be balanced at the speed at which the changes have subsided, i.e. that no more change in unbalance occurs from this speed to the service speed.

The shaft-elastic behaviour can also occur in rotors with other behaviours (see Chap. 6). The balancing procedures assigned to this shaft-elastic behaviour must be applied anyhow; the special procedure due to body-elasticity is then added. This superposition can make balancing significantly more complex and difficult.

10.4.10 Settling Behaviour Procedure

ISO 21940-12 does not assign a procedure letter to this behaviour (it is not a balancing procedure in the literal sense of the word, because no unbalance corrections are carried out), but describes the basic procedure.

The settling behaviour is characterised by a change in the unbalance state (at a certain speed, Sect. 6.2), which:

- Does not return when the speed is reduced.
- Only occurs once with a rotor (as long as its assembly state is not changed).

Procedure: the rotor is brought up to such a high speed that this change of unbalance state occurs *once* (e.g. a settling), then it can be balanced with a suitable procedure (A to I).

Description of the Balancing Task

11

Contents

11.1 Preliminary Note

How to arrange balancing tasks systematically today is shown in Sect. 4.1.5.

However, this is not yet sufficient to concretely describe the requirements by a rotor, e.g. for:

- The purchase of a suitable balancing machine.
- Choosing the right balancing machine from workshop stock.
- Outsourcing the balancing process to another company.

For this purpose, it seems useful to distinguish between rotors with rigid behaviour and rotors with flexible behaviour.

H. Schneider, *Rotor Balancing*, https://doi.org/10.1007/978-3-662-66049-2_11

The information compiled here relates to the ***balancing objectives.*** For the respective ***balancing process*** itself, some errors must be taken into account, as explained in principle in Sect. 4.7 and detailed in Chaps. 7 and 8.

are.

11.2 Balancing Rotors with Rigid Behaviour

In order to facilitate this step, with which many a company or person enters new territory, ISO 21940-21 contains a compilation of information in the appendix to describe the balancing task.

For low-speed balancing, a distinction is made between rotors with and without journals and the horizontal and vertical balancing machines commonly used for this purpose.

11.2.1 Rotor with Journals

If different rotor types are to be balanced, detailed information based on production drawings and assembly instructions of each type is best.

Drawings of the complete machine can provide important information, e.g. about the bearing and installation sequence, or about the operating situation.

In the case of individual rotors, the information on balancing is then added, e.g. on the mounting and drive on the balancing machine, balancing tolerance and tolerance planes, correction planes and type of correction (Fig. 11.1).

In case of mass-produced rotors, additional statistical data on the initial unbalance of a larger number of rotors is required (so that the distribution and maximum initial unbalance are known).

It is helpful to specify the desired distribution over the number of correction steps (one, two, or three), as well as the percentage of rotors that should be in tolerance after the intended correction (e.g. 97%).

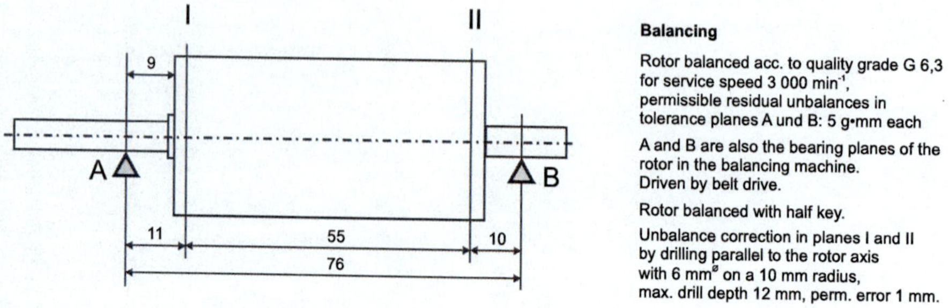

Fig. 11.1 Balancing instructions for a small electric armature

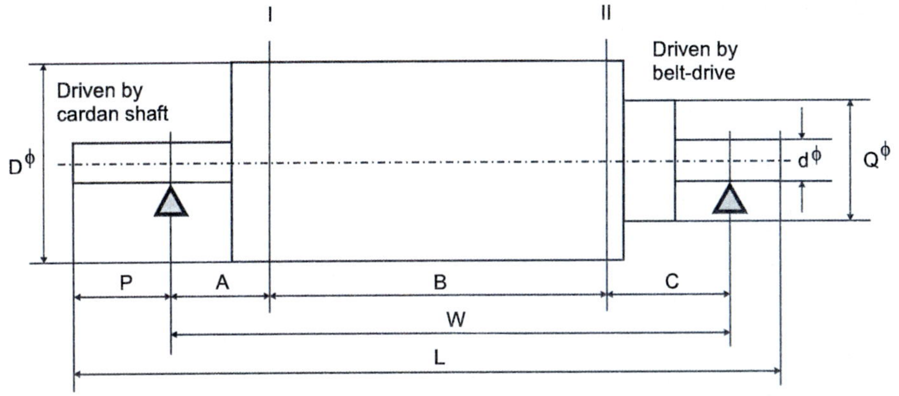

Fig. 11.2 Balancing a rotor with journals on both sides on a horizontal balancing machine: Abbreviated designations of the essential dimensions

11.2.1.1 Tabular Description of a Rotor Type

If many different types of rotors are to be balanced, the data essential for balancing must be collected.

The different sizes of a rotor type can be summarised in a table. For the dimensions of a rotor with journals on both sides, abbreviated designations are used in accordance with a dimension sketch, see Fig. 11.2.

For rotors that are supported overhung, abbreviated designations are used according to the dimensioning sketch, see Fig. 11.3.

An example of a series of medium size fans from 1.5 to 80 kg is shown in Table 11.1.

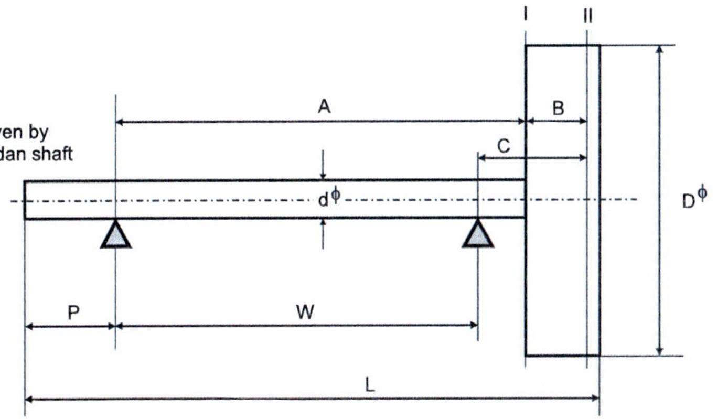

Fig. 11.3 Balancing of an overhung rotor on a horizontal balancing machine: Abbreviated designations of the essential dimensions

Table 11.1 Data for a horizontal balancing machine—table describing a series (rotor type with journals). Abbreviations according to Fig. 11.3

	Fans	Size 1	...	Size n	Unit
1	Mass (weight) *G*	1.5		80	kg
2	Batch size	20		1–3	Piece
3	Production rate at 100% capacity utilisation	15		———	Piece/h
4	Diameter *D*	250		1400	mm
5	Diameter for belt drive *Q*	Not applicable			mm
6	Length *L*	300		1600	mm
7	Bearing journal diameter *d*	20		60/80	mm
8	Distance between bearing planes *W*	265		1100	mm
9	Location of correction planes *I* and *II*				
	A	40		1200	mm
	B	170		350	mm
	C	55		450	mm
10	Shaft overhung *P*	25		50	mm
11	Service speed *n*	1500		900	min^{-1}
12	Critical speed[a]	Not applicable			min^{-1}
13	Moment of inertia mr_i^2 [b]	Not applicable			$kg \cdot m^2$
14	Aerodynamic power at service speed	0.8		45	kW
15	Max. initial unbalance	300		50,000	$g \cdot mm$
16	Quality grade G[c]	6.3		6.3	mm/s
17	Type of drive	Cardan shaft			
18	Type of correction	Addition of masses			

[a]Important e.g. for rotors that are originally supported in multiple bearings (crankshafts) but are only supported in two bearings in the balancing machine: the first flexural resonance speed in these bearing

[b]Compared to the aerodynamic power, the power needed for acceleration of fans can usually be neglected

[c]Quality grade G, or permissible residual unbalance per plane

11.2.1.2 More Tables

In the case described, motor armatures, pulleys and couplings are also to be balanced, for which corresponding tables are to be established.

11.2.1.3 Maximum Data

It may be useful to give maximum data beyond that of the current rotors to leave some reserve for the future. Example:

Mass $G_{max} = 800$ kg for electric armatures,
Diameter $D_{max} = 1{,}700$ mm for fans.

11.2.1.4 Additional Information on the Rotors

There are other characteristics of the rotors that are important for balancing, they can be described by a list, e.g.:

- In case of unusual geometry of the rotors, multiple correction planes or anomalous position of the correction planes, or in order to make certain details clearer, drawings are most advantageous.
- If the rotor is cantilevered (Fig. 11.3), indicate the negative (upward) load at A and the positive (downward) load at B.
- If we assume that the mass of the auxiliary shaft for the large fan in Table 11.1 is 20 kg (which is distributed approximately equally between the bearings), this results in a negative load of approximately 30 kg at A and a positive load of approximately 110 kg at B.
- If an axial force occurs during balancing (e.g. due to flow forces), its direction and presumed magnitude shall be indicated.
- Are the necessary holders (for mounting the rotors in the balancing machine) and connecting parts, e.g. drivers for the drive shafts, pulleys, auxiliary shafts, available or do they have to be procured?
- If rotors are to be balanced with their antifriction bearings, these are to be described, either by stating the standardized designation or by the type and outside diameter.
- If a specific balancing speed is desired, the reasons for this should be stated.
- Are there devices for correction?
- Are there any other special features of the rotor? For example, a rotating magnetic field, special aerodynamic effects?

11.2.2 Rotors Without Journals

Rotors without journals can be mounted on auxiliary shafts and then balanced on horizontal balancing machines the same way as rotors with journals, see Sect. 9.2.

However, vertical balancing machines are often used, with bearings and drive incorporated and having a flange with a centring device at the upper end. The rotor is mounted, usually with the help of an adaptor.

Compared to the rotor with journals, a new situation arises mainly due to the centring conditions. The parts may be centred by:

1. Two radial centring points (e.g. a long bore axis).
2. A radial and an axial centring on *one* end face.
3. A radial and an axial centring on *each* end face.

The vertical balancing machines can handle the first two cases well, but for the 3rd type of centring special solutions are necessary.

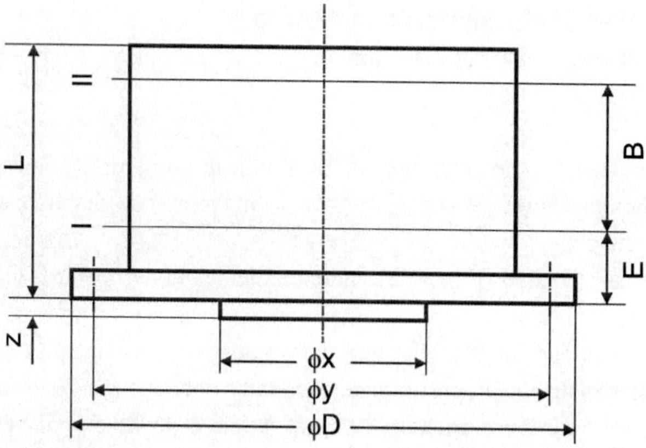

Fig. 11.4 Balancing on a vertical balancing machine: Abbreviated designations for the dimensioning of a rotor without journals, see Table 11.2 (for a vertical balancing machine). x, y and z are dimensions for mounting the adaptor

The essential dimensions of parts without journals can be given in accordance with ISO 21940-21 and the dimensioning scheme (Fig. 11.4). The dimensions x, y and z are dimensions for the mounting of the parts. Only for large series the machine flange is designed to accommodate the parts directly, mostly adapters are interposed.

For a series of a part type, the data are entered in a table (Table 11.2). Here too—as with the rotors with pivots—additional tables should be filled in for further series.

Table 11.2 Data for a vertical balancing machine: Table describing a series of rotors (without journals). Abbreviations according to Fig. 11.4

	Pulleys	Size 1	...	Size n	Unit
1	Mass (weight)	1.8		20	kg
2	Batch size	80		10	Piece
3	Production rate at 100% capacity utilisation	30		20	Piece/h
4	Diameter *D*	120		350	mm
5	Height *L*	70		250	mm
6	Locating hole	24		52	mm
7	Location of correction planes I and II				
	E	12		28	mm
	B	35		135	mm
8	Moment of inertia mr_i^2	$4.0 \cdot 10^{-3}$		0.35	$kg \cdot m^2$
9	Max. initial unbalance	220		1,900	$g \cdot mm$
10	Quality grade G [a]	6.3		6.3	mm/s
11	Type of correction	Drilling			

[a]Quality grade or permissible residual unbalance per plane

Certain maximum data may be of interest (see Sect. 11.2.1.3) and additional information on the rotors can describe the task more precisely (see Sect. 11.2.1.4).

11.3 Balancing Rotors with Flexible Behaviour

11.3.1 Low-Speed Balancing

Low-speed balancing is mainly used for rotors with rigid behaviour. However, there are a number of low-speed balancing methods (A-F) for rotors with shaft-elastic behaviour also, Sect. 10.4.

In principle, balancing machines of the same type are used, as described in Sect. 11.2. Sometimes, however, increased demands are placed on the balancing machines. For example, if the distance between the tolerance planes is small, but above all because of the sometimes increased tolerance requirements, see Sect. 8.4.

Because of the distribution of the total permissible unbalance tolerance over a large number of tolerance planes, these are the balancing methods:

C. Balancing individual components before assembly.
E. Sequential balancing during assembly.

In addition to the data for the rotor, the intended balancing procedure is also important here, with details of the process steps if necessary.

11.3.2 High-Speed Balancing

11.3.2.1 General

ISO 21940-12 describes three different high-speed balancing methods (G to I) for rotors with flexible behaviour, Sect. 10.4.

In the high-speed case, in addition to the typical data for a rotor with a journals (see Sect. 11.2), details on the behaviour of the rotor are added, i.e. (see Chap. 6):

- Shaft-elastic behaviour: Procedure G and H.
- Body-elastic behaviour: Procedure I.
- Plastic behaviour: Settling process.

Because of the speeds, which can go up to the service speed, or even to overspeed, balancing the rotors places much higher demands on the balancing machine and the necessary protective measures (see Chap. 17).

These requirements must be sufficiently specified.

11.3.2.2 Tabular Overview

For rotors with shaft-elastic behaviour, the following overview (by the author) lists some details (Table 11.3) that are important for a requirements profile.

The methods H and I use two planes for balancing, so here it is sufficient to specify the total permissible unbalance U_{per} and the speed to be used.

With method G, modal unbalances are also corrected for in addition to the low-speed unbalances—here a detailed specification of the unbalance tolerances is required. For this purpose it is best to use a representation as in Fig. 11.5.

Table 11.3 Data for a high-speed balancing machine—table to describe a series of rotors with shaft-elastic behaviour

Type	Designation	x	y	z
Drawing rotor	Designation			
Drawing machine	Designation			
Quantity	Pieces/a			
Main data				
Mass	kg			
Stat. load bearing drive side	N			
Stat. load bearing end side	N			
Total length	mm			
Bearing distance	mm			
Distance coupling—bearing	mm			
Diameter max.	mm			
Diameter bearing drive side	mm			
Diameter bearing end side	mm			
Rotordynamic data				
Service speed max.	min^{-1}			
Overspeed	min^{-1}			
1st Flexural resonance speed	min^{-1}			
2nd Flexural resonance speed	min^{-1}			
3rd Flexural resonance speed	min^{-1}			
Flexural modes	Drawings			
Correction planes	Number, location			
Balancing at several speeds				
1st Speed	min^{-1}			
2nd Speed	min^{-1}			
3rd Speed	min^{-1}			

(continued)

Table 11.3 (continued)

Type	Designation	x	y	z
4th Speed	min^{-1}			
Boundary conditions				
Drive				
Power	kW			
Torque	N·m			
Sleeve bearing				
Bearing drive-side				
Design	Designation			
Outer diameter	mm			
Width	mm			
High pressure oil	bar, °C, l/min			
Lubricating oil	bar, °C, l/min			
Bearing end-side				
Design	Designation			
Outer diameter	mm			
Width	mm			
High pressure oil	bar, °C, l/min			
Lubricating oil	bar, °C, l/min			
Special features				
Enclosure of the rotor				
Vacuum required?				
Protective structure				

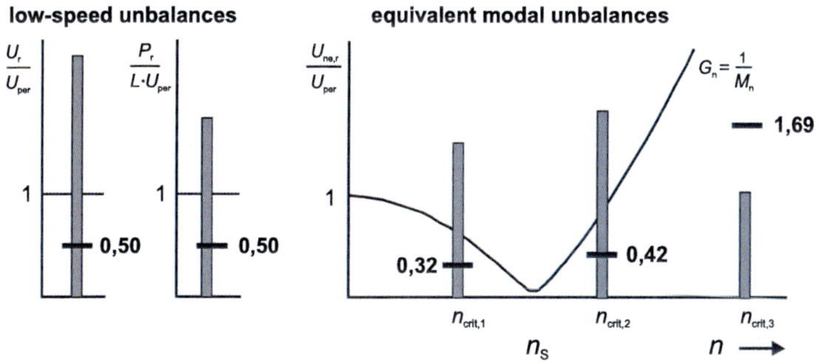

Fig. 11.5 Balancing on a high-speed balancing machine: The requirements due to the unbalance tolerances are best represented by such a diagram

Balancing Machines

12

Contents

H. Schneider, *Rotor Balancing*, https://doi.org/10.1007/978-3-662-66049-2_12

12.1 Preliminary Note

Each balancing task must be assigned to a suitable balancing machine. Since most balancing machines can accommodate a range of rotor masses and dimensions, different designs and sizes hold available in a machine pool are sometimes considered.

Trends are important when the balancing task is difficult in terms of the residual unbalance required, the correction step or the complexity of the process. For difficult situations:

- Balancing machine to be utilised by the rotor in the upper half of the load capacity (50–70%),
- Balancing speed (also for rotors in rigid condition) tends to be at the upper limit of the permissible range.

12.2 Quotation and Technical Documentation

It is not easy to interpret quotations on balancing machines and the associated technical documentation correctly. ISO 21940-21 gives some practical advice here, which also forms the basis for the following sections.

In order to improve clarity, this section is divided into horizontal and vertical balancing machines, non-rotating balancing machines and high-speed balancing machines.

The tables and lists are general, the figures in the sections are only to be taken as examples, and they do not represent any standardisation.

12.2.1 Horizontal Balancing Machines

Rotors with two bearing positions (or picked up from both ends) are usually balanced on horizontal balancing machines with two bearing pedestals (Fig. 12.1).

12.2.1.1 Limits for Rotor Mass and Unbalance

The information is best checked for completeness using the compilation in Table 12.1.

12.2.1.2 Efficiency of the Measuring Run

The criteria for the economic efficiency of the measuring run can be seen in Table 12.2 (see also Sect. 12.3.16).

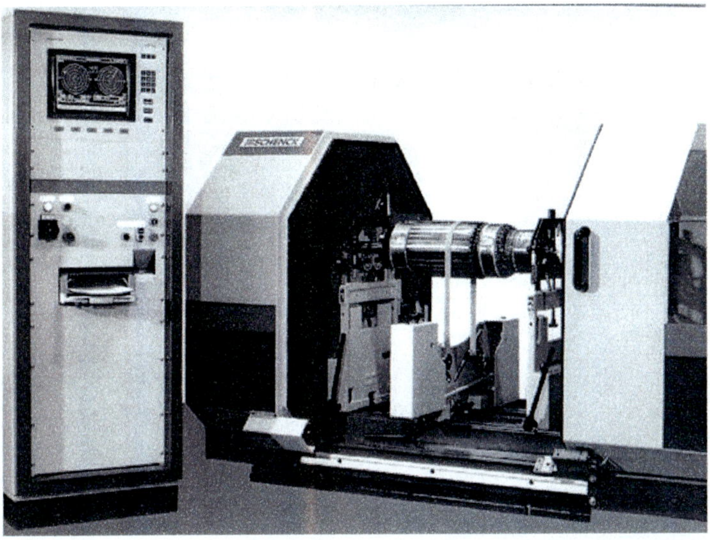

Fig. 12.1 General view of a horizontal balancing machine for balancing various rotor types

Table 12.1 Horizontal balancing machine; limits for rotor mass and unbalance

Manufacturer				Type			
1	Balancing speeds (or ranges)		min^{-1}	230	420	470	1250
2	Rotor mass	Max.	kg	— 750 —			450
		Min.	kg	Depending on balancing quality			
3	Occasional overload per bearing pedestal		kg	500	– none –		
4	Max. negative load per bearing pedestal		kg	— 50 —			
5	Max. moment of inertia of the rotor		$kg \cdot m^2$	4000	1200	400	140
6	Number of cycles (at max. moment of inertia)		Z/h	— 10 —			
7	Max. unbalance	Measurable	$g \cdot mm$	$2 \cdot 10^6$	$3 \cdot 10^5$	$2 \cdot 10^5$	$5 \cdot 10^4$
		Permitted	$g \cdot mm$	$2 \cdot 10^6$	$3 \cdot 10^5$	$2 \cdot 10^5$	$5 \cdot 10^4$
8	Minimum achievable residual unbalance U_{mar}		$g \cdot mm$	800	80	80	8
				However, not better than 0.5 $g \cdot mm$ per kg rotor weight per plane			

Table 12.2 Efficiency of the measuring run

Step		Time [s]	a	b	c	d
1	mechanical adjustment	180	•	•		
2	setting the measuring equipment	60	•	•		
3	preparing the rotor, e.g. add reference mark	1		•		•
4	mean acceleration time	2		•	•	•
5	taking readings	2		•	•	•
6	mean breaking time	2		•	•	•
7	transfer of readings to the rotor	10		•	•	•
8	others, e.g. activating a guard	-		•	•	•
Valuation and total times of the measuring run						
a	extremely important for single rotors: adjustment and setting	240 ← a				
b	time for the 1. measuring run of a new rotor	257 ← b				
c	time for additional readings of the same rotor	16 ← c				
d	important for series production: time for the next rotor of the same type	17 ← d				

12.2.1.3 Unbalance Reduction Ratio R_{UR}

The unbalance reduction ratio test is described in Sect. 13.5.

- For ISO test rotors type B: 95%
  (correction planes between the bearing planes, see Sect. 13.3.5)
- For ISO test rotors type C: 85%
  (correction planes outside the bearing planes, see Sect. 13.3.6).

12.2.1.4 Rotor Dimensions

The limitations in the contour of the rotor in a horizontal balancing machine are best described using the principle drawing in Fig. 12.2.

Restrictions in the contour of the rotor based on the example of a horizontal balancing machine:

- Rotor diameter max.:
  - Above bed: 1250 mm
  - Outboard, beyond the bed end: 2000 mm
- Maximum and minimum lengths:
  - Distance between bearing centres max.: 1540 mm
  - Distance between bearing centres min.: 100 mm
  - Max. distance between the coupling flange (the cardan shaft) and the centre of the distant bearing: 1540 mm
  - Min. distance between the coupling flange and the centre of the adjacent bearing: 0 mm.

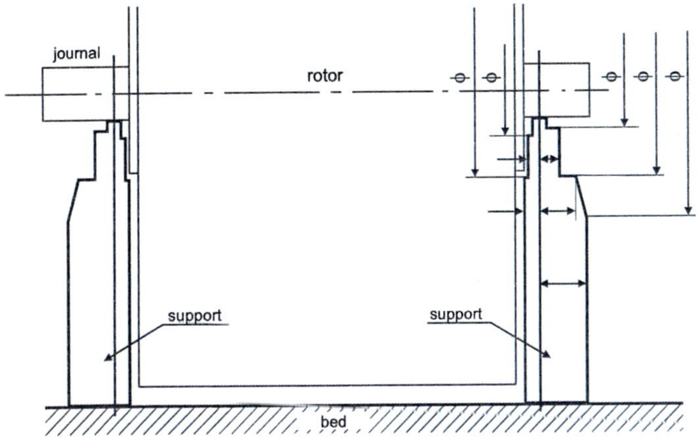

Fig. 12.2 Horizontal balancing machine: Principle drawing of bearing pedestal dimensions and the resulting limitation of the rotor contour

12.2.1.5 Bearing Journal

- Max. diameter: 180 mm
- Min diameter: 16 mm
- Max. permissible peripheral speed: 40 m/s

12.2.1.6 Setting Range of the Correction Planes

Calibrated setting: Ratio of bearing distance to correction plane distance max. 20:1. Position of correction planes arbitrary; also overhung. Permissible rotor mass centre position also outside the bearings. Setting possible either for two planes, or according to resulting unbalance and couple unbalance.

12.2.1.7 Drive

General explanations of the drive are given in Sect. 12.3.1. Here an example, Table 12.3, and details:

Details

- Starting torque, related to the nominal torque on the rotor: 120%.
- Torque peak, related to the nominal torque on the rotor: 120%.
- Transmission of the drive power to the rotor by cardan shaft, axially displaceable by ± 20 mm via gearbox.
- Motor
  - Type: Three-phase slip-ring motor
  - Rated power: 15 kW
  - Rated speed of the motor: 1450 min^{-1}
  - Connection data: 380 V, 50 Hz, three phases

Table 12.3 Example: Drive data of a horizontal balancing machine

Balancing speeds	Nominal torque on rotor	
min^{-1}	Star	Triangle
	N · m	N · m
230	220	650
420	120	360
740	70	200
1250	40	120

12.2.1.8 Brake

General explanations of the brake are given in Sect. 12.3.4. Here are details as an example:

- Type: countercurrent brake.
- Braking torque, related to the nominal torque: max. 100%.
- Can the brake be used as a locking brake? No.

12.2.1.9 Additional Information

Above data must be supplemented by a series of details relating to the following criteria:

- Transducer, functional principle of the balancing machine (Sect. 12.3.3),
- Display system (Sect. 12.3.2).
- Setting the measuring device (Sect. 12.3.6).
- Environmental influences (Sect. 12.3.14).
- Overall view of the machine, see Fig. 12.2.
- Plane separation (Sect. 12.3.6).
- Special accessories.
- Boundary conditions (Sect. 12.4).

12.2.2 Vertical Balancing Machines

If rotors do not have journals and are only picked up from one end (or via the bore), vertical balancing machines are usually considered (Fig. 12.3); in special cases, horizontal machines with a spindle are also used.

Many statements for the horizontal balancing machines (see Sect. 12.2.1) can be used with appropriate modifications, but in some places specific statements lead more quickly to the goal.

12.2.2.1 Limits for Rotor Mass and Unbalance

The information is best checked for completeness using the compilation in Table 12.4 (numerical values as an example).

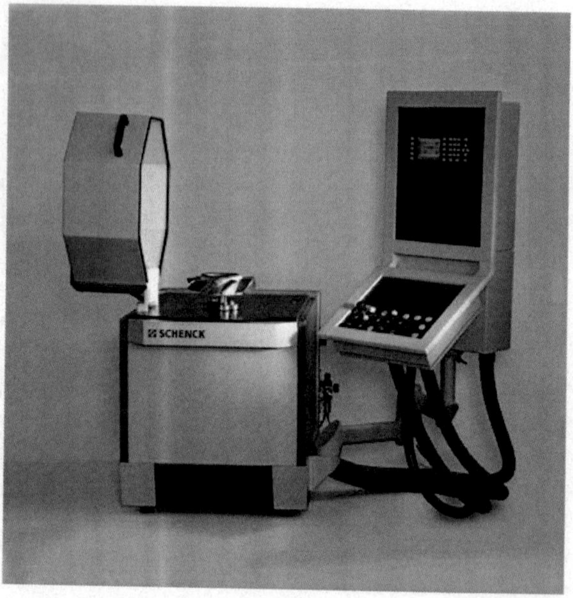

Fig. 12.3 Overall view of a vertical balancing machine for balancing various rotors without journals

Table 12.4 Vertical balancing machine, limits for rotor mass and unbalance

Manufacturer:			Type:		
1	Balancing speeds or ranges		min^{-1}	600	1200
2	Rotor mass (including adapter)[a]	Max.	kg	50	50
		Min.	kg	depending on balancing quality	
3	Occasional overload		kg	——	
4	Max. moment of inertia of the rotor (incl. adapter)		$kg \cdot m^2$	4.0	1.2
5	Number of cycles (at max. moment of inertia)		Z/h	30	30
6	Max. unbalance	Measurable	$g \cdot mm$	$5 \cdot 10^3$	$1 \cdot 10^3$
		Permitted	$g \cdot mm$	$2 \cdot 10^3$	$1 \cdot 10^3$
7	Minimum achievable residual unbalance U_{mar}		$g \cdot mm$	10	4
	However, not better than 0.5–2.0 $g \cdot mm$ per kg rotor mass				

[a]The following limit value should not be exceeded for the quality of the plane separation and calibration (when set according to the geometry) (h = height of the centre of gravity above the clamping plate): $G(h+60)^2\, n^2 = 10^{12}\ kg \cdot mm^2/min^2$

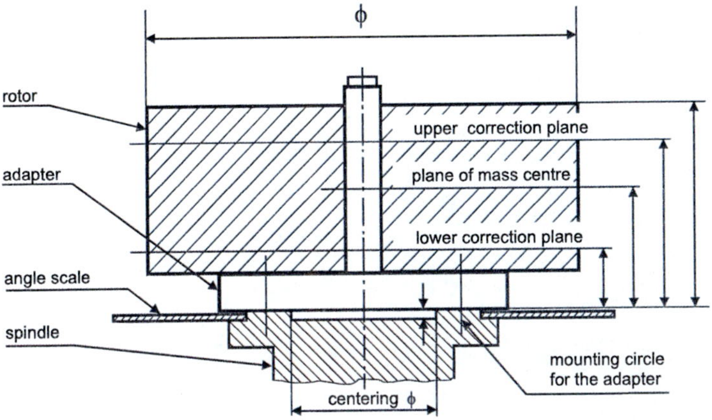

Fig. 12.4 Vertical balancing machine. Principle drawing of the spindle connection, an adapter, the limitation of the rotor contour, the position of the mass centre and the correction planes

12.2.2.2 Rotor Dimensions

The rotor dimensions suitable for the machine are best described by a drawing, (Fig. 12.4).

These include, above all, the connection dimensions of the spindle (or the adapter), the limitation of the rotor contour (e.g. by a protective cover or by correction units), the position of the mass centre and the correction planes.

Restrictions in the contour of the rotor:

- Rotor diameter max. (possibly limited by protective cover, or correction units): 350 mm
- Heights
  - Total height max. (possibly limited by protective cover): 240 mm
  - Max. height of mass centre (at max. mass)[1]: 170/55 mm

12.2.2.3 Influence of the Moment Unbalance

Most vertical machines are used for balancing in a single correction plane. Here, the influence of a moment unbalance on the indication of the resulting unbalance is interesting. The moment unbalance influence ratio (I_{MU}) is given in g · mm/g · mm^2. A test is described in Sect. 13.6.

[1] Data for 600/1200 min^{-1}; for infinitely variable drive, a curve is appropriate.

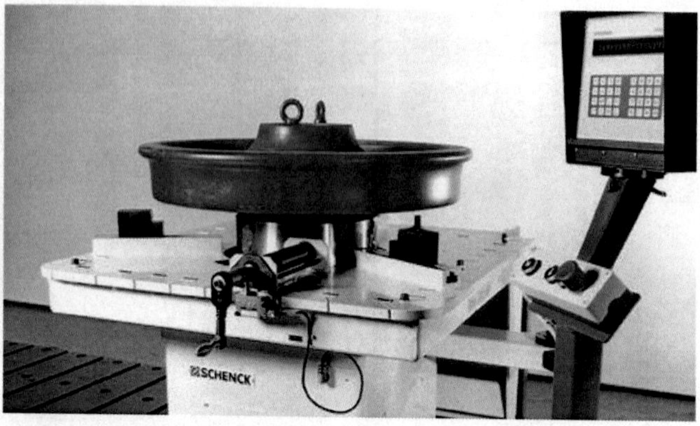

Fig. 12.5 Non-rotating balancing machine, here used for railway wheels

12.2.3 Non-Rotating Balancing Machines

According to the ISO definition (the old ISO 1925[2]), these are devices where the body to be balanced does not rotate, but which nevertheless provide a measured value for the unbalance vector, i.e. amount and angular position. They thus go far beyond simple balancing devices such as knife-edges or roller blocks.

The body is adapted with vertical axis (Fig. 12.5). The weight moment of its mass centre, which is outside the shaft axis by the eccentricity of the mass centre, is used for the measurement.

Only single-plane balancing can be carried out on non-rotating balancing machines, i.e. only the resulting unbalance can be measured and corrected (see Sect. 5.7.2), as the mass centre is already on the shaft axis in the case of moment unbalance.

The required data largely correspond to a vertical single-plane balancing machine, only all the data belonging to rotation are missing. A protective cover is normally not required also.

12.2.4 High-Speed Balancing Machines

For high-speed balancing machines (Fig. 12.6)—for balancing rotors with flexible behaviour—there is no ISO recommendation yet. However, experience has shown that some additional points need to be considered.

---

[2]This term is no longer included in the revised standard ISO 21940–2, but it is still used.

Fig. 12.6 High-speed balancing machine with plain bearing and vacuum tunnel, used for compressors of gear-driven compressors

Essential requirements for the balancing process of these rotors (see Sects. 10.4.7 to 10.4.10) are:

- An infinitely variable, correctly dimensioned drive.
- A suitable rotor support.
- A suitable bearing system.
- A measuring device that continuously informs the operator of the changing unbalance condition.
- A protective device for the maximum foreseeable case of damage.

12.2.4.1 Drive

The greater the speed dependence of the unbalance condition, the higher the accuracy has to be to use and control the drive.

The drive power is selected large enough that there is a power surplus for acceleration or braking in critical speed ranges or states. With large rotors, this would lead to very uneconomical drive sizes, so other solutions have to be sought.

12.2.4.2 Bearing Pedestals

Except for small rotors, only force-measuring solutions are offered today for high-speed balancing machines.

The risk for the rotors and the balancing machine can be reduced by using bearing pedestals with variable stiffness, which can be used to relocate and specifically bypass the critical speeds (see Sect. 6.4.2).

The bearing pedestals should be isotropic, i.e. equally stiff in all radial directions. In this case, all critical speeds occur only once.

If isotropy is not available, it may be that the flexural modes in the measuring direction only occur at higher speeds than perpendicular to the measuring direction. Then an additional measuring aid—a scanning of the bending line perpendicular to the normal measuring direction—is required.

12.2.4.3 Measuring Device

The measuring device shall display the bearing planes (or measuring planes) simultaneously and automatically and quickly follow the changed unbalance condition when the speed changes.

With resonance approximation or passages, the clarity of the vectorial display is most favourable (see Sect. 2.7.1 and Fig. 10.10).

12.3 Technical Details and their Assessment

Some technical terms and their meaning are now explained in alphabetical order. In accordance with ISO 1925[3], a detailed collection of vocabulary and definitions is compiled in Chapter 19 (see Sect. 19.3).

12.3.1 Drive

The drive has the task of bringing the rotor up to the desired balancing speed and then maintaining this with certain accuracy—usually required by the measuring device.

When balancing only *one* type of rotor in series, the focus is on simple coupling of the drive. However, if many *different* rotors are to be balanced, it must be possible to easily adapt the drive to the different requirements of the rotors.

With the balancing machines considered, measurement sensitivity and accuracy increase with speed. Universal balancing machines, which cover a large rotor range, therefore always have several rotor speeds (also for torque adjustment) (see also cardan shaft drive).

Each drive can be divided mentally into the electric motor with control and the mechanical part for transmitting the drive power to the rotor.

The coupling elements on the rotor and the parts that predominantly participate in the movement of the rotor should fully fulfil this transmission task, but at the same time be as light as possible, as they can influence the achievable balancing quality.

[3] The actual ISO 21940–2 has been significantly shortened in scope.

It is advantageous if the motor does not run at the same speed as the rotor in order to keep influences out of the unbalance measurement.

The electric motor shall be designed and controlled in such a way that during acceleration and braking it delivers a torque that is as uniform as possible and only slightly above the rated torque of the motor (a normal three-phase squirrel-cage rotor, for example, delivers about three times the rated torque when switched on directly; the coupling parts would become unnecessarily heavy in order to be able to transmit this torque peak).

Different motor systems are used for connection to the three-phase mains.

12.3.1.1 Squirrel-Cage Motor

The characteristics of these drives depend on the motor design and on additional starting aids with star-delta switching so that the starting and breakdown torque are not significantly higher than the nominal torque (Fig. 12.7). The ramp-up time is usually limited to a short time, e.g. 10 s, as otherwise unacceptable heating of the motor occurs.

Squirrel-cage motors are also used with pole-changing, e.g. with a speed ratio of 1:2, but the nominal torques are usually close to each other, so that there is an adjustment of the speed but not of the torque.

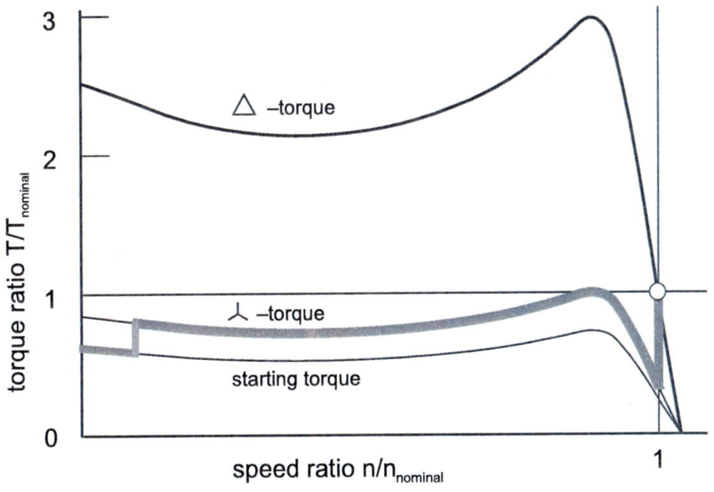

Fig. 12.7 Example of the torque curve of a three-phase squirrel-cage motor with: Start-up stage—star—delta connection

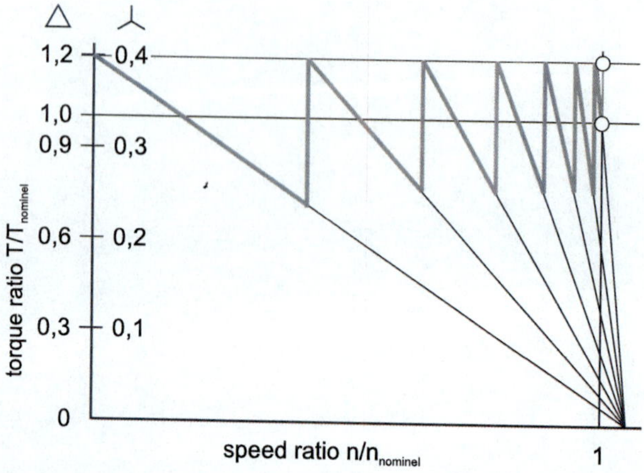

Fig. 12.8 Example of the torque curve of a three-phase slip-ring motor with seven-stage starter and star-delta connection

12.3.1.2 Slip Ring Motor

These drives have a multi-stage start-up circuit, which can be switched manually or automatically. In contrast to the squirrel-cage rotor, the torque on the rotor can be adjusted over the entire speed range by means of the star-delta switching. In both cases, the several starting stages can be switched so that the maximum torque is in a fixed ratio to the nominal torque (Fig. 12.8). In the case of the slip-ring motor with appropriate control, the run-up time can take considerably longer, e.g. 5 min.

For the motors mentioned so far, the speeds are coupled to the mains frequency due to their design. An adjustment is possible by selecting the number of poles.

12.3.1.3 DC Motor

If the speed is to be changed continuously, a DC motor (supplied by rectifier or inverter) is used, which solves the task most elegantly, because the torque can be kept at the desired value either manually or automatically during the entire run-up.

For motors without forced ventilation, run-up times of up to 5 min can be permitted; forced ventilation motors allow longer run-up times, but this is only economical for very large rotors.

12.3.1.4 Drive Power

The drive power is normally selected according to the moments of inertia of the rotors and the desired run-up times (see Sect. 20.4 and Fig. 20.9). With bladed rotors, on the other hand, the drive power must primarily be adapted to the ventilation; the moment of inertia is usually negligible.

The balancing speed has a significant influence on the air flow (see Sect. 20.4 and Fig. 20.10), possibly the air flow can be reduced by covering the rotor.

Fig. 12.9 Drive by cardan shaft when balancing a large pulley

Friction losses in the drive mechanics and the rotor bearing are not taken into account. Only in special cases do these factors have an influence on the choice of drive power (e.g. breakaway torque with plain bearings).

12.3.1.5 Cardan Shaft Drive

With the cardan shaft drive, the rotor is coupled axially via a cardan shaft (e.g. universal joint shaft), it is driven by it and held axially (Fig. 12.9).

The cardan shaft must be selected in such a way that the max. occurring torque is safely transmitted (see Sect. 20.4 and Fig. 20.6).

On the other hand, the cardan shaft includes a possibility of error (see Sect. 16.2), so it should be as light and small as possible. The maximum permissible cardan shaft is determined by the desired centre of gravity eccentricity (see Sect. 20.4 and Fig. 20.8).

With universally used balancing machines, the fulfilment of both requirements can encounter difficulties, example:

The transmission of 15 kW at 420 min^{-1} requires a cardan shaft for at least 350 N · m. However, if a residual unbalance of 10 g · mm/kg is to be achieved in normal workshop operation, only a cardan shaft for max. 250 N · m is permissible for a rotor of 30 kg. If the drive power (and thus the maximum occurring torque) cannot be reduced, a higher balancing speed must be used for this light rotor.

In universal balancing machines, the cardan shaft drive is therefore usually equipped with a multi-stage belt transmission or a manual gearbox.

Axial displacement of the cardan shaft facilitates coupling, especially with heavy rotors.

12.3.1.6 Belt Drive

With a belt drive, the rotor is driven either by a flat belt that wraps around the rotor (Fig. 12.10 and Fig. 12.11), or by a round belt that touches it tangentially (Fig. 12.12).

Fig. 12.10 Belt drive: Wrap-around belt drive for large rotors (paper machine roll)

Fig. 12.11 Belt drive: Hinged bracket belt drive for medium rotors (cutter)

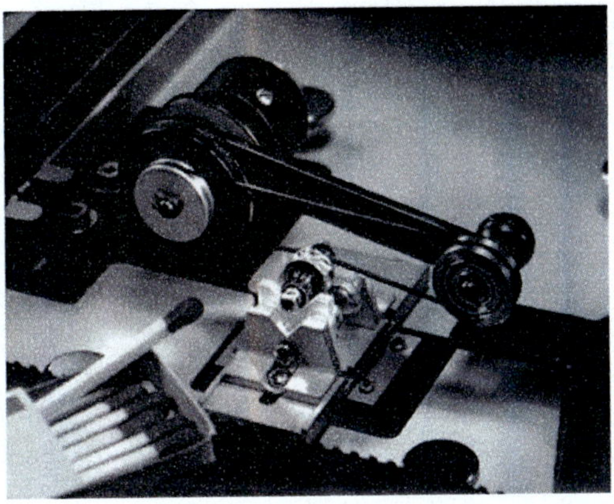

Fig. 12.12 Belt drive: Tangential drive by round belts for small rotors. Bearing in prisms (dental drill turbine)

The belt pulley on the motor and the motor speed determine the belt speed, the rotor speed results from the belt driven diameter on the rotor. Important technical data are the smallest and largest diameter of the belt on the rotor.

Since these support diameters depend on the possibilities on the rotor and cannot be chosen arbitrarily, several belt speeds are required, for example by pole-changing or speed-controlled motors and several belt pulleys.

Axial thrust stops prevent the rotor from moving axially, since the belt drive does not help.

12.3.1.7 Induction Field Drive

A squirrel-cage rotor can be driven by its own stator (see Sect. 12.3.1.8). If a series of armatures is to be accommodated in a stator, special arrangements must be made for quick insertion and removal. In the case of external rotor motors, simpler solutions are usually available (e.g. axial insertion into the bearing) than in the case of armatures with bearings on both sides, where an end shield would have to be removed.

Sometimes special induction field drives are used that are designed for high cycle numbers. For light squirrel-cage rotors, a radially open induction field can also be used, where the rotor is inserted from above (Fig. 12.13). However, only a limited diameter range can be covered with a single stator, since torque transmission deteriorates as the gap increases, causing starting and braking times to increase significantly.

Fig. 12.13 Open induction field drive for an X-ray rotary anode

12.3.1.8 Self-Drive

The self-drive of complete units, e.g. electric motors, usually does not cause any particular difficulties on the drive side, since the adaptation of the required drive power to the losses is given by the design. Though, the cycles required for balancing can overstrain the drive concept (example: gyros).

Of particular advantage is that no mechanical coupling elements are required and also the final bearing is automatically integrated (Fig. 12.14), so that in such cases extremely high balancing grades—G 1 or even G 0.4—can be achieved.

The sometimes very high service speeds can sometimes be reduced somewhat during balancing without loss of accuracy.

12.3.1.9 Compressed Air Drive

Drive by compressed air is still of some importance today. This type of drive is used with bladed rotors, e.g. with turbochargers or even with very light, smooth rotors (Fig. 12.15).

Fig. 12.14 Self-drive of a vacuum cleaner assembly

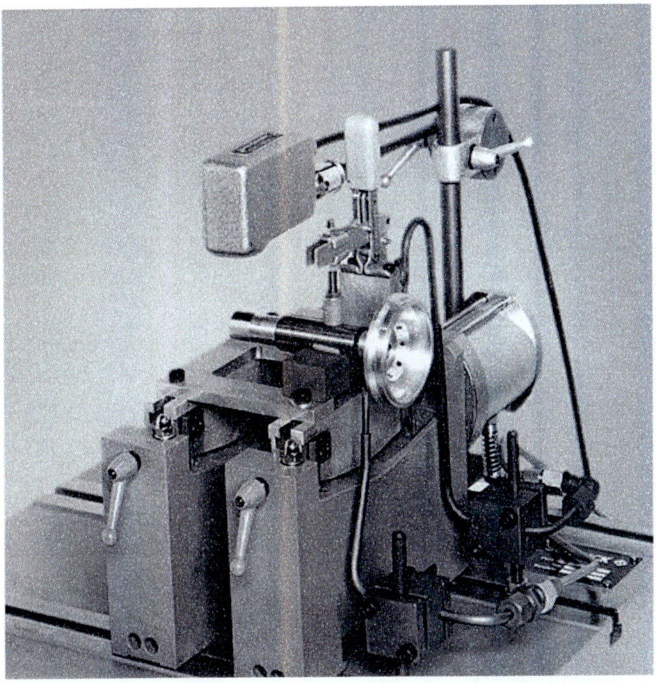

Fig. 12.15 Compressed air drive of a small textile rotor (spinning turbine)

Mainly for noise reasons, the compressed air drive is only used with small units; since in combination with a regularly split surface, a siren-like noise is produced. If no other type of drive can be used for balancing, attention must be paid to an appropriate noise protection on the machine.

If controlling is done by hand, quick accelerating, holding the speed to measure the unbalance, braking to standstill are problematic. But one can use quite reliable automated controls, whereby the biggest advantage of the air drive—its low price—is partly lost.

12.3.2 Display Systems

Balancing machines have display systems for the amount and angle of the unbalance or for correction. Since all newer measuring devices are based on computers, the typical displays are display panels designed from segment displays to various types of screens (ISO 21940-21 is somewhat outdated here).

In principle, screens allow all possible presentations of the measurement results (in reality in colour), e.g.: the vector display (Fig. 12.16), which allows a particularly quick view into the unbalance situation.

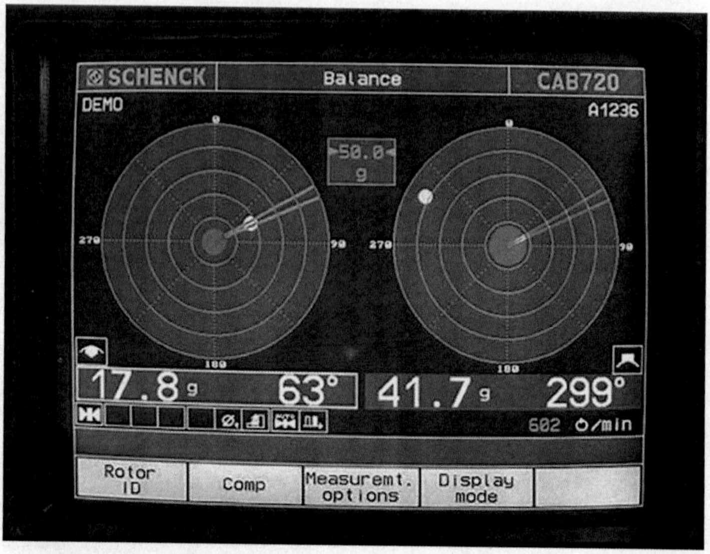

Fig. 12.16 Vector display on a screen, here combined with a numerical display, *on the left* indexed for correction

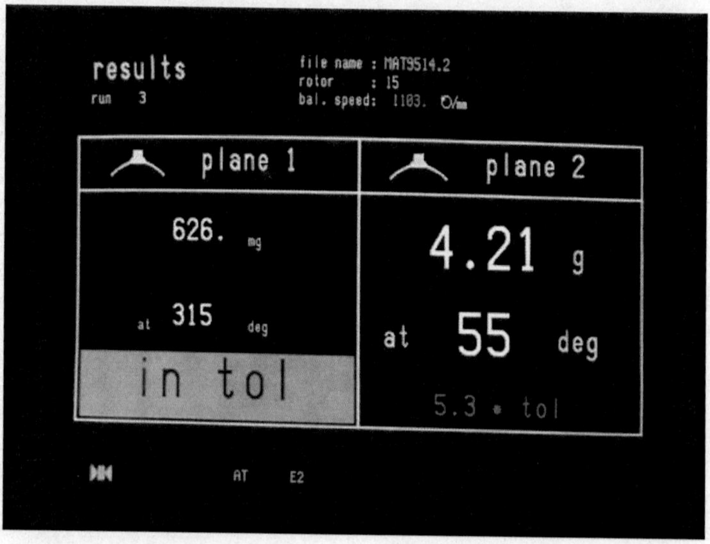

Fig. 12.17 Numerical display on a screen, *on the left* in tolerance, *on the right* out of tolerance

The statement is supplemented by a whole range of details, e.g. about balancing speed and type of unbalance (two planes) and whether the respective plane is in tolerance or not (with tolerance factor).

A full-format numerical representation shows Fig. 12.17: an unbalance correction by radial drilling for polar values Fig. 12.18.

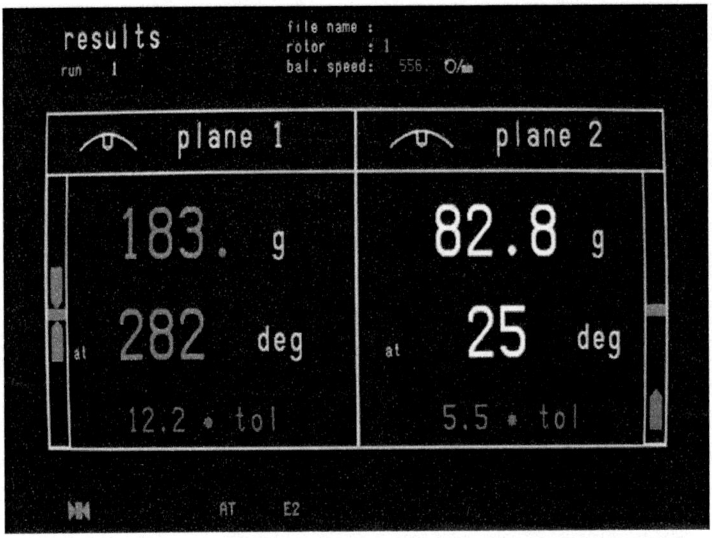

Fig. 12.18 Drilling instruction on a screen, polar, on the left already indexed for the correction

The question: which type of display is most appropriate, vectorial or numerical, can really only be answered in connection with the balancing task and the rotors, but there are typical tendencies:

The *vector display* gives the best impression of the vector character of the unbalance and the size and angle conditions of both correction planes. It is often combined with a numerical display that supports precise correction.

A full-format *numerical display* can be read well even from a distance for correction on large rotors, but in principle it has the disadvantage that the meaning of the displays is not recognised as quickly and that the angular uncertainty, which increases with decreasing unbalance size, is not made clearly visible.

12.3.3 Functional Principle of the Balancing Machine

Modern balancing machines consistently use mechanical–electrical transducers which convert the force, displacement, speed or acceleration of the vibrations into analogue electrical signals.

If these transducers alone are not sufficient to describe the machine system (e.g. force-measuring balancing machine + vibration velocity transducer), the mechanical element on which measurement is made (e.g. dynamometer) should be specified.

12.3.4 Brake

While for small rotors various types of friction brakes can be used, the drives for large rotors preferably contain an electric braking circuit. In some drive systems, the braking energy can be fed back into the grid.

An assessment is best made by braking time and number of cycles, as well as by the braking torque compared to the nominal torque of the motor (see Sect. 12.2.1.8).

12.3.5 Speed

There is always much debate about the correct balancing speed.

Even if one tries to take the phrase to heart: *as low as possible, as high as necessary,* it is unfortunately not quite easy to determine the speed, because it depends on three important blocks:

- The rotor.
- The balancing machine.
- The targeted economic efficiency.

These blocks are broken down further (Fig. 12.19, Fig. 12.20 and Fig. 12.21). From these partly contradictory influences, one tries to find a workable compromise in each individual case. The weighting of the various components plays a major role here, so that different solutions can arise depending on the point of view and available experience.

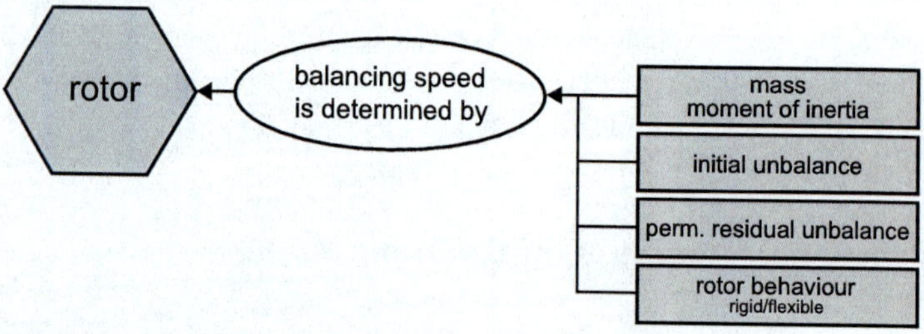

Fig. 12.19 Balancing speed; criteria of the rotor

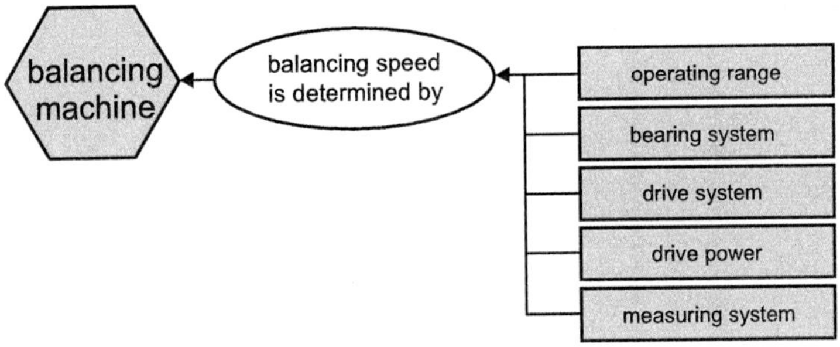

Fig. 12.20 Balancing speed; criteria of the balancing machine

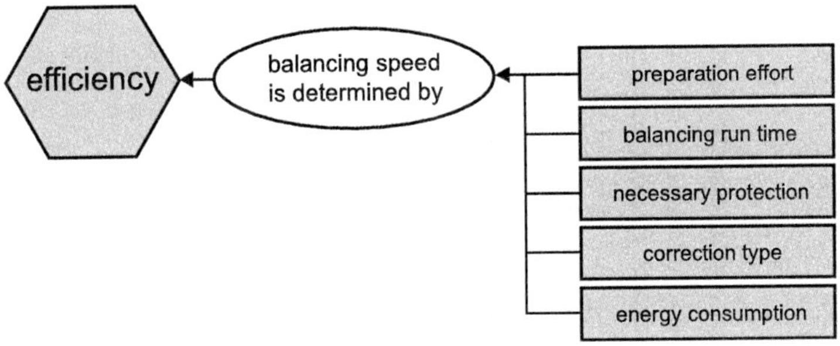

Fig. 12.21 Balancing speed; criteria of economic efficiency

12.3.6 Calibration and Setting of the Measuring Device

This includes all work required to enter the information on calibration weights, correction planes, bearing planes, balancing radii and, if necessary, balancing speed into the balancing machine's measuring device.

The transducers on almost all machines measure in the planes of the bearing pedestals, often identical to the bearing planes. However, the bearing reactions are normally influenced by the unbalances in both correction planes. A calculation is used to separate the planes so that each reading is only determined by the assigned correction plane.

In addition, the display is converted to units, e.g. g correction mass, mm drilling depth, mm material length, etc., which support the balancing process as well as possible.

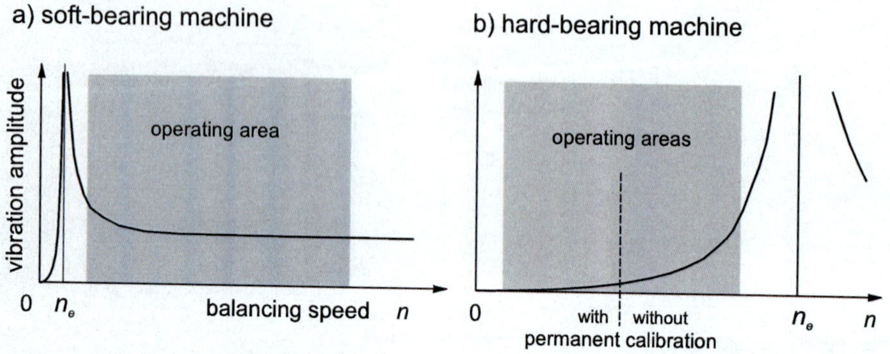

Fig. 12.22 Working areas of the balancing machines: (**a**) above the resonance of rotor/bearing system for the soft-bearing balancing machine and (**b**) below for the hard-bearing balancing machine

12.3.6.1 Soft-Bearing Balancing Machine

In soft-bearing balancing machines, the signals coming from the transducers are not only dependent on the unbalances and their position in relation to the correction planes, but also on the rotor (its masses and moments of inertia) and the bearing support. They operate in the supercritical range (see Sect. 2.7.1.3) of the rotor/bearing system (Fig. 12.22).

The correct setting is therefore found by applying known unbalances (calibration masses) in the correction planes. With analogue measuring instruments, this procedure is made much easier by a compensation device (to suppress the display of the original unbalance); with computer instruments, the system calculates the influence values and stores them.

There are soft-bearing balancing machines where vibration exciters are used in the bearing planes to perform calibration when the rotors are at standstill. However, it seems to be difficult to achieve good unbalance reduction ratios (see Sect. 12.3.15).

An assessment of the calibration of the measuring device for soft-bearing machines is possible on the basis of the following questions:

- In what way is the measuring device for the first rotor of a new type calibrated?
- How many runs (time) are required to adjust the measuring device for two planes?
- How accurately must the speed be maintained during adjustment and subsequent balancing?
- If a compensation device is available: What is the permissible residual unbalance?

12.3.6.2 Hard-Bearing Balancing Machine

Hard-bearing balancing machines operate in the subcritical range (see Sect. 2.7.1.1) of the rotor/bearing system (Fig. 12.22).

If the speed range reaches close to the resonance area (or if the rotor already deforms too much), the calibration of the measuring device is carried out the same way as described for the soft-bearing balancing machine.

If the speed range used is sufficiently far from resonance, the mass properties of the rotors (and support) can be neglected when setting the measuring device. The machine can be calibrated permanently.

Thus the plane separation—i.e. in this case the conversion of the bearing reactions to the correction planes—is only dependent on their axial distances.

These balancing machines are carefully calibrated *once* using a test rotor. All subsequent rotor types that fall within the working range can then be adjusted by entering the corresponding lengths and correction radii when the machine is at a standstill (Fig. 12.23); they show the unbalance already in the first run.

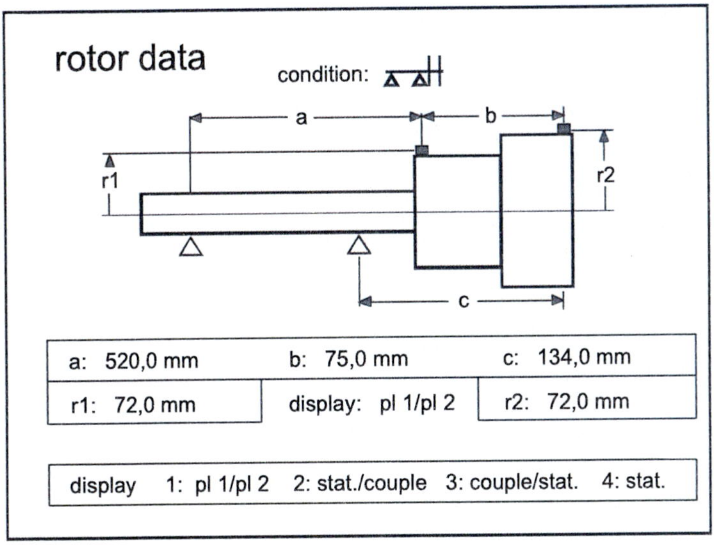

Fig. 12.23 Setting diagram of a hard-bearing balancing machine with calibrated setting, for overhung correction planes on the right hand side

The questions for assessing the measuring equipment are:

- Does the measuring device have a calibrated setting or does it have to be recalibrated for other speeds and masses in each case?
- How accurately must the speed be maintained during calibration and subsequent balancing?
- How many runs are required to calibrate the measuring device for two-plane balancing?

12.3.7 Foundation

The specified properties of the balancing machine are only ensured through correct installation and connection to the ground.

It is important to identify which type of foundation is required, as the effort involved can vary widely. For example; what is needed:

- A work table (workbench)?
- A reinforced hall floor?
- A special concrete foundation?

Sometimes the balancing machine shall be installed not on the ground floor but on an upper floor. Only a thorough analysis protects against negative consequences.

12.3.8 Minimum Achievable Residual Unbalance U_{mar}

The minimum residual unbalance that can be achieved with a balancing machine is given in g · mm, or as specific unbalance in g · mm/kg.

These specifications are of interest for the entire working range of the balancing machine, whereby the influence of the mass and the speed play a role above all. They are supplemented by the absolute limits of the drive means (e.g. cardan shaft, belt) or the measuring device.

U_{mar} is influenced by the measuring method, the size and angle display, the plane separation, the drive, the bearing, etc.

The indicated value primarily refers to the balancing machine, but the value can be worsened by out-of-round pivots of the rotor, or balancing accessories with errors. U_{mar} is checked by a test, preferably with a special test rotor (see Sect. 13.4).

12.3.9 Bearing Support

In horizontal balancing machines, the bearing supports the rotor, allows it to rotate and usually must allow different journal diameters to be accommodated (while maintaining the centre height).

12.3.9.1 Twin-Roller Bearing

The easiest way to support rotors is a twin-roller carriage on which the bearing journal is placed (Fig. 12.24).

They are adjusted in height for different journal diameters in order to maintain the height of the shaft axis. A sliding piece or roller from above the journal prevents the rotor from jumping out of the bearing if the centrifugal forces are too high.

Fig. 12.24 Twin-roller bearing, height adjustable, used for medium to heavy rotors. It is supplemented by a radial counter bearing (from above)

12.3.9.2 V-Block Bearing

For very light rotors, V-block bearings—made of different materials—are also wear-resistant enough (Fig. 12.25).

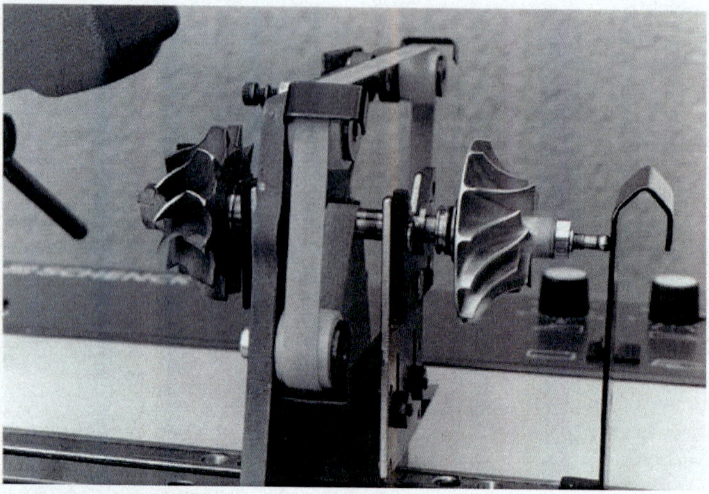

Fig. 12.25 V-block bearing for light rotors (small turbochargers)

12.3.9.3 Sleeve-Bearing

For high speeds and heavy rotors, sleeve-bearings are used, e.g. hydrostatic (Fig. 12.26) or hydrodynamic bearings (Fig. 12.29).

Bearings similar or identical to the operating condition are used, but they are sometimes run with lower oil flow.

Until now, there have been no universal or easily adaptable solutions; plain bearings must be precisely tailored to the bearing journals.

Lightweight rotors with sensitive bearing surfaces can also be accommodated in aerostatic bearings.

Fig. 12.26 Hydrostatic bearing

12.3.9.4 Spindle Bearing

Bearing systems with antifriction bearings or with sleeve-bearings are used as spindle bearings for balancing rotors without journals. They are used in vertical balancing machines, and in horizontal balancing machines, for example, for adapting drive shafts (Fig. 12.27).

Fig. 12.27 Spindle bearings for a cardan shaft

12.3.9.5 Service Bearings

In addition, systems are important that can accommodate the rotor with its service bearings, e.g. with its antifriction bearings (Fig. 12.28).

For high-speed and very heavy rotors, hydrodynamic bearings are mostly used, ranging from simple circular bearings to tilting pad bearings (Fig. 12.29).

Fig. 12.28 Service bearings in the form of antifriction bearings on the compressor of a jet engine

Fig. 12.29 Service bearings in the form of sleeve bearings: balancing and overspeeding system for large generators

12.3.9.6 Special Bearing Systems

As described in Sect. 9.3, certain assemblies must be balanced as complete units. In all cases where the bearing housings have to be included—or even complete assemblies—a special support system is required for this task (Fig. 12.30).

Fig. 12.30 Special support system for complete units (e.g. a motor spindle)

12.3.10 Mass Moment of Inertia, Number of Cycles

In order to be able to assess the performance of the drive and its control, the maximum permissible mass moment of inertia of the rotor is given for each speed that can be accelerated in a specified time.

In addition, there is information on the number of cycles. This is the number of runs (acceleration and deceleration) that the machine can perform in 1 h without thermal problems. It is assumed that a rotor is balanced with the respective maximum permissible moment of inertia (depending on the speed).

12.3.11 Measuring Principle

In addition to the unbalance signal, the signal from the transducers usually contains a wealth of other vibration signals from the rotor, the bearing, the drive and the environment (Fig. 12.31).

From these *disturbing vibrations* - which are sometimes several powers of ten larger than the *useful signal* - the unbalance signal must be filtered out so carefully that it represents the unbalance in size and direction sufficiently accurate.

In computer-based measuring instruments, the display will be digital in any case, the conversion from analogue signal (of the transducer) to digital information can take place at different points between the transducer and the display.

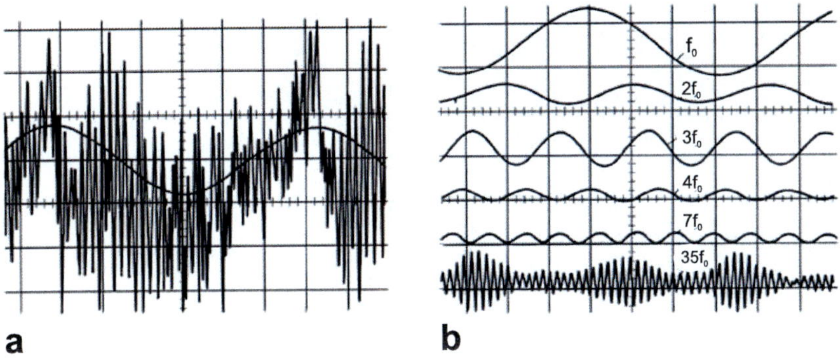

Fig. 12.31 Signal of the transducer: (**a**) with reference signal (sine), (**b**) some of the oscillation components, separated by Fourier analysis

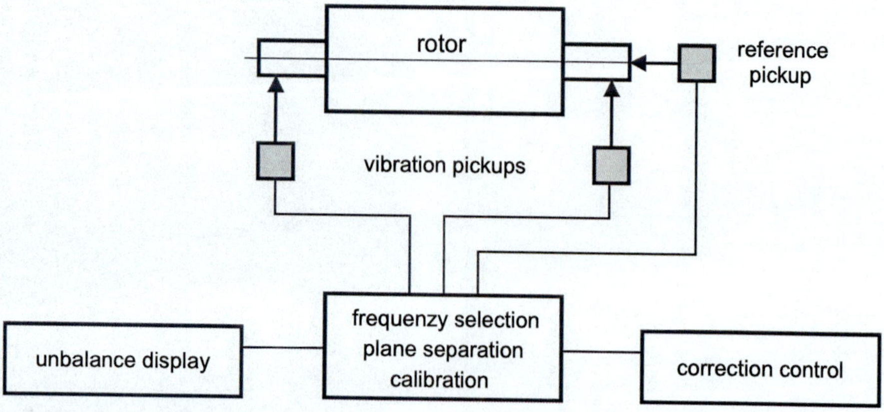

Fig. 12.32 Schematic of the measuring chain of a balancing machine

The quality of the respective design determines how well the interfering vibrations are filtered out and thus the measuring accuracy of the size and angular position of the unbalance. It is best determined by the tests for the minimum achievable residual unbalance U_{mar} (see Sects. 12.3.8 and 13.4) and the unbalance reduction ratio R_{UR} (see Sects. 12.3.15 and 13.5).

For a realistic test, the boundary conditions (e.g. roughness and run-out errors of the bearing journals, bearing type and its condition as well as environmental influences) must also be similar to service conditions. A schematic of the entire measurement chain is shown in Fig. 12.32.

12.3.12 Test Rotors, Test Masses

ISO 21940-21 describes a series of tests (see Sect. 13.4 ff.) based on precisely defined test rotors and test masses.

12.3.12.1 Test Rotors

There are three different types of test rotors for different tasks, the systematics are shown in Fig. 12.33 (for details on the test rotors, see Sect. 13.3).

For horizontal balancing machines, the inboard rotor (type B) is the usual case. The size and test speed should be selected to suit the particular application.

If the user has special requirements, e.g. balancing planes close together or even overhung, then special tests may be useful. For this purpose, ISO has developed a series of suitable outboard rotors (type C), each consisting of a special shaft in combination with a test rotor for vertical balancing machines (type A).

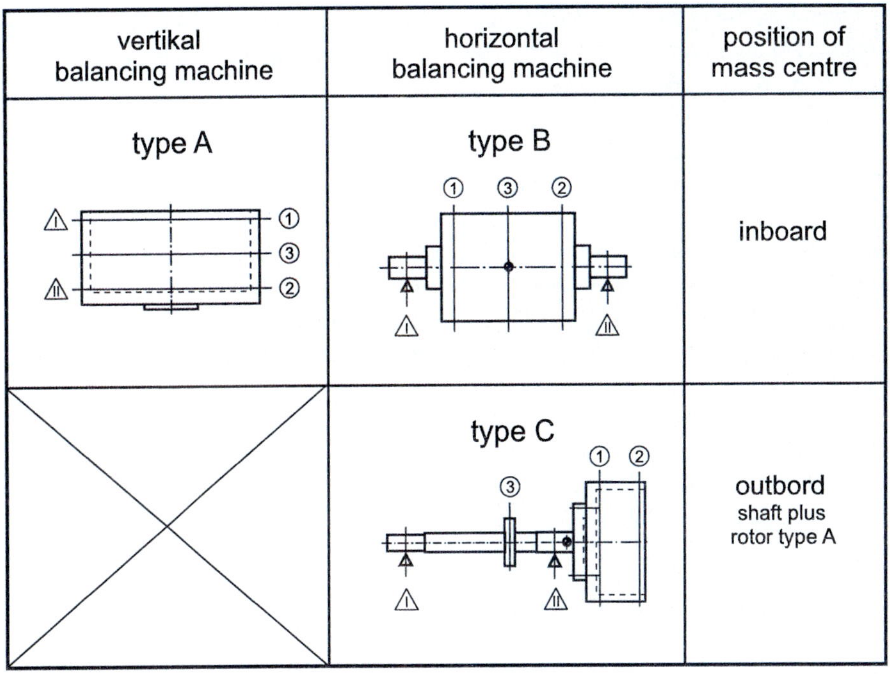

Fig. 12.33 Systematics of test rotors according to ISO 21940-21

In addition to the standardised test rotors mentioned above, the user can also develop and use his own test rotors as long as the principles and test routines set out in the standard are adhered to. This idea will be emphasised more strongly in the next edition of ISO 21940-21.

12.3.12.2 Test Masses

The test masses can usually be screwed on, their unbalance (mass centre radius times mass) must be known exactly.

See Sect. 13.3 for all necessary details.

12.3.13 Overload

An occasional overload by the rotor mass is only of interest for the lowest speed and is also only indicated for this area. This maximum mass can be carried by the balancing machine without instantaneous damage. However, continuous exposure can lead to a deterioration of the balancing machine's performance.

Overloads can of course also occur due to: too high a balancing speed (Gn^2 or $Gh^2 n^2$ -value), too high bearing loads (with outboard rotors also negative loads), too high torques on the transmission means (e.g. cardan shaft, driver, adapter).

12.3.14 Environmental Influences

Some external conditions may affect the guaranteed performance of the balancing machine, e.g. temperature, humidity, fluctuations of the mains in voltage and frequency.

Of interest is the range in which the specified performance is maintained.

In all cases where the stated performance relates to another condition, state whether the values change significantly, e.g. due to:

- Use of roller bearings on the rotor journals,
- Axial run-out errors of the plane where the axial counter bearing works (e.g. in the case of belt drive).

12.3.15 Unbalance Reduction Ratio R_{UR}

The unbalance reduction ratio is the change in unbalance achieved with *a single* correction, divided by the initial unbalance; it is given as a percentage.

Unfortunately, this statement is not very clear; a statement about the correction step—ratio initial unbalance/final unbalance—would be better (Table 12.5).

Example

An increase in R_{UR} from 80 to 85% only implies an improvement in the correction step from 5:1 to 7:1 (a factor of 1.4), while the same step in a different percentage range has a different significance: An increase in R_{UR} from 93 to 98% implies an improvement in correction step from 14:1 to 50:1 (factor 3.6)—a much larger, and therefore much more significant advantage. ◀

When stating the unbalance reduction ratio, it is assumed that the addition or removal of unbalances is carried out without error and that the balancing machine is operated with normal care.

The test for checking the unbalance reduction ratio is described in Sect. 13.5.

Table 12.5 Unbalance reduction ratio: The specification of the correction step is much clearer

R_{UR} (%)	80	85	90	93	95	97	98
Correction step	5:1	7:1	10:1	14:1	20:1	33:1	50:1

12.3.16 Economic Efficiency

This is the ability of the balancing machine to assist the operator in balancing a rotor in such a way that the rotor can be balanced to a specified residual unbalance in the shortest possible time.

Efficiency is best determined with a test rotor (Sect. 13.3). This is based on a test rotor that is in the lower third of the balancing machine's weight range. If a specific rotor type is of particular interest, a test rotor with similar data should be used in deviation from ISO.

The following designations are used:

Measuring run: It consists at most of the steps (see Table 12.2):

1. Setting up the machine including accessories, adapters etc.
2. Setting the measuring device.
3. Preparing the rotor for the balancing run.
4. Mean acceleration time.
5. Obtaining the measuring values.
6. Average braking time.
7. Time for transferring the measured values to the rotor.
8. Other procedures, e.g. according to safety regulations.

Points 1 and 2 are of particular importance for balancing *single rotors.*

Steps 1–8 are necessary for the *first* run of a rotor, 4–8 for the *following* runs of the same rotor.

For the *first* run of a *recurring* rotor type (series production), steps 3–8 are required.

Balancing run: In addition to the measuring run, which can have different scopes (see above), it also includes the subsequent unbalance correction for this measuring run.

Floor-to- loor time: It is the total time for all balancing and control runs necessary to achieve the required residual unbalance, as well as for loading and unloading. The number of balancing runs depends on the typical original unbalance, the tolerance and the unbalance reduction ratio.

Production rate: The production rate is very often used. It is the reciprocal value of the floor-to-floor time expressed in units/h and is sometimes supplemented with the utilisation rate (e.g. 95%).

Cycle time: If there is sometimes or always more than one rotor in the machine (balancing machine and correction station), the cycle time is to be indicated, i.e. the time that elapses from the output of one rotor to the output of the next.

12.4 Boundary Conditions

In addition, there are a whole range of boundary conditions determined by the intended installation site of the balancing machine, as well as issues relating to preliminary acceptance at the manufacturer's works, assembly and handover, final acceptance and maintenance.

Administrative issues (transport, transfer of risk, insurance) and technical details also need to be clarified (based on ISO 21940-21).

- Power supply: Single-phase or three-phase alternating current, voltage with maximum deviation, frequency, neutral lead available, loadable? What regulations must the electrics comply with?
- Is tropical insulation required?
- Is compressed air available, at what pressure and with what pressure fluctuations?
- Is the hall floor where the machine is to be installed stiff, i.e. comparable to a concrete slab lying on compacted earth? How thick is the concrete of the hall floor?
- Are there any vibration generators nearby, e.g. hammers, heavy vehicles, etc.? If so, indicate the frequency and duration of the disturbance.
- Who should inspect or accept the balancing machine, and where? According to which conditions?
- What language should the operating instructions have, or what other languages would be acceptable?
- Which legal regulations must be observed (e.g. safety, explosion protection)?
- Is a service engineer needed to install the machine and adjust it?
- Should a service engineer brief the staff?
- Should the operator be sent to the balancing machine manufacturer for training?
- Is a maintenance contract for the balancing machine of interest?

Some of these issues are usually not essential from the beginning, but many of them become important over time and are easily overlooked.

If the balancing machine manufacturer has all the necessary information, he can submit an offer that fully covers the specific problem.

Above all, it is important that the documentation is detailed and clear so that all important aspects for the practical use of the balancing machine can be read out.

Tests on Balancing Machines

13

Contents

H. Schneider, *Rotor Balancing*, https://doi.org/10.1007/978-3-662-66049-2_13

13.1 Preliminary Note

Balancing machines are measuring devices for unbalances and must therefore be checked regularly for quality assurance purposes. The type of inspection and its frequency depend on the respective balancing task and the risks that arise in the event of a malfunction.

If a product is manufactured in series, large quantities of measured values can be obtained and used as a basis. Statistical means are usually applied for evaluation.

Very different criteria can be discussed (Fig. 13.1), e.g.:

- How repeatable is a measured value? Here we ask about the scatter (Fig. 13.1).
- How accurate is the measurement? Here we check how far the measured mean deviates from the target, e.g. a gauge block (Fig. 13.1).
- How comparable are measurements, for example with different systems? Here, influences of the different measuring conditions and measuring equipment become visible (Fig. 13.1).

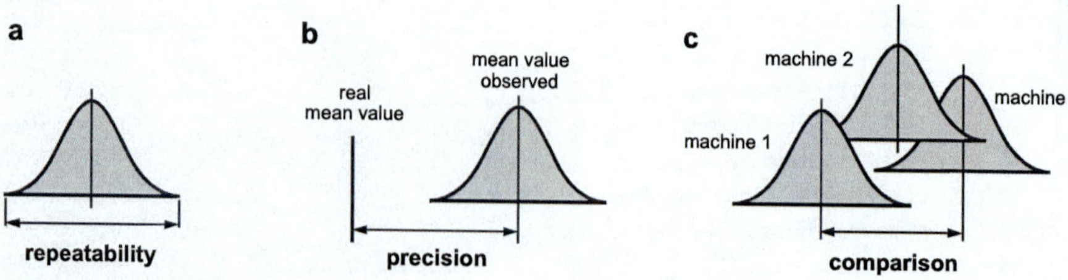

Fig. 13.1 Checking measured values by statistical means; examples: repeatability (**a**), precision (**b**) and comparison (**c**)

This also applies to balancing machines used for mass production. For unbalances, however, there is no independent standard, no "standard meter". Even "masters" that are sometimes used have their limits.

In industrially manufactured products, most measured variables are one-dimensional, e.g. measurements of length. They are normally distributed; the scatter is much smaller than the nominal value.

Unbalance, on the other hand, is a two-dimensional quantity whose nominal value generally is zero. As a result, some special features occur in the frequency distribution and in the statistical evaluation, which must be taken into account. For general multidimensional statistical problems see e.g. K. STANGE.

Unfortunately, there are neither international nor national technical standards for statistics in this demanding balancing area.

Some of the basic problems are pointed out in Sect. 13.2. Detailed statements on the use of statistical means in balancing technology can be found in the relevant balancing literature, e.g. H. SCHÖNFELD.

For balancing machines that are primarily used for individual rotors, the statistics are only applicable to a limited extent. Here ISO 21940-21 suggests various test rotors and a series of tests for checking essential performance characteristics of the balancing machines (see Sects. 13.3 to 13.7).

13.2 Statistics with Unbalances

The measurement of the unbalance of a *collective* of rotors of the same type results in a scattered field in the two-dimensional representation (Fig. 13.2).

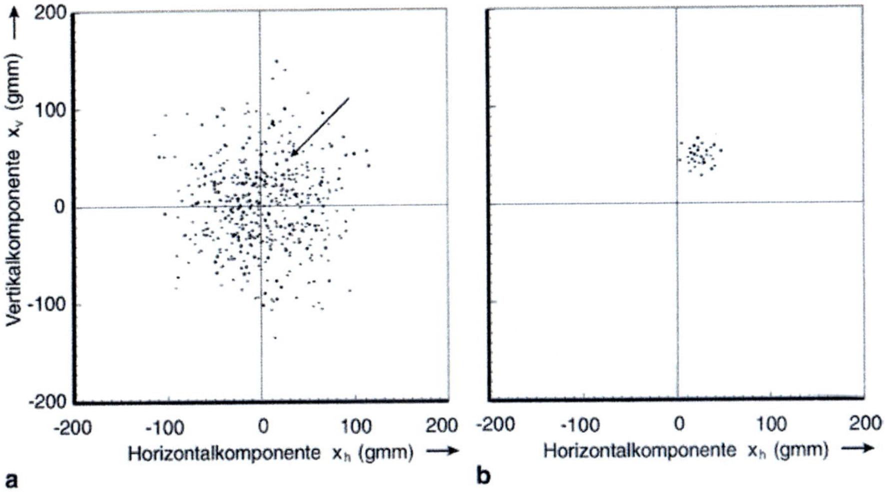

Fig. 13.2 Unbalance measurements of a collective of rotors (**a**) and repeated measurements of a single rotor (**b**), of a single value in a, marked with an arrow

If the measurement is repeated several times for a *single* rotor, a scattered field shows up also (Fig. 13.2b), but it is typically smaller because some deviations are omitted.

The unbalance measurements of a *collective* not only contain a statement about the unbalance, but also on systematic and random errors. These errors occur at various points in the overall system: the balancing machines, the rotor and the interface between the rotor and the balancing machine (see Chap. 16).

Sometimes the type of reference setting (the basis for the angular position) is also added, so that very different shapes and positions of scattered fields can become visible.

Depending on what is evaluated, whether unbalance components or amounts of the unbalance vectors, different distribution types also result.

13.2.1 Circular Scatter Field

A circular scatter field can either be centric (Fig. 13.3a), or eccentric due to a systematic error—fix in amount and angle (Fig. 13.3b).

If one evaluates the *unbalance components*, one obtains in both cases typical (one-dimensional) normal distributions (Fig. 13.4a):

- Without a systematic error, the mean values of the components are zero.
- With a systematic error, the mean values are different from zero.
- Mean and standard deviation are independent of each other.
- The usual calculation rules for (one-dimensional) normal distributions apply.

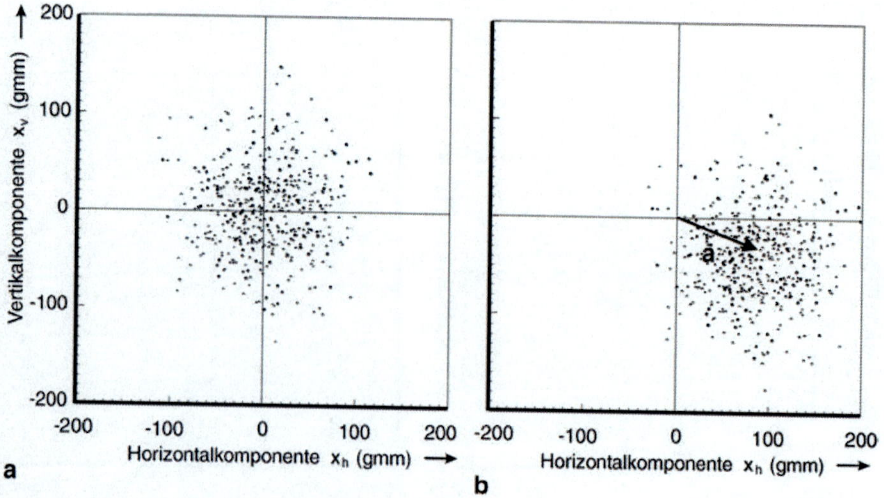

Fig. 13.3 Position of the scatter field of unbalances, centric (**a**) and eccentric by the mean value vector *a* (**b**)

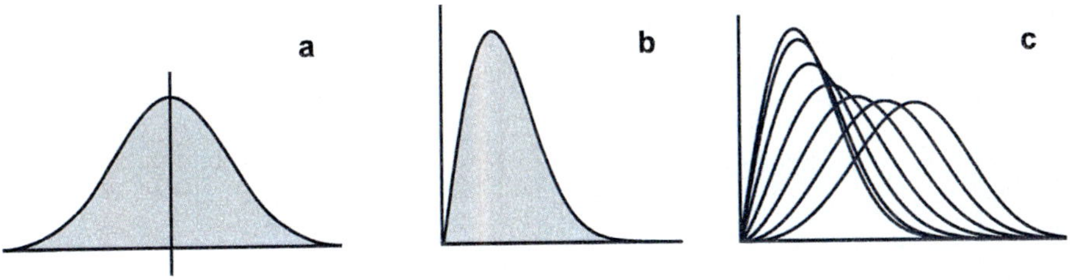

Fig. 13.4 Frequency distributions of the unbalances of Fig. 13.3. One-dimensional normal distribution (**a**), two-dimensional normal distribution (one-sided distribution) for the centric scatter field (**b**), which approaches the one-dimensional normal distribution with increasing eccentricity (**c**)

If, on the other hand, one starts from the *polar* representation and evaluates the *amounts of* the unbalance vectors, a different picture emerges (Fig. 13.4b):

- The distribution is one-sided limited, because there are no negative amounts of the vectors.
- It develops a special form, the (one-sided) two-dimensional normal distribution.
- The standard deviation is usually smaller than with the one-dimensional normal distribution.
- The mean value depends on the standard deviation.
- Special calculation rules apply.

As the eccentricity increases, the shape changes and approaches the one-dimensional normal distribution (Fig. 13.4c).

13.2.2 Ring-Shaped Scatter Field

A ring-shaped scatter field occurs when a systematic error is contained in the measured values, which is defined in amount, but whose angle can take on arbitrary values. This can be a systematic error that is *not* assigned to the reference mark in a fixed angular position.

13.2.2.1 Example 1

The package of an electric armature has a typical unbalance due to the manufacturing process. The mounting of the shaft is not determined angularly, during balancing the shaft groove is scanned as angular reference.

The scatter ring can again lie centric (Fig. 13.5a), or eccentric by an additional systematic error *with* fixed angular reference (Fig. 13.5b).

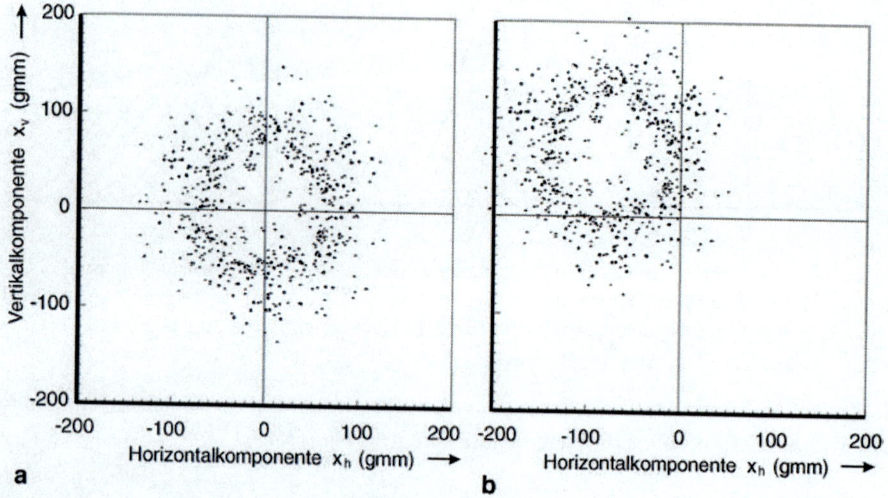

Fig. 13.5 Position of the scatter ring of unbalances, centric position (**a**) and eccentric (**b**)

13.2.2.2 Example 2

The above mentioned electric armature is a squirrel cage rotor; it is cast in a machine with a fixed relation to the shaft groove, where a systematic error occurs. In addition, there is another systematic error that does not correlate with the shaft groove.

If one evaluates the *unbalance components*, one obtains in both cases distributions with two maxima and a saddle in between (Fig. 13.6a).

If we start from the *polar* representation and look at the amounts, the following picture emerges:

- The centric scatter ring leads to a two-dimensional distribution (Fig. 13.6b), which looks similar to that for the normal scatter circle with eccentricity (Fig. 13.4b).
- The eccentric scatter ring leads to a deformed one-sided distribution (Fig. 13.6c), which with increasing eccentricity becomes more similar to the distribution of the components with centric or eccentric scatter ring (Fig. 13.6a).

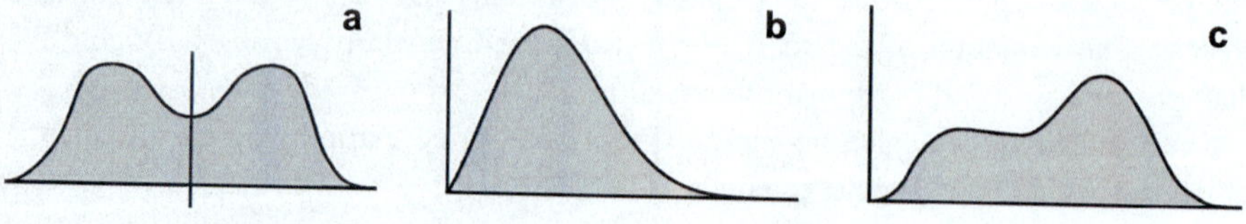

Fig. 13.6 Frequency distributions of the unbalances of Fig. 13.5. Components for the centric and the eccentric scatter ring (**a**), amounts of the unbalance vectors for the centric scatter ring (**b**), for the eccentric scatter ring (**c**)

Such different measurement conditions must be recognised and avoided at all costs. The usual statistical evaluations would otherwise lead to erroneous statements. It is not always as easy to understand as in the example shown.

13.2.3 Characteristics of One- and Two-Dimensional Normal Distributions

For both distributions (Fig. 13.4a and Fig. 13.4b), a mean value $\overline{x}$ and a standard deviation S can be calculated the usual way.

Typical differences become visible:

- The mean value $\overline{x}_B$ of the two-dimensional distribution is always larger than the mean value $\overline{x}_K$ of the one-dimensional distribution (components).
- The standard deviation S_B of the two-dimensional distribution is always smaller than the standard deviation S_K of the one-dimensional distribution (up to a factor of 0.655).

As the eccentricity measure a/S_K increases—with the vector mean a (Fig. 13.3b) and the standard deviation S_K of the components—the characteristics of the two-dimensional distribution approach those of the one-dimensional distribution asymptotically; from a measure of about 5 they can be regarded as equal.

However, if one wants to form a frequency sum from the mean and standard deviation, different rules also apply for the lower and upper limits.

Limits for frequency sums:

- One-dimensional normal distribution (symmetrical): lower limit $\overline{x}_K - qS_K$, upper limit $\overline{x}_K + qS_K$.
- Two-dimensional normal distribution (one-sided limited): lower limit 0, upper limit $\overline{x}_B + qS_B$.

Whereby the factors q are different and, in the case of the two-dimensional normal distribution, also change with the measure of eccentricity. For a measure of eccentricity $a/S_K = 1$ (residual unbalances in series operation are typically between 0 and 1.5) the values for q are stated in Table 13.1.

Table 13.1 Factors q for one- and two-dimensional normal distribution for different frequency sums and an eccentricity measure $a/S_K = 1$

Frequency sum (%)		95	99	99,73
q	One-dimensional normal distribution	1,96	2,58	3,00
	Two-dimensional normal distribution	1,85	2,68	3,19

13.2.4 Further Special Features

Scatter fields of unbalance can also occur in elliptical form, with a clear difference between the two axes, if unbalance characteristics of the rotors are different in these axes (e.g. with crankshafts). A prerequisite for recognising the situation and evaluating it correctly is here that the reference for the angle is clearly assigned to these axes.

Further problems lie in the fact that most rotors are balanced in two planes, so that sometimes the correlation of the unbalances of the two planes is also an interesting criterion, or the correlation of the initial unbalance to the residual unbalance.

13.2.5 Spot Checks or a Hundred Percent Check

Most of the machining steps are carried out under identical conditions for each part (e.g. with the same NC programme), because the goal is to have identical parts of one type. The check of important features is done with the help of spot checks, with different factors being calculated depending on the sample size.

With balancing, on the other hand, each rotor is treated individually, its unbalance is measured and corrected, the result often is checked and, if necessary, a further correction step follows. This way, it is a hundred percent check.

13.2.6 Key Figures

Statistical evaluationss are also used to determine key figures for the balancing machines, e.g. regarding:

- Uncertainty of measurement
- Machine capability
- Process capability.

In balancing, it is important to understand the particular circumstances and the current situation (see Sect. 13.2.2 to 13.2.5).

In addition, the requirements must be sensibly adapted to the respective boundary conditions. For example, different deviations are observed depending on how the individual steps for balancing are organised:

- Measurement, correction and check take place in the same station.
- Measuring and correction take place in different stations, and the rotor returns to the measuring station for checking.
- Measurement, correction, check take place in three different stations.

13.2.7 Rejects

In mechanical production a missed target (a set point with tolerance) often means that the part is scrap.

In rotor balancing, the target value is zero with a tolerance. If the tolerance is missed in the first machining step, it can usually be achieved in another step—on the same machine or possibly on a different one.

There is therefore normally no scrap in balancing; only in rare cases when the material intended for correction has been used up—e.g. balancing of crankshafts—can scrap also occur here.

13.3 Test Rotors and Test Masses

13.3.1 General

ISO 21940-21[1] defines three different series of test rotors for testing balancing machines. Each series is intended to characterise a typical case that occurs in practice.

13.3.2 Test Rotors Overview

The test rotors and balancing tasks are described as follows:

Type A	Rotors without journals, mostly balanced on a vertical balancing machine (or balanced on a horizontal machine with integrated spindle), balanced in one or two correction planes
	Service bearing planes may be anywhere; i.e. one on each side, or both on one side of the main body
	However, for the tests described in this standard it is assumed that one bearing is at each side of the rotor
Type B	Inboard rotors with journals, balanced on horizontal balancing machines, mostly with two balancing planes between the bearings
	Service bearings are positioned on either end of the rotor
Type C	Outboard rotors with journals, balanced on horizontal balancing machines, with two overhung correction planes
	Service bearing positions are one-sided, similar to those on the proving rotor

[1] ISO 21940: Mechanical vibration—Rotor balancing—Part 21 (2019): Description and evaluation of balancing machines.

For all test rotors and their use, ISO 21940-21 sets out a number of basic rules; the most important are quoted here:

- Each rotor has three test planes. The same test rotors and test masses are used for one-plane tests and two-plane tests.
- Tests with rotor type C are to be specifically agreed on.
- For universal balancing machines, a test rotor with a mass in the lower third of the machine's load capacity should generally be used.

These series of test rotors are presented with their data in the next sections.

13.3.2.1 Individual Test Rotors

However, instead of these standardised test rotors, individual test rotors can also be used, which may be created from production rotors, for example—because of the proximity to the task or for cost reasons[2].

However, the errors caused by them during balancing (see Chap. 16) must be negligible. The prerequisites to be mentioned are:

- Reproducibility of measurements: it may be necessary to glue fits, cast windings, replace flexible elements.
- Precision of the test unbalances: requires accurate masses and precise positioning.

13.3.3 Test Masses

High demands are generally placed on the test masses.

The test masses can usually be screwed on, their unbalance (centre of gravity and mass) must be known exactly. The following test masses are required:

- For the test for the minimum achievable residual unbalance U_{mar}:
  - A test mass corresponding to 10 times the claimed minimum achievable residual unbalance $U_{mar,cl}$.
- For the unbalance reduction ratio R_{UR} test:
  - Two identical test masses with 20 to 60 times the claimed minimum achievable residual unbalance $U_{mar,cl}$ for the stationary unbalances.
  - Two equal test masses with 5 times the stationary unbalances, for the travelling unbalances.

[2] This point will be emphasised more in the next issue of the standard.

The mass must be accurate to 0.1 (100%—R_{UR}). For an unbalance reduction ratio of 90%, this means to 1%.

The position of the test masses on the rotor must have the appropriate accuracy in all coordinates:

- The correction plane distance with the percentage calculated above.
- The balancing radius with the same percentage.
- The angular position with the same percentage, related to 1 rad (e.g. 1% of 1 rad = 1% of 57.3° ≈ 0.6°).

13.3.4 Test Rotors in Detail

13.3.4.1 Type A Test Rotors

These test rotors represent rotors without journals, for balancing machines with a spindle, to accommodate rotors at one end (Table 13.2 and Fig. 13.7). Most of these machines have a vertical axis, but there are also versions with a horizontal axis.

The test rotor consists of a cylinder open at the top; the three test planes lie on its mantle.

Table 13.2 Test rotors type A. Recommended data according to ISO 21940-21, for dimensions see Fig. 13.7

Rotor No.	Rotor mass	Major diameter	Minor diameter	Hight	Details										Highest test speed[c]
	m	D	d	H	X	Y^a	Z^a	F	G	I^b	J^b	K^b	R^b	T	
			≈ 0,9D	≈ 0,5D	≈ 0,075D	≈ 0,175D	≈ 0,175D	≈ 0,06D							
	kg	mm	mm	mm	mm	mm	mm	mm		mm	mm	mm	mm	mm	min^{-1}
1	1,1	110	99	55	8	20	20	6,5	M3	50,8	0,4 x 45°	4,2	76,2	6,6	20.000
2	3,5	160	144	80	12	30	30	9,5	M4	50,8	0,4 x 45°	4,2	76,2	6,6	14.000
3	11	230	206	127	19	45	45	13	M5	114,3	0,4 x 45°	4,2	133,35	10,3	10.000
4	35	345	310	170	25	60	60	20	M6	114,3	0,4 x 45°	4,2	133,35	10,3	6.000
5	110	510	460	255	38	90	90	30	M8	114,3	0,4 x 45°	4,2	133,35	10,3	4.000

All tolerances and residual unbalances shall be in accordance with test aims.
Test rotors from SAE ARP 4162A may be used instead, with test masses modified to suit the ISO-tests.

[a] Except for *Y* and *Z*, dimensions may be varied.

[b] Interface (spigot) dimensions shall comply with the requirements of SAE ARP 4162A proving rotors (Rotor Nos. 2 to 5).

[c] Refers to rotors. Test mass design can limit highest speed.

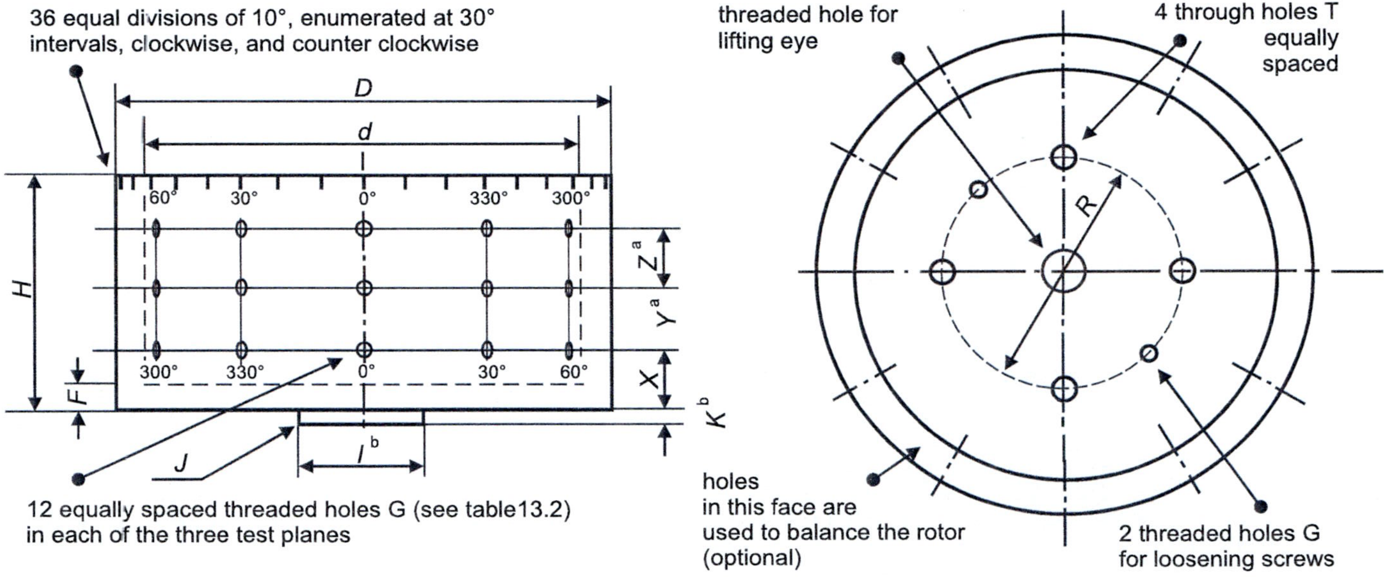

Fig. 13.7 Type A test rotors for vertical balancing machines. Dimensions and notes see Table 13.2

13.3.4.2 Test Rotors Type B

Type B test rotors represent rotors with journals, for balancing machines with two bearing positions, i.e. accommodating rotors at both ends (Table 13.3 and Fig. 13.8). Most of these machines have a horizontal axis, but there are also special cases with a vertical axis.

The test rotor consists of a compact roll with journals on both ends; with test planes on the cylinder surface. Thus it represents the typical case of an inboard rotor, whose centre of gravity lies between the bearings, as do the correction planes.

The maximum and critical speeds given in Table 13.3 refer to rigid bearings, with flexible bearings these limiting speeds decrease accordingly (see Sect. 6.4.2).

If old ISO test rotors with only 8 holes per test level still exist, they can be modified to the new dimensions.

SAE ARP 4162 also specifies roll-shaped test rotors, but they are longer. They can also be used for the ISO tests, but other test masses must be prepared according to the respective ISO requirements.

For balancing tasks where the correction planes are overhung on one side or where the centre of gravity is outside the bearing pedestals, use the type C test rotors.

Table 13.3 Test rotors type B. Recommended data according to ISO 21940-21, dimensions see Fig. 13.8

Rotor No.	Rotor mass	Major diameter	Overall length	Shaft diameter	Bearing distance	Details									Resonance speed [c]	Highest test speed [d]
	m	D	L	D	$A+B+C$	A[a], C[a]	B[a]	E	F	P_1	H[b]	K[b]	P_2[b]	N	$\approx 7{,}6\ 10^6/D$	$\approx 7{,}6\ 10^5/D$
			$\approx 2{,}5D$	$\approx 0{,}3D$[b]	$\approx 2D$	$\approx 0{,}5D$	$\approx D$	$\approx 0{,}25D$	$\approx 0{,}5D$							
	kg	mm	mm	mm	mm	mm	mm	mm	mm	mm	mm	mm	mm		min^{-1}	min^{-1}
1	0,5	38	95	11	76	19	38	9,5	19	31	-	-	-	M2	200.000	20.000
2	1,6	56	140	17	112	28	56	14	28	46	-	-	-	M3	140.000	14.000
3	5	82	205	25	164	41	82	20,5	41	72	-	-	-	M4	95.000	9.500
4	16	120	300	36	240	60	120	30	60	108	4	7	30	M5	65.000	6.500
5	50	176	440	58	352	88	176	44	88	160	1,4	30	47	M6	45.000	4.500
6	160	260	650	78	520	130	260	65	130	240	1,8	42	62	M8	30.000	3.000
7	500	380	950	114	760	190	380	95	190	350	2,2	57	84	M10	20.000	2.000

All tolerances and residual unbalances shall be in accordance with test aims.
Test rotors from SAE ARP 4162A may be used instead with test masses modified to suit the ISO-tests.

[a] Dimensions A, B, and C may be varied, provided they meet the requirements: $A \approx B/2$, $C \approx B/2$

[b] End-drive interface dimensions comply with typical drive shafts.

[c] Resonance speeds are calculated for rotors running in rigid bearings.

[d] Refers to rotors. Test mass design can limit highest speed.

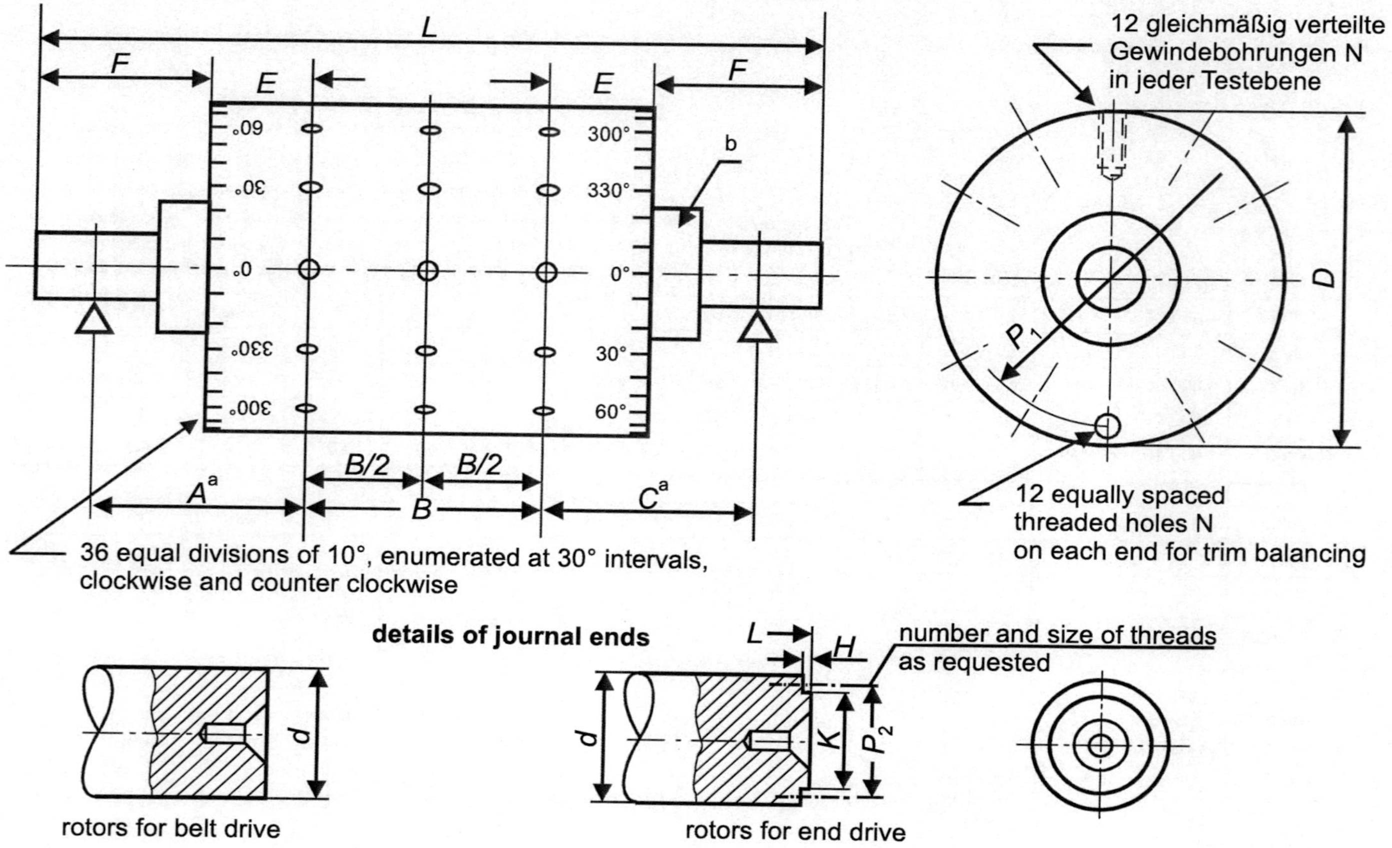

Fig. 13.8 Type B test rotors for horizontal balancing machines and tests for inboard rotors. Dimensions and notes see Table 13.3

13.3.4.3 Test Rotors Type C

Type C test rotors represent rotors with an overhung mass centre, for balancing machines with two bearings. This means an overhung accommodation of rotors and thus, for horizontal machines, a negative load on the remote bearing (Table 13.4 and Fig. 13.9). Many horizontal balancing machines have the ability to balance these rotors. However, as such rotors are rare; a test with a type C test rotor should be specially agreed between the manufacturer and the user of the balancing machine.[3]

The maximum and critical speeds given in Table 13.4 refer to rigid bearings, with flexible bearings these limiting speeds are reduced accordingly (see Sect. 6.4.2).

The test plane for the minimum achievable residual unbalance U_{mar} is located on the shaft in such a way that this results in unbalances in the bearing planes in the ratio 1/3 and 2/3 (see Sect. 7.5.2.2).

The test planes for the unbalance reduction ratio R_{UR} are located on the cylinder mantle of test rotor A. Type C test rotors thus represent the typical case of an outboard rotor, whose mass centre lies outside the bearings, as do the correction planes.

[3]The data for the test rotors C are actually being discussed in the ISO committee: please use the current edition as a basis.

Table 13.4 Test rotors type C. Recommended data according to ISO 21940-21, dimensions see Fig. 13.9

Component		Assembly														
Shaft No.	Test rotor type A	Test rotor type C	Mass	Bearing load on rotor [a]		Details					Major diameter	Bearing distance	Distance of correction planes		Resonance speed [d]	Highest test speed [e]
			m	A	B	Y [b]	d_1 [c]	d_2	d_4	N [b]	d_6	L	A	B		
	No.	No.	kg	N	N	mm	mm	mm	mm		mm	mm	mm	mm	min^{-1}	min^{-1}
1	1	1	2,2	-3	24	20	17	21	50	M3	110	164	41	40	25.000	4.000
2	2	2	6,2	-8	70	30	25	30	72	M4	160	240	62	60	17.000	2.800
3	3	3	19,5	- 25	220	45	36	45	106	M5	230	352	93	90	14.500	1.900
4	4	4	60	- 75	670	65	58	65	156	M6	345	520	140	120	8.000	1.300
5	5	5	190	- 230	2100	95	78	95	230	M8	510	760	203	180	5.500	900

All tolerances and residual unbalances shall be in accordance with test aims.
Test rotors from SAE ARP 4162A may be used instead of test rotor Type A, with test masses modified to suit the ISO-tests.

[a] For indication see Abb. 13.9.

[b] Dimensions may be varied, provided the centre of mass stays outboard with the same overhung Y, and the position of holes N between bearings is maintained.

[c] End-drive interface dimensions for Nos. 3 to 5 meet the requirements described for test rotors Typ B, Nos. 4 to 6.

[d] Resonance speeds are calculated for rotors running in rigid bearings

[e] Refers to rotors. Test mass design can limit highest speed.

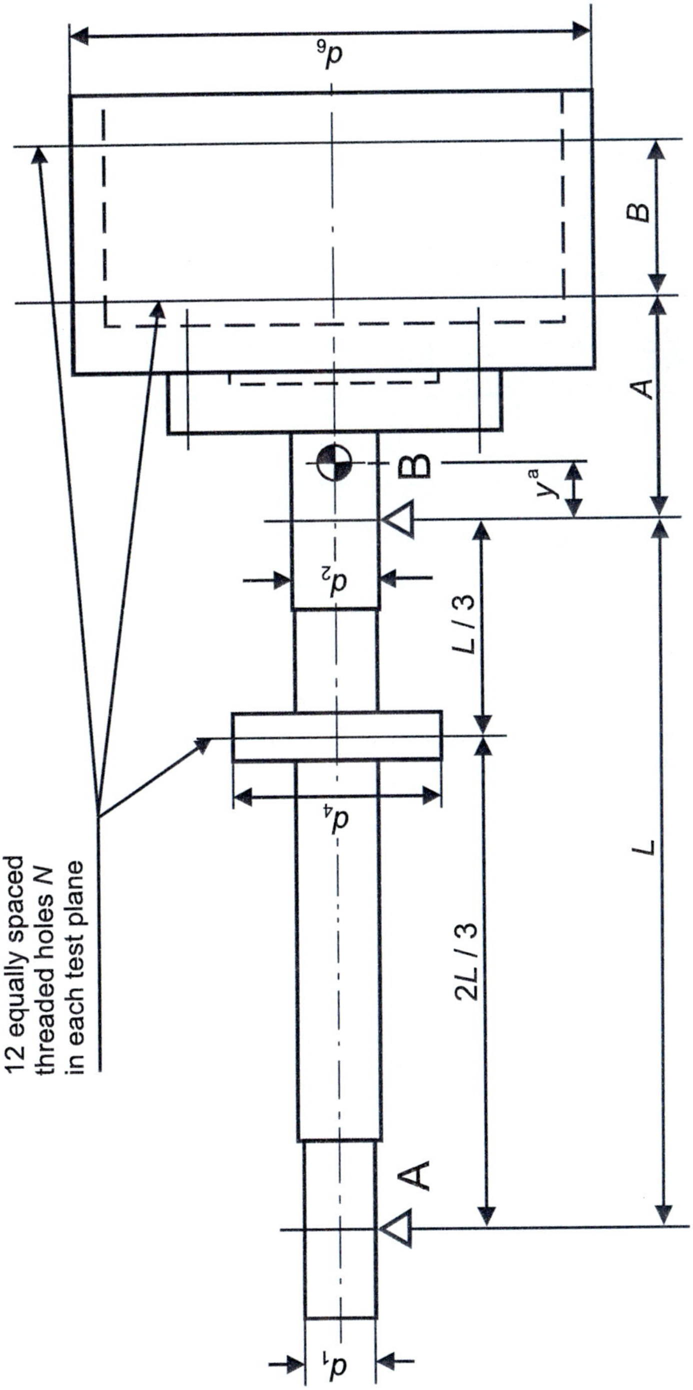

Fig. 13.9 Type C test rotors for horizontal balancing machines and tests for outboard rotors. Dimensions and notes see Table 13.4

13.3.5 Test Conditions

The most important test conditions for the tests for the minimum achievable residual unbalance U_{mar} (see Sect. 13.4) and for the unbalance reduction ratio R_{UR} (see Sect. 13.5) for the different test rotors are shown in Table 13.5, 13.6 and 13.7 respectively.

Table 13.5 Overview of test conditions for test rotor A

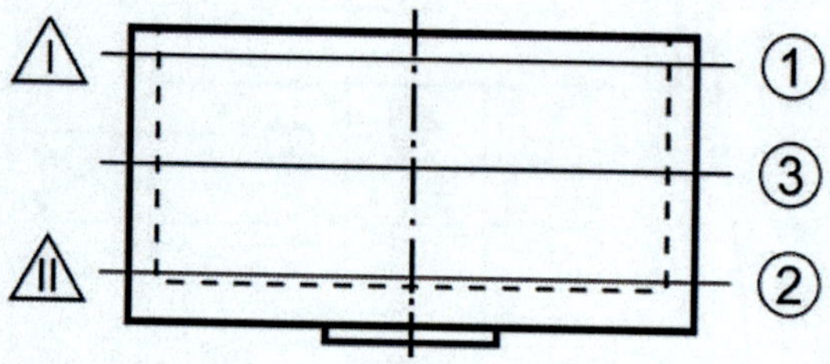

	Test: minimum achievable residual unbalance U_{mar}, see 13.4	Test: unbalance reduction ratio URR, see 13.5
	Balancing with setting: resultant unbalance	
single-plane	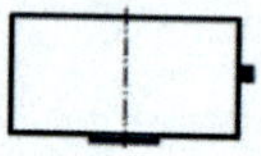 test mass in plane ③ corresponds to 10 x U_{mar} measuring: resultant unbalance	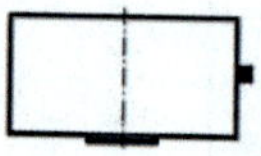 test mass in plane ③ $U_{station}$ = 20 to 60 x U_{mar} U_{travel} = 5 x $U_{station}$ measuring: resultant unbalance
	Balancing with setting: two correction planes near to ①,②	
two-plane	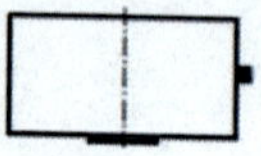 test mass in plane ③ corresponds to 10 x U_{mar} measuring: planes △I, △II	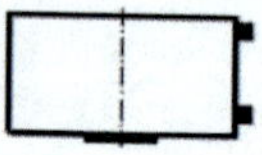 test masses in planes ①, ② $U_{station}$ = 20 bis 60 x U_{mar} U_{travel} = 5 x $U_{station}$ measuring: planes ①, ②

Table 13.6 Overview of test conditions for test rotor B

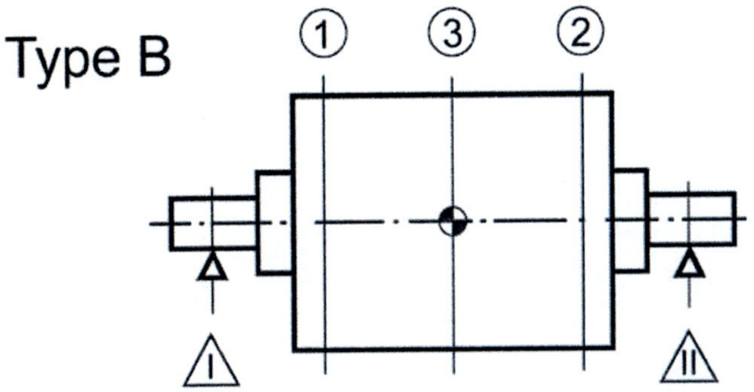

①, ②, ③ test planes   Ⅰ, Ⅱ measuring planes for U_{mar}

	Test: minimum achievable residual unbalance U_{mar}, see 13.4	Test: unbalance reduction ratio URR, see 13.5
	Balancing with setting: resultant unbalance	
single-plane	 test mass in plane ③ corresponds to 10 x U_{mar} measuring: resultant unbalance	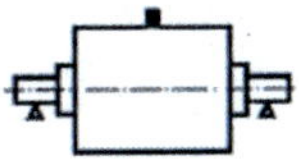 test mass in plane ③ $U_{station}$ = 20 to 60 x U_{mar} U_{travel} = 5 x $U_{station}$ measuring: resultant unbalance
	Balancing with setting: two correction planes near to ①,②	
two-plane	 test mass in plane ③ corresponds to 10 x U_{mar} measuring: planes Ⅰ, Ⅱ	 test masses in planes ①,② $U_{station}$ = 20 to 60 x U_{mar} U_{travel} = 5 x $U_{station}$ measuring: planes ①,②

Table 13.7 Overview of the test conditions for test rotor C, Variant 1: (**a**) Test with single unbalances, Variant 2 (**b**): Test with resulting unbalance and couple unbalance

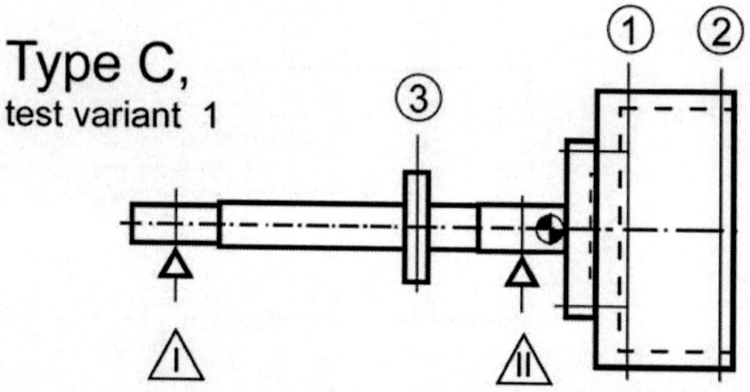

①, ②, ③ test planes △I, △II measuring planes for U_{mar}

	Test: minimum achievable residual unbalance U_{mar}, see 13.4	Test: unbalance reduction ratio URR, see 13.5 variant 1: dynamic unbalance
	Balancing with setting: resultant unbalance	
single-plane	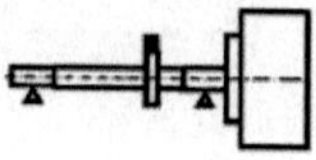 test mass in plane ③ corresponds to 10 x U_{mar} measuring: resultant unbalance	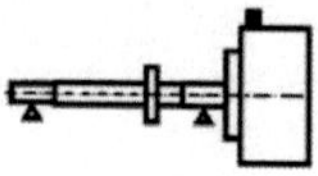 test mass in plane ① $U_{station}$ = 60 to 100 x U_{mar} U_{travel} = 5 x $U_{station}$ measuring: resultant unbalance
	Balancing with setting: two correction planes near to ①,②	
two-plane	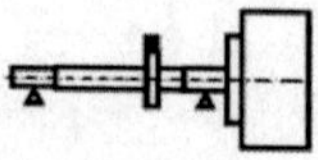 test mass in plane ③ corresponds to 10 x U_{mar} measuring: planes △I, △II	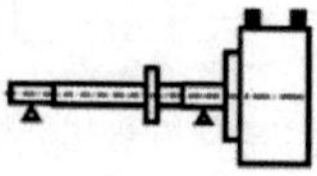 test masses in planes ①,② $U_{station}$ = 60 to 100 x U_{mar} U_{travel} = 5 x $U_{station}$ measuring: planes ①,②

Table 13.7 (continued)

Type C,
test variant 2

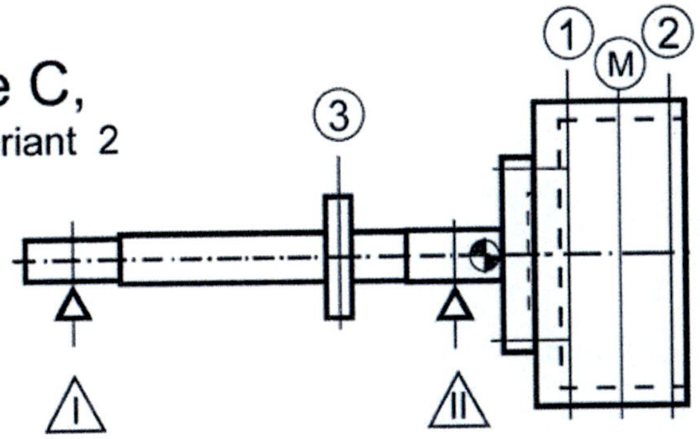

①, ②, ③, Ⓜ test planes △I, △II measuring planes for U_{mar}

	Test: minimum achievable residual unbalance U_{mar}, see 13.4	**Test: unbalance reduction ratio URR, see 13.5** **variant 2: result./couple unbalance**
Balancing with setting: resultant unbalance		
single-plane	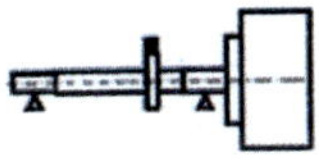 test mass in plane ③ corresponds to 10 x U_{mar} measuring: resultant unbalance	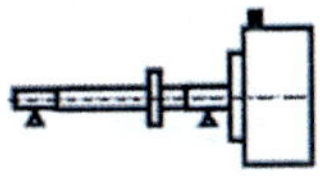 test mass in plane ① $U_{station}$ = 60 to 100 x U_{mar} U_{travel} = 5 x $U_{station}$ measuring: resultant unbalance
Balancing with setting: two correction planes near to ①,②		
two-plane	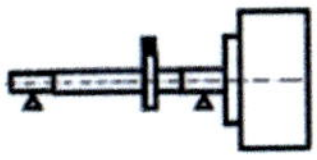 test mass in plane ③ corresponds to 10 x U_{mar} measuring: planes △I, △II	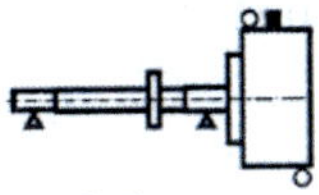 resulting unbal.: test mass in plane Ⓜ $U_{station}$ = 20 to 60 x U_{mar} U_{travel} = 5 x $U_{station}$ couple unbalance: test mass in plane ①,② $U_{station}$ = 20 to 60 x U_{mar} U_{travel} = 5 x $U_{station}$ measuring: planes ①,②

13.4 Test of the Minimum Achievable Residual Unbalance U_{mar}

For the entire rotor U_{mar} is the minimum achievable residual unbalance For easier handling, it is better to split this value between both reference planes for the tolerances (see Sect. 7.2.1). The two rules in Sect. 7.5.2 are used for distribution to the planes.

These tolerance planes in Table 13.5, 13.6 and 13.7 for the different types of test rotors are marked with triangles. The test involves a number of steps that must be carried out very carefully.

13.4.1 Start Condition

The mechanics and the measuring device are set for the selected test rotor (planes see Table 13.5, 13.6 and 13.7). Make sure that the residual unbalance of the rotor does not exceed 5 times the specified minimum achievable residual unbalance per plane.

Two artificial unbalances are applied near the correction planes, corresponding to 10–20 times the specified minimum achievable residual unbalance per plane.

These unbalances do not need to be of equal size and precisely known; however, care should be taken that they are:

- Not in the same plane.
- Not in one of the correction planes.
- Not at the same angle.
- Not be offset by 180°.

13.4.2 Correction

The rotor is balanced by the normal balancing process (with the usual operatorl care) by correction in both correction planes, a maximum of four corrections are allowed. The conditions and measurement results are recorded in a log (Table 13.8).

Table 13.8 Measured values of the test rotor during balancing (based on ISO 21940-21).

Location:	Date:	Test person:	Signature:		**Balancing for tests**
Balancing machine	Type:		Manufacturer:		
Test rotor	Type:			No.:	
Mass:		kg;	Test speed:		min^{-1}
$U_{mar\ 1}$:		g·mm;	$U_{mar\ 2}$:		g·mm
10 x $U_{mar\ 1}$:		g·mm;	10 x $U_{mar\ 2}$:		g·mm
Test mass 1:		g;	Test mass 2:		g
Effective radius 1:		mm;	Effective radius 2:		mm

Plane			**1**		**2**	
Reading Run No.	Correction	Rotor condition	Amount multiple $U_{mar\ 1}$	Angle °	Amount multiple $U_{mar\ 2}$	Angle °
1	no	initial unbalance				
2	1. step	after 1. correction				
3	2. step	after 2. correction				
4	3. step	after 3 correction				
5	4. step	after 4. correction: residual unbalance				
6	no	after 60° reference change				

After correction, the angle reference is shifted by 60° relative to the rotor. In the case of drive shaft machines by rotating the drive shaft relative to the rotor. In case of other drives e.g. by offsetting the scanning mark. This makes errors of the drive system, the spindle bearing and certain errors of the measuring device measurable in full size.

After measurement runs 5 and 6, no further correction steps are allowed.

If the residual unbalance in run No. 5, but especially after changing the reference in run No. 6, is not below (or at least close to) the claimed minimum achievable residual unbalance per plane, the reasons should be clarified first, as under these circumstances the U_{mar} test will probably not be passed.

13.4.3 Test Runs with Test Masses

In test plane 3 (previously test masses were used in two planes), a test mass with 10 times the value of U_{mar} is applied, in any order in all positions (12) available.

Measurements are taken in reference planes I and II—in Table 13.5, 13.6 and 13.7 marked with triangles. The amount indication for each angular position of the test mass is noted separately for each reference plane (Table 13.9). **Remark**: Make sure that the display really rotates around the zero point.

Table 13.9 Data sheet for the test on U_{mar} (based on ISO 21940-21).

Location:	Date:	Test person:	Signature:	**U_{mar} test**
Balancing machine	Type:		Manufacturer:	
Test rotor	Type:		No.:	
Mass:	kg;	Test speed:		min^{-1}
10 x U_{mar}:	g·mm;			
Test plane 3:	g;			
Effective radius 3:	mm;			

Test mass angular position	**Amount of unbalance**	**Amount of unbalance**	**Multiples of mean values**	**Multiples of mean values**
°	plane I	plane II	plane I	plane II
	g·mm	g·mm	—	—
0				
30				
60				
90				
120				
150				
180				
210				
240				
270				
300				
330				
Sum			—	—
Mean value			—	—

13.4.4 Evaluation of the U_{mar} Test

With the help of Table 13.7 the sum of all measured values is first calculated separately for each plane and afterwards the mean values are calculated (twelve measured values, i.e. divisor 12).

By subsequently dividing each measured value by the respective mean value, specific values are produced, i.e. multiples of the mean values. These values are noted in the two right-hand columns of Table 13.9 and plotted in a general diagram (Fig. 13.10)—separately for both planes.

The limit lines (dashed lines) in this diagram are ± 12% away from the mean value 1. They correspond to the claimed minimum achievable residual unbalance per plane plus a small margin (20%) for errors when performing the test.

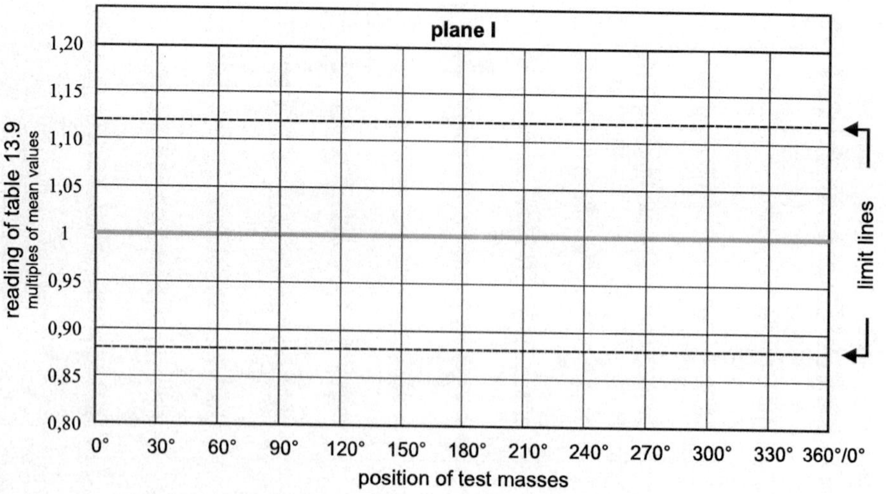

Fig. 13.10 Diagram for the evaluation of the U_{mar} test. Identical for plane I and II

The machine has passed the U_{mar} test if *all* measured values in the two planes—with *one* exception per plane—are within the limit lines.

If the measured values form a clear sine line, one can assume that a systematic error has become visible (e.g. due to the angular shift of the reference by 60°), which would remain undetected during the normal balancing process. If the test was nevertheless passed, it can be concluded that the achievable minimum residual unbalance is still below the claimed value.

The corresponding test for single-plane balancing machines runs accordingly, the same tables and images are used for evaluation.

13.4.5 Abbreviated U_{mar} Test

For repetitions, ISO 21940-21 allows shortening the test. The preparation remains identical, but when setting test masses every second position is skipped, so that instead of 12 test runs there are only 6 runs.

To calculate the mean value, the readings are divided by 6 and the data entered in the diagram (Fig. 13.10).

In this abbreviated test, *all* measured values must lie *within* the dashed borderlines to pass the test.

13.5 Unbalance Reduction Ratio Test R_{UR}

This test checks the overall accuracy of the amount indication, the angle indication and the plane separation (for single plane balancing machines only amount and angle indication is checked).

For test rotor type C, instead of a test in two planes, Table 13.7, Variant 1, a test with resulting unbalance and moment unbalance may be useful; the details are explained in ISO 21940-21 and are fixed in Table 13.7, Variant 2.

In the past, two tests were made with different unbalance levels; today it is only one test, well above the minimum achievable residual unbalance.

13.5.1 Start Condition

The initial state is the rotor well balanced according to Sect. 13.4.2, after moving the reference by 60°, the machine shows the heavy spot.

Test and measurement planes are planes 1 and 2, in Table 13.5, 13.6 and 13.7 marked with circles.

In each plane, a test mass with the unbalance $U_{stationary}$ is now attached at a freely chosen (but not the same) angle for the duration of the test (each test mass means 20 to 60 times U_{mar}). These starting conditions—together with the data of the machine and the rotor—are entered into a scheme for the R_{UR} test (Table 13.10).

13.5.2 Test Runs with Test Masses

A travelling test mass (corresponding to exactly fivefold $U_{stationary}$) is placed in each plane in one of the free positions, the angular positions is entered in Table 13.10.

In the first run, the unbalances of both correction planes are measured and also noted in Table 13.10.

For the following runs, the travelling test masses are each placed one pitch further, in one plane in *one* direction of rotation, in the other plane in the *opposite* direction. The positions occupied by the stationary test masses are left out, resulting in a total of 11 test runs.

13.5.3 Evaluation of the R_{UR} Test

Dividing the measured values by the unbalance of the stationary test masses results in multiples of $U_{stationary}$ (Table 13.10). Thus it is possible to work with a standardised diagram that is always applicable. In this diagram, the stationary value 1 is entered in a fixed angular direction, the travelling value 5 in a total of eleven 30° jumps (Fig. 13.11).

Table 13.10 Data sheet for the R_{UR} test (based on ISO 21940-21)

Location:	Date:	Test person:	Signature:	**URR test** Type A, B
Balancing machine	Type:		Manufacturer:	
Test rotor	Type:	No.:		
Mass:	______	kg;	Test speed: ______	min^{-1}
Claimed e_{mar} :	______	g·mm/kg;	Claimed U_{mar} : ______	g·mm
$U_{station}$ = 20 to 60 x U_{mar}:	______	g·mm;	U_{travel} = 5,0 x $U_{station}$: ______	g·mm
Test mass stationary:	______	g;	Test mass travelling: ______	g;
Effective radius stationary:		mm;	Effective radius travelling:	mm;

Run No.	**Test mass angles**				**Reading**		**Reading** plane 1 divided by $U_{station}$	**Reading**		**Reading** plane 2 divided by $U_{station}$
	plane 1		plane 2		plane 1			plane 2		
	stationary	travelling	stationary	travelling	amount	angle		amount	angle	
	angle °	angle °	angle °	angle °	g·mm	°	multiple of $U_{station}$	g·mm	°	multiple of $U_{station}$
1										
2	Test mass stays as in run 1		Test mass stays as in run 1							
3										
4										
5										
6										
7										
8										
9										
10										
11										

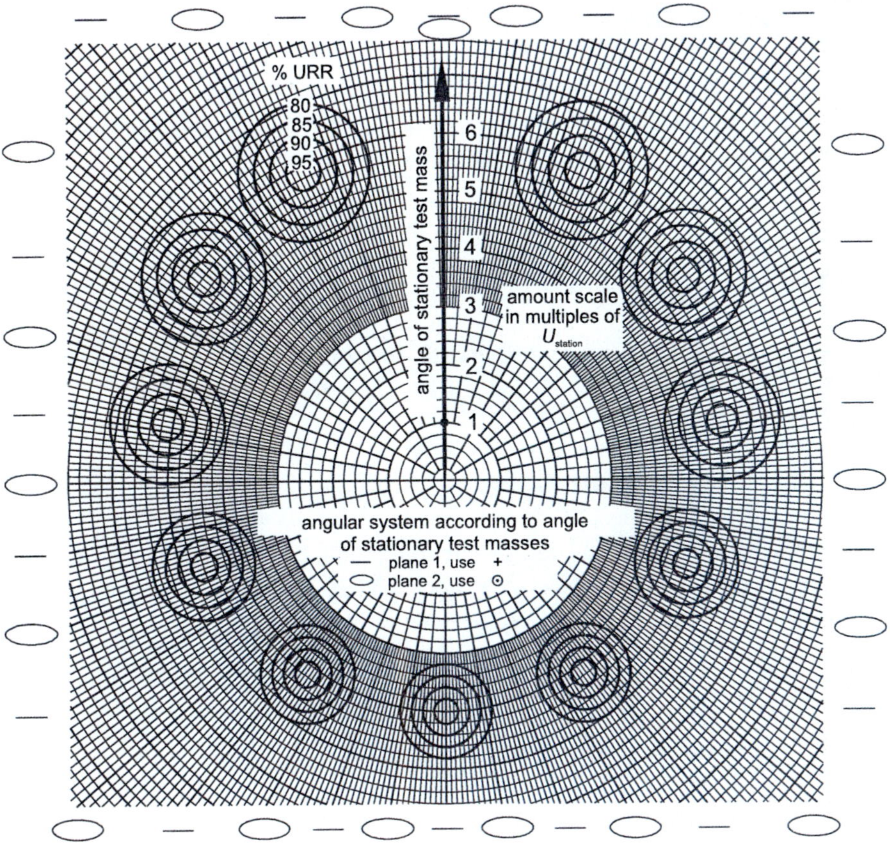

Fig. 13.11 Diagram for the evaluation of the R_{UR} test for two planes. For one level an adapted diagram applies

The random angular positions of the stationary test masses are recorded by entering a separate angular scale for each plane. Concentric circles mark areas of certain R_{UR} values around the nominal values of the indications.

After entering all (specific) measured values, the achieved R_{UR} value can be read off. A certain R_{UR} value is considered to have been reached if only *one* value in each plane is outside.

13.5.4 Abbreviated R_{UR} Test

For repetitions, ISO 21940–21 allows a shortening of the test. The start condition remains identical, but when setting test masses, every second position is skipped. The travelling test masses must therefore start in such a way that they do not collide with the stationary ones.

Instead of 11, only 6 test runs are made and the data entered in the diagram (Fig. 13.11).

Here, *all* measured values must lie *within* the relevant R_{UR} circles during the assessment to pass the test.

13.6 Test of the Moment Unbalance Influence Ratio I_{MU}

Most vertical balancing machines are used for balancing in *a single* correction plane. Here, the influence of a moment unbalance on the display of the resulting unbalance I_{MU} is interesting and must be specified for the balancing machine.

$$I_{MU} = \frac{U_r}{P} \quad \mathrm{g \cdot mm/g \cdot mm^2} \tag{13.1}$$

13.6.1 Starting Conditions

The test rotor is balanced as described in Sect. 13.4.2 including changing the reference by 60°.

13.6.2 Test Runs with Test Masses

Then two unbalances of equal size (e.g. the *travelling* test masses of the R_{UR} test, see Sect. 13.5.2) are placed in the two outer planes (1 and 2), offset from each other by 180°.

This moment unbalance is advanced three times by 90° each time and the measured values A for the resulting unbalance are noted.

13.6.3 Evaluation of the I_{MU} Test

None of the four measured values A may exceed the value of the claimed minimum achievable residual unbalance plus the product of the applied moment unbalance with the claimed influence value:

$$A \leq U_{mar,cl} + P\, I_{MU} \tag{13.2}$$

Example

Claimed minimum achievable residual unbalance: $U_{mar,cl} = 5$ g · mm; claimed moment unbalance influence ratio: $I_{MU} = 1$ g · mm/400 g · mm^2. The test masses each have an unbalance of $U_t = 100$ g · mm (assumed), the plane spacing is $l = 40$ · mm.

Indication $A \leq 12$ g · mm. ◀

Question

Is the indication A in tolerance? ◀

Solution

The moment unbalance results in $P = lU_t = 4,000\ \mathrm{g \cdot mm}$. The maximum permissible measured value is then:

$$A_{zul} = 5\ \mathrm{g \cdot mm} + 4,000\ \mathrm{g \cdot mm}^2 \frac{1}{400} \frac{\mathrm{g \cdot mm}}{\mathrm{g \cdot mm}^2} = 15\ \mathrm{g \cdot mm}$$

Conclusion

The test was passed.

13.7 Compensation Test for the Index Process

For index balancing (see Sect. 9.2.2) today mostly aids in the measuring instruments are used. The purpose of the test is to check the function and accuracy of these aids.

If the rotor would really be indexed on its mounting, difficulties arise with reproducibility, which is often in the same order of magnitude as the test aim.

To avoid these problems, the rotor remains firmly connected to its mounting during the test. The index is simulated with test masses, one group of which is held in place while the other group is moved.

13.7.1 Start Condition

The initial condition is the well balanced test rotor according to Sect. 9.2.2 or a test rotor whose unbalance in each plane is less than 5 times U_{mar}.

13.7.2 Test Runs with Test Masses

The test masses of the R_{UR} test are used, the *stationary* ones represent the unbalance of the mounting side, the *travelling* ones the unbalance of the rotor side:

- In plane 1: $U_{stationary}$ at 30° and $U_{travelling}$ at 150°.
- In plane 2: $U_{stationary}$ at 150° and $U_{travelling}$ at 30°.

With these unbalances—which represent the initial state—the *first measurement* run is made and the first step of compensation is carried out according to the manufacturer's specifications.

Now the two large test masses are shifted by 180° to simulate the indexing:

- In plane 1: $U_{travelling}$ is shifted from 150° to 330°.
- In plane 2: $U_{travelling}$ is shifted from 30° to 210°.

With these unbalances, the *second measurement* run is made and the second step of compensation is carried out according to the manufacturer's specifications.

Now the two large test masses ($U_{travelling}$) are removed so that the rotor no longer appears to be unbalanced.

The remaining unbalance of the mounting ($U_{stationary}$) should be compensated by the process of indexing, i.e. a *new measuring* run (third) should ideally result in a display of zero, in reality the limits of the compensation device become apparent.

13.7.3 Evaluation of the Compensation Test

The compensation passes the test if the remaining reading in both planes is less than or equal to 0.02 times $U_{stationary}$.

14 Unbalance Correction

Contents

H. Schneider, *Rotor Balancing*, https://doi.org/10.1007/978-3-662-66049-2_14

14.1 Preliminary Note

Balancing a rotor with rigid behaviour means to change the unbalance condition in such a way that the mass inertia axis coincides sufficiently exactly with the shaft axis (see Sect. 5.4.2) (Fig. 14.1).

In the case of rotors with shaft-elastic behaviour, the correction of modal unbalances additionally are to be taken into account.

14.2 Types of Correction

One type of correction is to shift the shaft axis towards the inertia axis. This procedure – called mass centring (centring by balancing) – is used in few cases in single-plane balancing and in two-plane balancing. Here, for example, the rotor is picked up in an (arbitrary) initial position, the unbalance about the thus given shaft axis is determined. The rotor is displaced during rotation so that the two axes coincide and the (new) centring holes are made (Fig. 14.2).

If parts of the rotor are subsequently machined (fits, bearing journals, etc.), the resulting new unbalance must be taken into account (bias) (Fig. 14.3).

The much more common way, however, is to shift the mass axis of inertia towards the shaft axis by removing, shifting or adding material. For this purpose, all processes that are technically feasible can be used.

Very large unbalances can only be corrected economically by adding material; for normal and small unbalances there are no priorities in principle.

The type of correction chosen in each individual case depends above all on the rotor and the required correction times as well as the costs. A short overview should clarify the variety of possibilities.

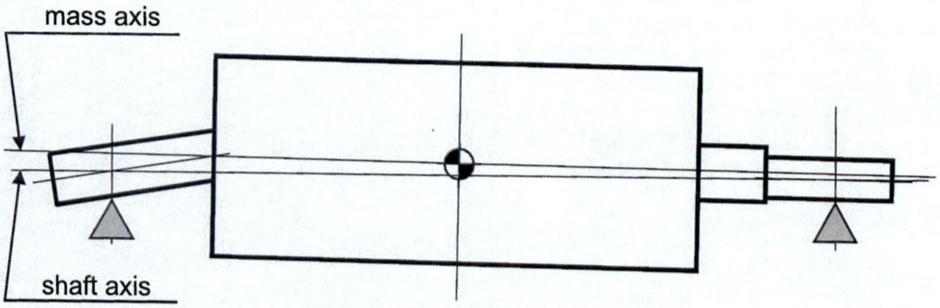

Fig. 14.1 Balancing a rotor with rigid behaviour. Mass axis and shaft axis must be brought into congruence

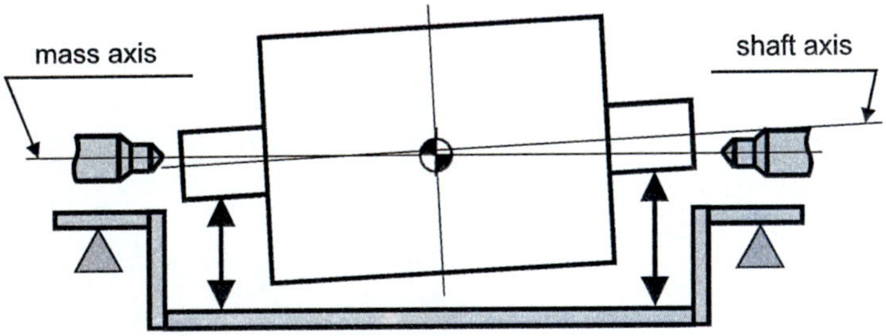

Fig. 14.2 Correction of a rotor with rigid behaviour. Changing the shaft axis by moving the rotor in a frame so that its mass axis coincides with the axis of rotation. Then new centring holes are applied according to the mass axis: This is called mass centring

Fig. 14.3 Mass centring of a cast crankshaft

14.2.1 Material Removal

All conceivable methods of material removal are possible, e.g.:

- Drilling
- Milling
- Eccentric turning
- Grinding
- Planing
- Cutting
- Ablation by a beam, e.g. a laser (Fig. 14.4).

Fig. 14.4 Removal of material with laser during rotation (on a gyroscope)

In many cases, already existing material of the rotor is removed, e.g. when drilling or milling into the package of an electric armature.

Sometimes, however, additional material is provided for balancing, in the form of rings, cams, or pivots, which are then shortened or removed.

In very rare cases, a material is used here that is not determined by the basic function of the rotor, but is only chosen for correction.

Important criteria for the design of correction by removal of material are compiled in Fig. 14.5.

Removal of material

Criteria

- ☐ **applicable for**
  - rotor material
- ☐ **adjustable for**
  - initial unbalance
  - correction step
  - balance tolerance
- ☐ **permissible error of**
  - magnitude
  - angle
- ☐ **necessary time for**
  - setup
  - correction step
  - each rotor (average)
- ☐ **expenditure for**
  - equipment
  - disposal
  - coolant / lubricant
- ☐ **preparation of**
  - rotor
- ☐ **stress of**
  - rotor
  - operator
  - environment

Fig. 14.5 Important criteria for the correction design by removal of material

14.2.2 Relocating Material

These procedures are only applied in special cases, e.g. if the correction has to be repeated often.

All methods of relocating material are eligible, e.g.:

- Radius change, e.g. by screwing in or out, for correction in components.
- Counter-indexing (amount) and co-indexing (angle) of two (equal) eccentric masses at each end of the rotor.
- Counter-indexing (amount) and co-indexing (angle) displacement of two (or more) masses of equal size, e.g. in an annular groove (Fig. 14.6).

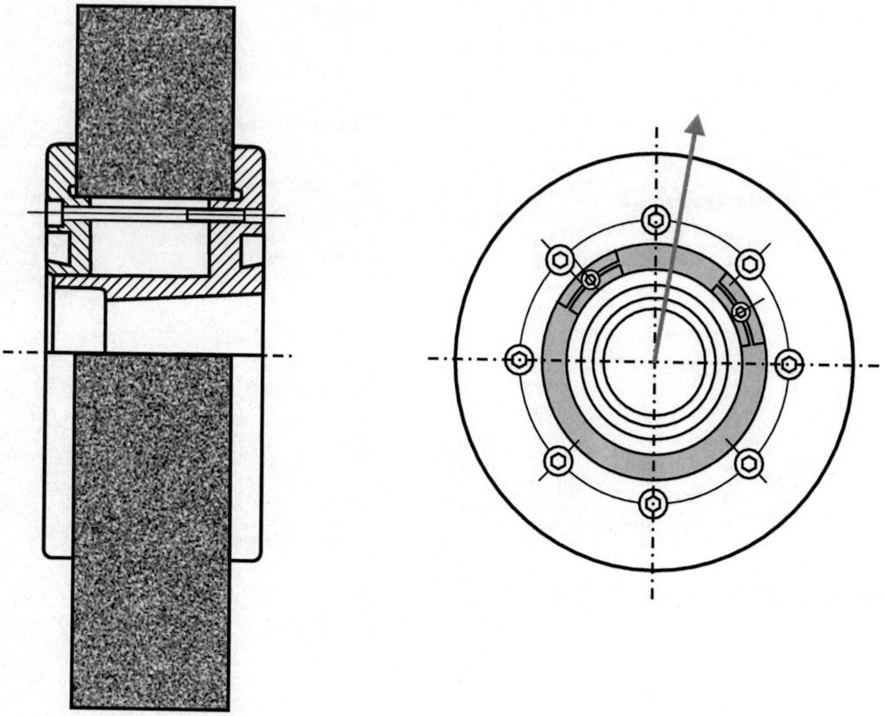

Fig. 14.6 Correction by shifting material, e.g. for grinding wheels: Shifting of groove stones in dovetail grooves. Two to three stones are used in one or two correction planes each. The spread is used to set the size of the correction; the vector sum is used to set the angular position

14.2.3 Adding Material

This is the most extensive type of correction, because the material can often be freely selected when adding. Mostly, material with a high specific density is chosen – up to heavy metals – but sometimes relatively light material is used in order to be able to dose small values better.

Various methods of adding material, e.g.:

- Inserting notched pins.
- Adding screws.
- Pulling in round steel.
- Screwing on correction masses.
- Riveting of correction masses.
- Insertion of sheet metal strips.
- Putting on clips.
- Putting rings on pivots.

Addition of material

Criteria

- **adjustable for**
  - initial unbalance
  - correction step
  - balance tolerance
- **permissible error of**
  - magnitude
  - angle
- **necessary time for**
  - setup
  - correction step
  - each rotor (average)
- **expenditure for**
  - equipment
  - material
  - tool for attaching correction material
- **preparation of**
  - rotor
  - material
- **stress of**
  - rotor
  - operator
  - environment

Fig. 14.7 Important design criteria for the correction by adding material

- Inserting slot nuts.
- Weld on.
- Solder on.
- Spraying on.
- Application of putty (curing).
- Bonding weights to car wheels (to rims).

There are also cases where material removal and addition are combined: if the surface is to be even after correction, it is possible, for example, to drill into steel and then fill the hole with plastic or aluminium.

Important criteria for the design of the correction by adding material are compiled in Fig. 14.7.

14.3 Correction Time

The correction time determines the time per balancing run to a large extent, especially with modern, high-performance balancing machines.

A negative example: after measuring the unbalance, the operator calculates the size of the correction masses, and then gets the corresponding correction masses – e.g. from flat material – sawn, drilled and deburred (or makes it himself). Then a hole is drilled in the rotor at the measured angular position (at a given radius), a screw and nut and the necessary tools are brought in and the correction mass is attached. The machine is not used for half an hour. If the required residual unbalance reading has not yet been achieved, the same procedure is repeated once or twice more.

It is also typical that in many cases, that there is no suitable scale available to weigh the correction masses or that the fastening material (screws, washers and nuts, or welding material) is not properly taken into account.

14.3.1 Organisation of the Correction

Balancing can almost always be organised reasonably, even with changing rotors, e.g. by:

- The indication displays the correction directly (Fig. 12.18), or tables are used for correction (see Sect. 20.4).
- The preparation of stepped correction masses.
- A correction at fixed points is provided and the holes are already machined during the mechanical production.

Screwing on the correct weights (or set of weights) is then a matter of a few minutes.

14.3.2 Automation of the Correction

It is not possible to classify the type of correction as a function of the time required, but for each type of correction three essential levels can be specified, which generally mean increasing machine costs as the balancing times are reduced:

- Hand-operated correction.
- Semi-automatic correction.
- Fully automatic correction.

In order, this means:

- Transferring the measured values and performing the correction manually.
- Transferring the measured values manually, correction is performed automatically (or vice versa).
- Transferring the measured values and correction is performed automatically.

These boundaries are fluid and thus cannot be captured exactly, as there are always additional aids that build a bridge to the next level.

In addition, two measured values must be transmitted for the correction in each plane: the amount and the angular position. There are cases in which, for example, the angular position is automatically turned in but the amount has to be drilled out manually, others in which the angular position is turned in manually and the amount is automatically drilled out.

The addition of well-prepared weights usually takes less time than the removal of the material (at the same financial effort).

The correct criterion to use when selecting the optimal one from the abundance of possibilities is the floor-to- floor time, which also includes the time for loading and unloading.

14.4 Correction Errors

When correcting the measured unbalance, errors are unavoidable in practice; this leads to the fact that a reduction of the initial unbalance according to the unbalance reduction ratio of the balancing machine is not possible.

The correction mass is never completely accurate in size, is not placed on the right radius and not at the right angle. In addition, the influence of the always varying volume is sometimes not correctly assessed. The setting of the measuring device then refers to an assumed (averaged) value, so it may not be correct in individual cases. Errors therefore occur in:

- The correction masses.
- The location of the correction planes.
- The correction radii.
- The angles of the correction masses.

The size of the individual error is very much dependent on the type of correction, so it cannot be stated separately. However, typical examples will be used to show the conditions to be observed.

14.4.1 Correction Masses

The error can be kept small if defined masses are added or removed (0.1–1%). The error becomes extremely large when material is removed by hand over a wide area (20–50% depending on the experience).

14.4.2 Correction Planes

When correcting into the axial direction (e.g. axial drilling from the end faces), the centre of gravity of the correction mass shifts with size.

The plane separation therefore does not apply in general, but only for a certain drilling depth. In all other cases there is an influence on the display of the other side.

The percentage influence is directly proportional to the deviation from the target correction plane. However, it is not the percentage influence that is decisive for the disturbance due to the incorrect plane separation, but the absolute influence, i.e. the false indication in the other plane.

The absolute influence is the product of the correction mass and the percentage influence. The absolute influence – and thus this error – is averaged when set to about 2/3 of the average axial correction depth.

With computer devices, a correction of this error can be carried out with a special software.

14.4.3 Correction Radii

The correction radius can be very accurate when correcting axially, for example (about 0.1–1% error), but the error can become very large when correcting radially or on the circumference.

Modern computer devices can also support the various balancing methods here with special software and keep the deviations within acceptable limits.

The following correction methods lead to errors in the radius.

14.4.3.1 Radial Correction

In case of radial correction (e.g. by drilling), the error depends on the radius of the surface from which drilling is done and on the drill depth (see Sect. 20.4, Fig. 20.23).

The measuring device is set to half the average drill depth; the relative error in the correction of this plane is then the same in both directions – with smaller or greater drill depth.

14.4.3.2 Correction on the Circumference

If the correction mass is distributed over a larger angle on the radius r (e.g. milling a bead ring), the effective radius decreases with a larger angle, i.e. the correction unbalance no longer increases linearly with the angle (Sect. 20.4, Fig. 20.25).

The measuring device is best set to the mean value of the effective radii at maximum angle and angle 0° (not to the effective radius of the mean angle). This makes the relative error the same for extreme corrections.

Task

Maximum angle 150°, effective radius 0.74 r; minimum angle 0°, effective radius r.

Solution

Mean radius (0.74 r+r)/2=0.87 r. (Radius of the mean angle of 75° would be 0.93 r).

14.4.3.3 Correction by Spreading Two Correction Masses

By spreading two equally sized correction masses on the radius r by the angle α any unbalance between 0 and $2U$ can be corrected (see Sect. 20.4, Fig. 20.24).

The measuring device is best set to the radius of a single mass. Display for angle and magnitude separately with a non-linear magnitude division so that the angle of spread can be read directly.

With computer devices, a correction of this deviation can be carried out with special software.

14.4.4 Correction Angle

The angular position of the correction masses can be very accurate with defined material addition or removal (1–3% deviation), but the deviation rises sharply with manual surface correction (20–30%).

Also important are the coupling of the angular reference system, its accuracy and freedom from backlash, as well as the precise transfer of the measured angular value to the rotor.

14.4.5 Permissible Correction Errors

The fundamental considerations of Sect. 14.4.1 to 14.4.4 are supplemented by situational conditions:

- If the unbalance is far from the tolerance, very precise work must be done, the permissible errors in size and angle are very limited (Fig. 14.8),
- If the unbalance is only just outside the tolerance, approximate correction is sufficient; the permissible errors in size and angle are very large (Fig. 14.9).

These large differences in the necessary precision are sometimes not recognised correctly, so that work is always done with the same accuracy. Correction takes longer and thus becomes more expensive than necessary.

Vector display is a great help here. If it is then combined with a tolerance display (see Fig. 12.16), all the necessary indications are actually given – they only have to be used.

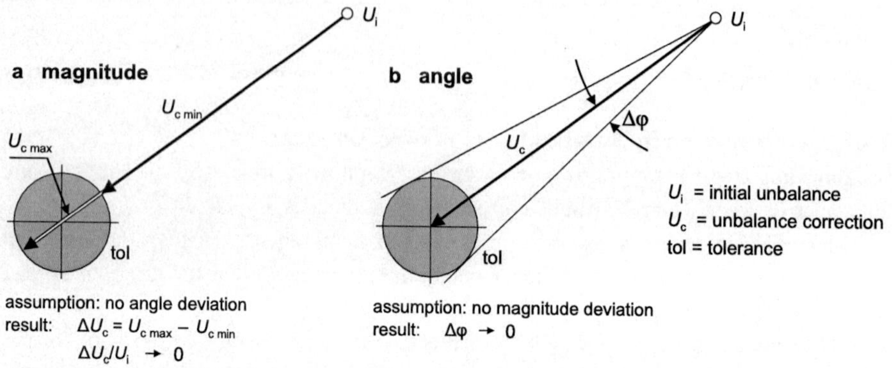

Fig. 14.8 Permissible errors for large unbalances compared to the tolerance. The permissible specific amount error (**a**) and the permissible angle error (**b**) are each considered separately, the combination of errors in size and angle reduces the permissible errors even more

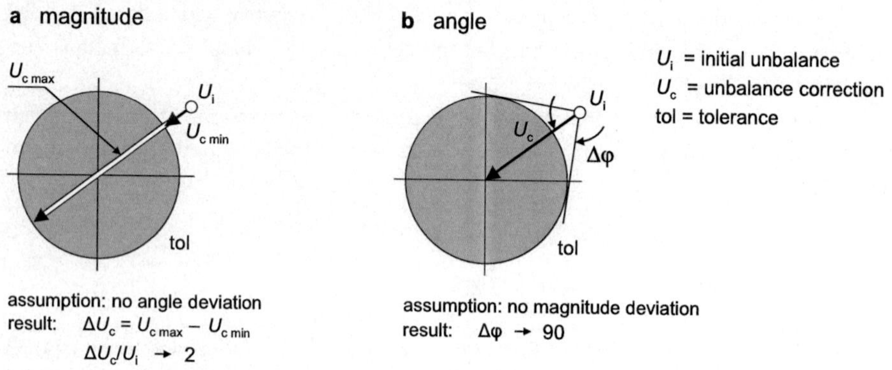

Fig. 14.9 Permissible errors with small unbalances compared to the tolerance. The permissible specific amount error (**a**) and the permissible angle error (**b**) are each considered separately, the combination of errors in size and angle reduces the permissible errors even more

14.5 Unbalance Reduction Ratio

14.5.1 General

The ratio between the initial unbalance and the residual unbalance, in conjunction with the unbalance reduction ratio, determines the size of the permissible error during correction.

14.5.2 Small Correction Step

With a wanted ratio of 5:1 and an unbalance reduction ratio of 90% (which theoretically means 10:1 reduction), a maximum error of about 10% is permissible during correction. Then the tolerance can still be reached with one step.

14.5.3 Large Correction Step

If the ratio of initial unbalance to residual unbalance is larger, but still smaller than the unbalance reduction ratio, then only a very exact correction in *one* step leads to the goal (in borderline cases, the correction errors can be reduced to approx. 1% – with a carefully selected and organised method). In such a limit range, however, it must be checked whether the time (and cost) required for a one-step correction does not exceed that for a two-step correction.

If the ratio of initial unbalance to residual unbalance is greater than the unbalance reduction ratio, it is no longer possible to balance in one step.

14.5.4 Series

These statements given above only apply to single rotors. For large-volume products, the magnitude of the unbalance of a larger number of rotors is statistically evaluated. The correction is designed in such a way that a certain (large) percentage can be balanced within the required tolerance (see e.g. Sect. 15.4). Rotors with extreme unbalances, for which the tolerance cannot be achieved this way, are e.g. sorted out or finished manually on another balancing machine.

Preparation and Execution of Rotor Balancing

15

Contents

15.1 Preliminary Note

If balancing is properly planned, it is only necessary to ensure that the operator balances according to instructions. The more details are put into the decision of the balancing personnel, the more clearly the fact becomes visible that other departments of the company have not completely fulfilled their tasks.

One central point is often not sufficiently regarded: the proper introduction of the difference in balance quality between the ***balancing target*** and the ***balancing process***, see Sect. 4.8.

This chapter describes, among many other tasks, how the chosen balancing process has an impact on the permissible *unbalance readings* to be observed in this process.

H. Schneider, *Rotor Balancing*, https://doi.org/10.1007/978-3-662-66049-2_15

Or the other way round: How shall the balancing process be adjusted to achieve the permissible unbalance readings for this balancing process and finally to secure the originally chosen balancing target?

15.2 Causes for Unbalances

If one realizes that today almost all rotors are balanced, the question arises, where these unbalances come from and whether they cannot be avoided.

Unbalance is always present when the mass distribution of the rotor is not symmetrical with respect to its shaft axis, i.e. one of its central principal axes of inertia is not sufficiently coincident with the shaft axis (see Sect. 14.1).

In the case of welded fan wheels, this asymmetry can be very pronounced, it is somewhat smaller in the case of simple, assembled bodies, e.g. wound electric motor armatures, becomes very small in the case of turned parts, can be negatively influenced by mounted roller bearings – but it is so noticeable even in the case of high-speed parts with plain bearings and ground on all sides that balancing cannot be waived, because in most cases the manufacturing tolerances are not as fine as the balancing tolerances. This can best be assessed by comparison with the permissible eccentricity e_{per} of the centre of gravity (see Sect. 7.3.2).

The main causes can be summarized in three groups (Fig. 15.1). Many of these causes can be influenced in magnitude, but generally they cannot be avoided to such an extent that balancing would not be necessary. Today, unbalance tolerances—expressed in displacement of the centre of gravity—are typically one to two powers of ten smaller than the associated manufacturing and assembly tolerances.

And in those cases where the errors could be reduced to the necessary extent, it has been found that avoiding rotor balancing does *not* lead to a more economical solution.

Errors occur in following areas

❑ **design**

- parts not rotationally symmetrical
- surfaces not machined
- run-out

❑ **material**

- blowholes in casted parts
- changing density
- changing dimensions (sheet metal)
- antifriction bearing run-out

❑ **manufacturing and assembly**

- form errors during welding and casting
- clamping error, e.g. journal eccentric, tilted
- permanent deformation
- distortion, deformation by welding, shrinking
- differences in material e.g.screws, washers

Fig. 15.1 Causes of unbalance from three different areas

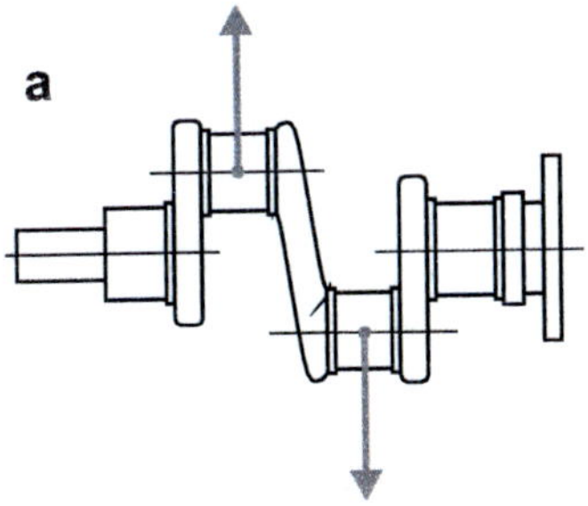

Fig. 15.2 Calculational balancing: Two-cylinder crankshaft (**a**), with counterweights (**b**). (In the case of two-cylinder crankshafts, a portion of the connecting rod is usually also balanced, i.e. the crankshaft itself is not in tolerance even with counterweights)

15.3 Design Guidelines and Drawing Specifications

Already during the design of a rotor, care must be taken that all asymmetrically lying masses are balanced as far as possible (Fig. 15.2)—in the case of flexible rotors even near the same plane—and that the required number of balancing planes is created at the right places (see Sect. 4.6).

In addition, it is necessary to consider (usually in conjunction with the work planning) which type of correction (see Sect. 14.2) is to be used. Material is to be added accordingly, holes or grooves etc. are to be provided.

The condition in which the rotor is to be balanced must be indicated on the production drawing—e.g. with rolling bearings fitted, or with half key.

The correction planes and the bearing planes should be marked and dimensioned, as well as the correction radii, if they are not arbitrarily selectable. Furthermore, clear information shall be given on the correction type, the tool to be used, any existing limits in the correction (e.g. maximum permissible drilling depth) and the permissible residual unbalance per plane (see Fig. 11.1).

15.4 Layout of the Correction

The frequency distribution of the unbalance plays an important role for the most suitable balancing process of mass-produced rotors (see Sect. 13.2). The unbalance of a number of rotors is measured and shown in a bar graph (Fig. 15.3).

The unbalance axis is scaled in multiples of the tolerance T, and a multiple V of the tolerance is entered in this diagram. V is the maximum unbalance that can be corrected within one step. This value—assumed here to be 10:1—is derived from the unbalance reduction ratio R_{UR} of the balancing machine and the correction intended.

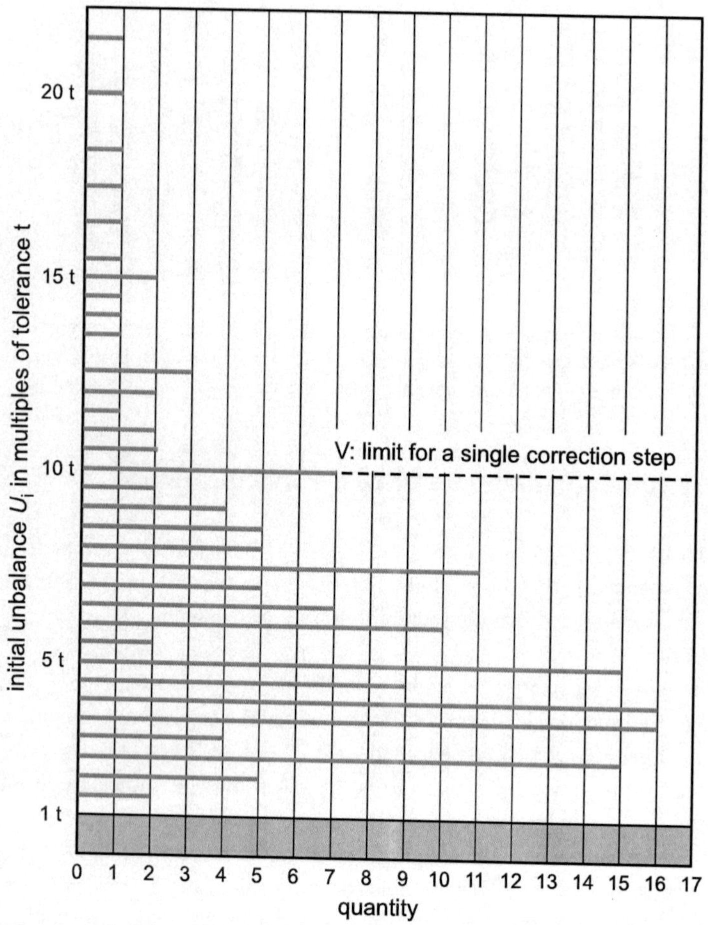

Fig. 15.3 Typical distribution of the initial unbalance amount (centric frequency distribution) of a large-volume product

Using V, it is clear which percentage of rotors can probably be balanced within one step. In Fig. 15.3 this is about 87%. All rotors with unbalances greater than V require more than one steps.

The measured rotors show 22 T as the largest unbalance, but in a few cases larger unbalances can also occur. Therefore, it must be determined for which max. unbalance the correction is to be designed and whether larger unbalances are subject to special treatment or whether they are rejects.

The correction for all rotors with unbalances *greater than* V only needs to ensure that values smaller than V are achieved *after the* first correction. For stepped masses, for example, two masses (15 T and 25 T) would be sufficient to always achieve results better than 5 T (0.5 V) for initial unbalances between 10 and 30 T (1 V and 3 V).

For unbalances *smaller than* 10 T (initial unbalances, or intermediate results after the first step), steps between 1.5 T and 2 T are recommended (depending on the accuracy of the balancing machine and the precision of the correction). For 1.5/3/4.5/6/7.5/9 times T six different masses are necessary, for 1.8/3.6/5.4/7.2/9 times T only five masses (but then there is less scope for errors).

The gradation of the steps and the correction masses must be optimized for each rotor in conjunction with the intended correction method in order to achieve minimum unit costs.

15.5 Work Planning

Like any other production process, rotor balancing must be planned correctly by the work planning department to make it economical. I.e. the possibilities of the balancing machine and the type of correction should be adapted properly to the respective rotor.

Semi-automatic and fully automatic machines are usually tailored to very specific rotors and all individual times and the piece time are already discussed when selecting the most suitable machine. With changing rotor types and extensive manual operation the difficulty of estimating the times correctly increases. But it is precisely here that a systematic approach is worthwhile: while in the case of fully automatic machines, further refinements in individual cases can sometimes reduce the cycle time by 5 to 10%, in the case of previously unplanned manual balancing, it is often possible to achieve time savings of 50 to 90% without difficulty.

Work preparation normally leads via work studies to work plans for individual processes. This also applies, of course, to the balancing process of individual rotors. Here also, it is possible to draw conclusions for similar, newly arising rotors from work plans for work already carried out.

In any case, the step from the balancing tolerance of the *balancing target to* the permissible unbalance reading of the *balancing process* must be made here, see Sect. 4.8.

15.5.1 Rotor Condition During Balancing

The balancing target describes the permissible residual unbalance for the delivery, i.e. what is noted on the production drawing of a rotor to be balanced Sect. 15.4.

If the rotor is to be balanced in a different condition—e.g. without rolling bearings, or not completely assembled—this condition shall be stated, but also the magnitudes of the resulting errors shall be specified by the designer (e.g. in Fig. 11.1). Only then the errors of the balancing process, together with all other errors of the process itself can be properly taken into account.

15.5.2 Permissible Unbalance Readings for the Balancing Process

The permissible unbalance readings for the balancing process result from the tolerances for the balancing target (and the errors that may be specified there, see Sect. 15.5.1) taking into account all errors associated with the balancing process itself, see Chap. 16.

ISO 21940-14 describes procedures on how systematic errors can be reduced and the remaining errors can be determined. In this book, the problem is explained in Chap. 16.

Realistic planning of the permissible unbalance readings for the balancing process is a basic prerequisite for achieving the desired balancing target. Nevertheless, it seems that balancing people do not like to deal with this—at least the author has not been able to get his hands on any acceptable practical example.

15.5.2.1 Commonly Practiced Approach

In practice, a deduction as a percentage is usually made for the balancing process: in the range between 20 and 50%; sometimes even a better quality grade is used—but not with the usual factor of 2.5, but with the intermediate factor of 1.6.

If problems nevertheless arise in operation (or during acceptance) with the used markdown, the percentage normally is increased. Only if problems remain and a further increase of the markdown would lead to difficulties, the causes will be investigated. In this case, Sect. 15.5.2.2 comes into play.

The so-called "unbalance tolerance" selected for the balancing process sometimes is even used externally (for a customer), but this is not correct.

15.5.2.2 Current Approach of the Standards

A systematic approach is generally recommended, because the balancing process only this way can be completed with a minimum of effort and with a good chance of success.

The examples in Sects. 4.8, 7.8 and 8.4 show that it is possible to obtain indications for the optimisation of the balancing process with relatively simple means. The prerequisite is that some information is at hand about:

Influences of the Balancing Machine

- Limit values of the balancing machine, e.g. smallest achievable residual unbalance U_{mar} and unbalance reduction ratio R_{UR} for all speeds of interest.
- Limits for the compensation of measured values.
- For single-plane balancing machines: moment unbalance influence I_{MU}.
- Influences of the available drives at all speeds of interest.
- Mass data of drive shafts, drivers, adapters.

Influences of the Rotor

- Support: Deviations from the service condition,
- Runout errors at all relevant points,
- Data of the fits used,
- Data of the keys used and their systematics,
- Special influences, e.g. hydraulic or electromagnetic forces, constraint forces,

- Display disturbances (e.g. due to ventilation)—can may be reduced by longer measuring time and averaging,
- Scatter from run to run (e.g. due to moving blades)—can be reduced by averaging over several runs.

If many errors are to be considered, the quantities are not simply added, but according to ISO 21940-14, they are averaged according to the rule of root mean squares, see Sect. 16.4.

15.6 Automation

The correction by material removal on small electric armatures is used as example, in order to discuss different automation levels, different processes and cycle times.

Examples

a) Drilling in a separate station, manual transport: Table 15.1, Fig. 15.4,
b) Drilling in the balancing machine: Table 15.2 and Fig. 15.5,
c) Milling in a separate station, manual transport: Table 15.3 and Fig. 15.6,
d) Milling in a fully automatic balancing machine: Table 15.4, Fig. 15.7.

The tables are structured as follows: In column 1 the activities of the operator are listed, in column 2 the commands and in column 3 the functional steps. ◀

From top to bottom, the tables represent the respective time sequence. The process starts e.g. in Table 15.1 with 3 steps of the operator:

- Pick up a rotor
- Put on scanning ring
- Insert the rotor into the balancing machine

Then he gives the order:

- Balancing

The balancing machine then performs:

- Run up
- Measuring
- Locking the indication,
- Braking

This is how it systematically continues.

From Tables 15.1 to 15.4 it is obvious how the operator becomes less important until a fully automatic machine operates.

Table 15.1 Universal balancing machine with manual drilling machine standing next to it (see Fig. 15.4)

Operator	Command	Process steps
Pick up a rotor Add scanning ring Insert rotor into balancing machine	Balancing	Acceleration Measuring Indication locked Braking
Transfer rotor to drilling machine Index for plane 1 Drill Turn around rotor Index for plane 2 Drill		
Transfer rotor to balancing machine	Checking	Acceleration Measuring Indication locked Braking
Pick up rotor Remove scanning ring Put down rotor		

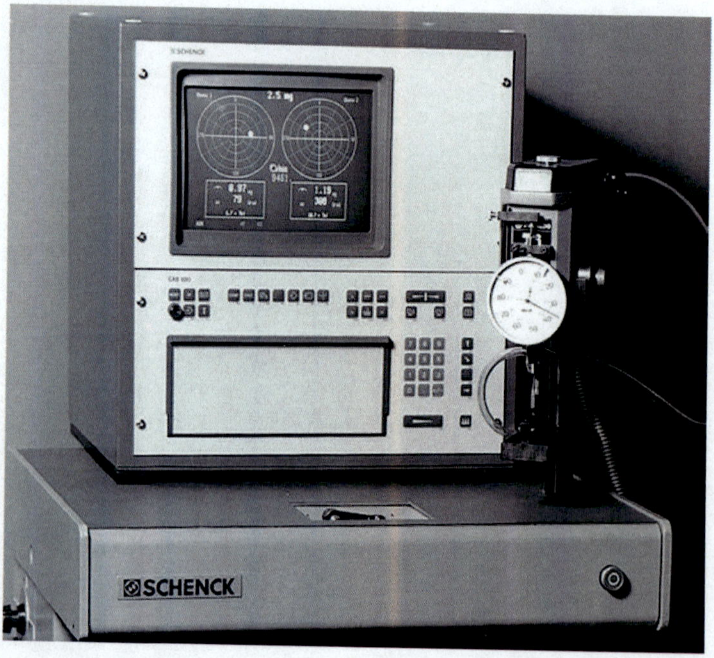

Fig. 15.4 Balancing machine for small electric armatures in table-top design with separate manual drilling station (two stations design). The steps for the balancing process are compiled in Table 15.1 the time schedule is shown in Fig. 15.8

Table 15.2 Balancing machine with two attached drill units (s. Fig. 15.5)

Operator	Command	Process steps
Pick up a rotor Insert rotor into balancing machine	Balancing	Acceleration Measuring Locking readings Braking
Index for plane 1 Clamp Drill Unclamp Index for plane 2 Clamp Drill Unclamp		
	Checking	Acceleration Measuring Locking readings Braking
Remove rotor Put down rotor		

Fig. 15.5 Balancing machine with two drilling units attached (one station). The steps for balancing are compiled in Table 15.2 the time schedule is shown in Fig. 15.8

Table 15.3 Balancing station with automatic correction station at side (see Fig. 15.6). The rotors are balanced in an overlapping cycle

Operator	Command	Process steps
Pick up a rotor Insert rotor into balancing station	Balancing	Acceleration Measuring Locking readings Braking
Remove rotor Index rotor to a given position Insert rotor into correction station	Correction	Clamp Corrections in planes 1, 2 Unclamp
Remove rotor Insert rotor into balancing station	Checking	Acceleration Measuring Locking readings Braking
Remove rotor Put down rotor		

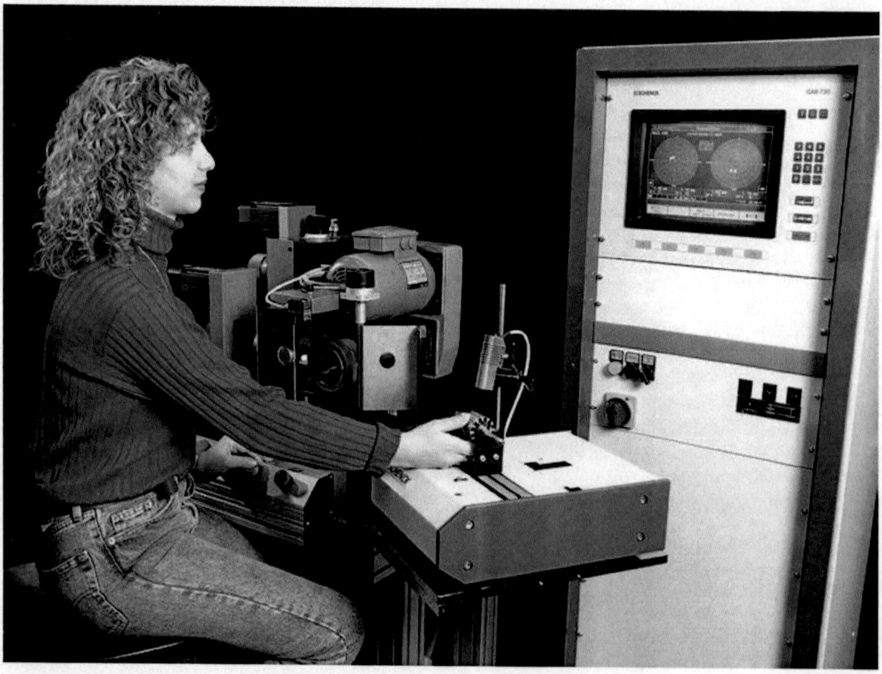

Fig. 15.6 Balancing station with automatic correction station at side. The steps for balancing are compiled in Table 15.3 and the time schedule is shown in Fig. 15.8

Table 15.4 Fully automated balancing machine with rotary transfer (see Fig. 15.7)

Operator	Position, movement	Process steps
	Rotor in feeder Transport to measuring station 1	Acceleration Measuring and storing readings Braking, indexing for plane 1
	Transport to correction station 1	Clamping Correction in plane 1 Unclamping
	Transport to measuring station 2	Acceleration Measuring and storing readings Braking, indexing for plane 2
	Transport to correction station 2	Clamping Correction in plane 2 Unclamping
	Transport to control station	Acceleration Measuring Braking
	Transport to output conveyors	Putting down; good/bad separately

Fig. 15.7 Fully automated balancing machine with 6-station rotary transfer. The steps for balancing are compiled in Table 15.4 the time schedule is shown in Fig. 15.8

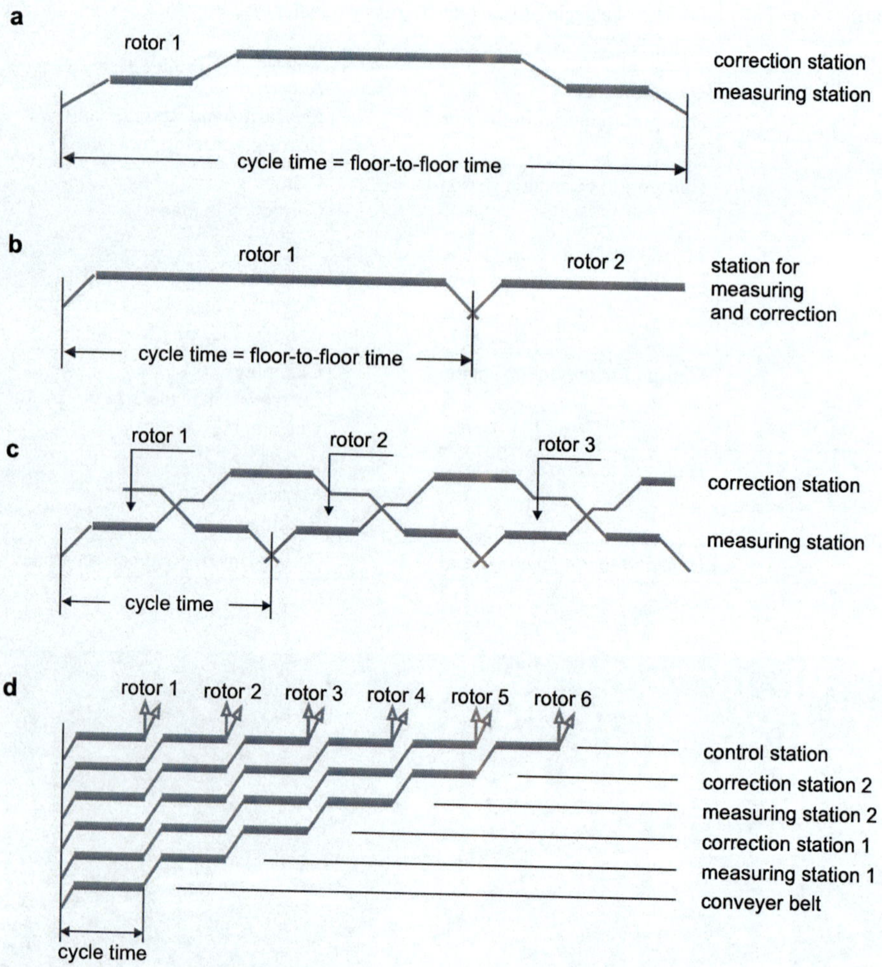

Fig. 15.8 Schedules for the machines of Figs. 15.4, 15.5, 15.6 and 15.7. The times are not to scale (comparing **a–d**), but they show important tendencies

Figure 15.8 shows the schedules (not to scale in the comparison of **a–d**) of the different solutions. The individual stations are represented by different horizontal straight lines, and the motion of a single rotor is shown by a polyline.

If there is more than one rotor in the process, the cycle time is only a fraction of the floor-to-floor time.

a) Two stations: One rotor runs through the machine, then the next.
b) One station: Here measuring *and* correction are done in the same set-up.
c) Two stations, overlapping cycle: While one rotor is in the correction station, the rotor that has just been corrected is checked and then the next rotor is measured in the balancing station.

d) Fully automatic machine with rotary transfer: In all stations, the processes run in parallel, i.e. there are always six rotors in the machine.

15.7 Loading and Unloading

Loading and unloading must be carefully considered and organized, especially with:

- Series production.
- Correction outside the balancing machine.
- Very valuable rotors or those, easily destructible.
- Unhandy and heavy rotors.

As a general rule:

- Transport routes are to be kept as short as possible.
- The movement should be mainly horizontal.
- Mechanical aids (hoist, crane, lifting device) should relieve the operator of heavy physical work, be precisely controllable and be available at all times.

There is also a whole range of automation options for transport.

In the case of ***manually operated*** balancing machines, all loading operations are also performed by the operator (Fig. 15.9).

Fig. 15.9 Manually operated balancing machine. Balancing of turbine disks of a jet engine

In the case of ***semi-automatic*** balancing machines, the rotor is only *transported* by hand (possibly also between the stations of the machine) and is removed again at the end of the working cycle (Fig. 15.10).

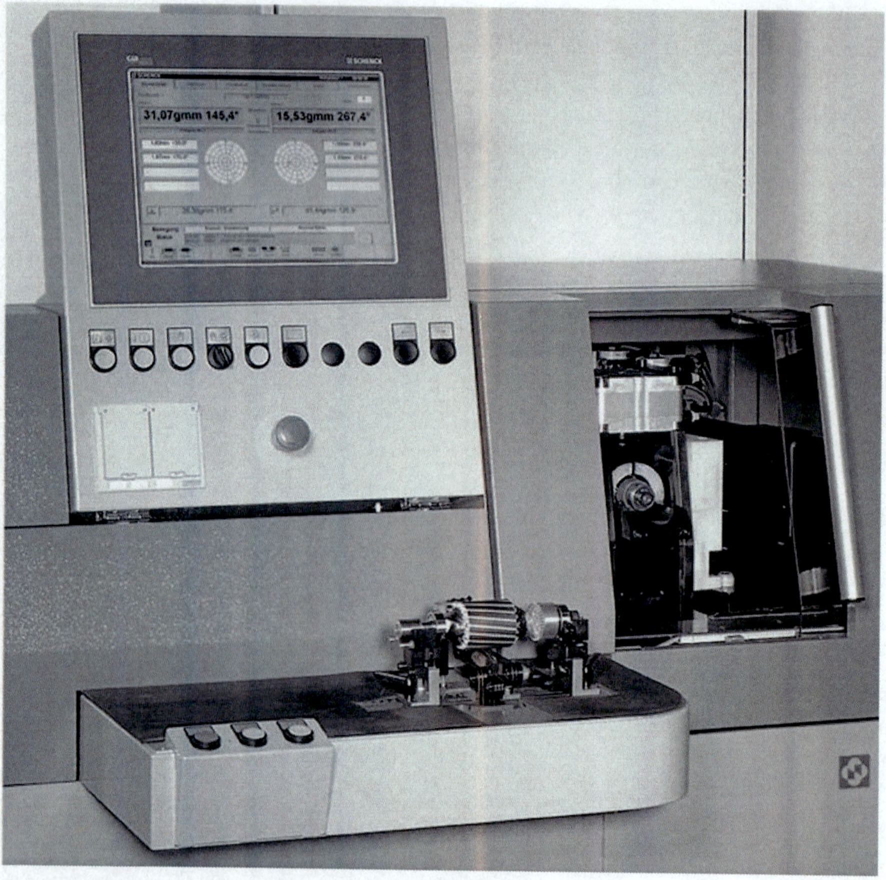

Fig. 15.10 Semi-automatic machine, balancing station *on the left*, correction station *on the right.* The transport is done by hand

In the ***fully automatic*** balancing machine, the rotor is conveyed by a transport system through the various stations, balanced in the process and discharged again. A transport system, e.g. a robot, can thereby turn a semi-automatic machine into a fully automatic one (Fig. 15.11).

Automatic loading and unloading only makes sense for large series. In the case of small series, the savings do not even outweigh the additional set-up time required, so it is more advantageous to work with manual aids here.

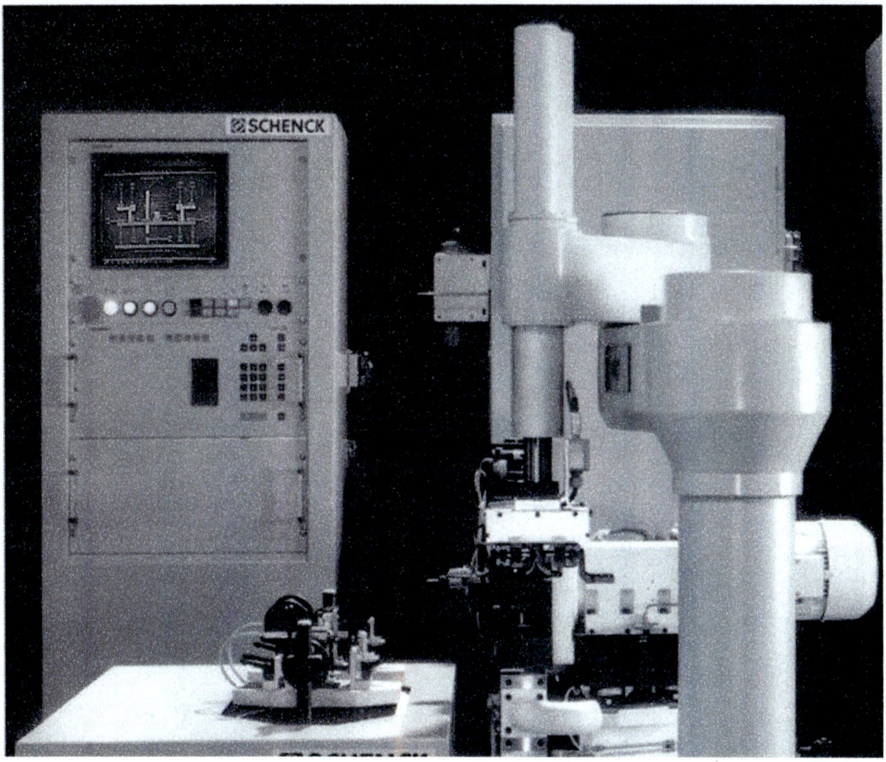

Fig. 15.11 Fully automatic machine, balancing station on the left, correction station in the background, linked by a robot handling two rotors simultaneously

15.8 Preparations on the Rotor

In order to balance a rotor, it must be mounted into a balancing machine, driven and provided with an angular reference. Not all rotors fulfil these requirements by themselves; they must be specially prepared.

It is recommended to include all preparations for balancing in the standard production process, e.g.:

- Assemble the service antifriction bearings for support in the balancing machine.
- Produce the centring and the threaded holes for the drive shaft connection,
- Drill holes for the unbalance correction at fixed locations in the correct angular position to the whole pattern of the drive shaft connection.

Often there are only small things in the production process, but that are of great importance for balancing. It is advantageous, e.g., to have a small cast-on cam that is used as a reference mark for the angle during balancing, which is also used later to index to the correct angle for the correction.

The optimum solution depends so much on the rotor, its design, the manufacturing, the balancing machine with its measuring system and the balancing process that it is certainly not easy to find. However, the problem is very interesting in an engineering sense, and its correct answer contributes significantly to short balancing times and thus to an economical balancing.

15.9 Production Cycle Rotor Balancing

If balancing is properly planned and organized, it is only necessary that the operator balances according to instructions.

Conclusion

The more details are put into the decision of the "balancer", the more clearly the fact becomes visible that other areas of the company have not completely fulfilled their tasks.

Errors in the Balancing Process

16

Contents

H. Schneider, *Rotor Balancing*, https://doi.org/10.1007/978-3-662-66049-2_16

16.1 Preliminary Note

According to ISO 21940-11[1] the unbalance tolerance (usually the quality grade) describes the unbalance permissible for the *delivery*, thus representing an important contractual detail for manufacturer and customer.

When balancing, the limits have to be tightened, on the one hand because the measuring equipment exhibits errors, and on the other hand because a different situation usually exists during balancing compared to the delivery condition—or is deliberately created for economical or practical reasons—e.g. by:

- A different bearing.
- A different assembly condition.
- Balancing accessories, e.g. adapters, drivers, parallel keys etc.
- By physical effects, such as windage, constraining forces, axial thrust, or magnetic fields.

[1] ISO 21940 Mechanical vibration—Rotor balancing, part 11 (2016): Procedures and tolerances for rotors with rigid behaviour.

How these errors—especially the unintentional ones—can be detected and how they can be reduced is described in ISO 21940-14[2] and in this chapter.

Note

Some important details of ISO 21940-14 have been superseded by more recent specifications in ISO 21940-11. Therefore, a correction of part 14 is in progress, and the standard is also to be completely revised.

This chapter therefore has a different structure (compared to the 8th edition) and a modified presentation of the issues.

The remaining individual errors ΔU_i—after possible corrections—are quantified and expressed as combined error ΔU.

Target

The permissible reading $U_{\text{reading per}}$ for the residual unbalance is to be smaller than U_{per} by ΔU, see Fig. 16.1.

Many appliers of balancing technology therefore reduce the permissible unbalance reading for the balancing process by a certain percentage (or a "small" quality level), but this is, so to speak, a "blind flight": if ΔU is really small, the effort is unnecessarily large, and if ΔU is large, it may not be sufficient.

With the aid of a careful ***determination*** of the possible balancing errors, the intended balancing process can be checked and, if necessary, improved by determining whether:

- The balancing machine is suitable.
- The balancing speed is correctly selected.
- The degree of assembly is sufficient.
- The errors are permissible for the parts assembled later.
- The provided drive and the drivers are sufficient.
- The required adaptors meet the requirements.

If ΔU is large, the permissible unbalance reading $U_{\text{reading per}}$ for the balancing process can become unrealistically small, or even negative (i.e. unrealistic) values.

[2] ISO 21940 Mechanical vibration—Rotor balancing, part 14 (2012): Procedures for assessing balance errors.

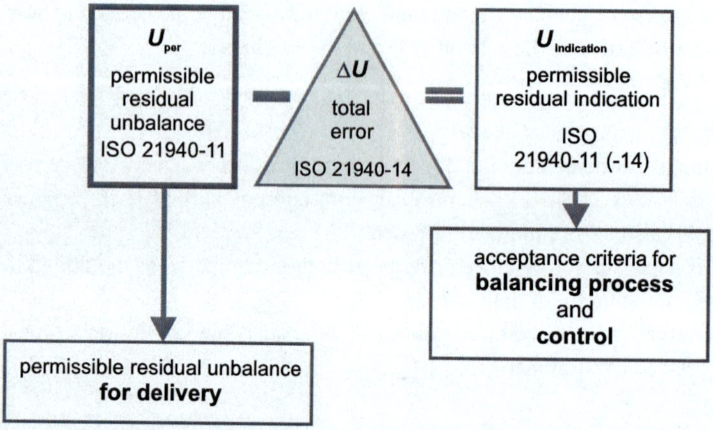

Fig. 16.1 The combined error of the balancing process is subtracted from the permissible residual unbalance for the delivery and results in the permissible reading for the balancing process and for a control

16.2 Causes for Errors

16.2.1 General

Any error in the rotor between the delivery condition and the condition during balancing can lead to an error which is not negligible, see the following lists in Sects. 16.2.2 to 16.2.5.

In addition, the measurement of the unbalance results in errors due to limitations of the measuring chain, Sect. 16.2.6.

16.2.2 Missing Parts

The rotor is not as completely assembled as it will be delivered: one or several parts are missing, e.g.:

- Antifriction bearings.
- Parallel keys.
- Cooling fans.
- Blades.
- Stages.

16.2.3 Additional Parts or Effects

Additional parts are connected to the rotor: to make balancing possible, easier, or by accident, e.g.:

- Drive pulley.
- Auxiliary shaft.
- Balancing spindle.
- Adaptors.
- Enclosed liquids or solids.

16.2.4 Changed Rotor State

On the balancing machine, the rotor has a changed unbalance state depending on: the running conditions, the quantity and the type of the medium conveyed, the bearing planes or the bearing type, for example:

- Speed.
- Direction of rotation.
- Missing or different housing.
- Covers.
- Conveyed medium.
- Support condition.

16.2.5 Rotor Behaviour is Not Reproducible

The unbalance indications change from run to run, or are dependent on certain environmental influences or time, e.g.:

- Movable blades.
- Thermal effects.
- Intermittent bending of the rotor.

16.2.6 Unbalance Measurement

The measurement of the unbalance in the balancing machine includes errors, e.g. due to.

- Unbalance reduction ratio R_{UR}.
- Minimum achievable residual unbalance U_{mar}.

16.3 Handling of Errors

16.3.1 Error Types

Three different types of errors are of interest for balancing, they differ in origin and appearance.

For these types of errors a few examples (based on ISO 21940-14, but arranged differently), they will be explained later.

16.3.1.1 Systematic Errors

Errors with constant amount and angle, they can be determined by calculation or measurement—possibly with special methods.[3] Examples are:

- Unbalance of the drive shaft of the balancing machine.
- Run-out of the pulley on the rotor shaft axis.
- Run-out of balancing bearings.
- Unbalance of an auxiliary shaft.
- Run-out of the rotor seat on the auxiliary shaft.
- Play of the rotor on the auxiliary shaft (when loosened and systematically reassembled).
- Misalignment between bearing journal centres in service and used surfaces for balancing.
- Parallel keys and grooves that do not correspond to the nominal condition.
- Magnetism in the rotor.
- Errors caused by renewed, unsystematic assembly.
- Permanent bending of the rotor after balancing.
- Play of the rotor on the auxiliary shaft (special procedure, see Sect. 16.6.2, note 4).

16.3.1.2 Random Errors

Errors whose magnitude and angle change in an unpredictable manner over a series of measurements (under the same conditions), e.g. due to:

- Loose fitted parts,
- Liquids or solids in cavities.[4]

---

[3]The prerequisite is that the angular reference of the balancing machine is at a fixed angle to the systematic error. If not, the systematic error appears as a scalar error, see Sect. 16.3.1.3.

[4]A random error only occurs if the foreign bodies remain in *varying* amounts at *alternating* positions. If they always collect *completely* in the *same* place, they act as a systematic error, if they collect at *any* angle but *completely*, they act as a scalar error.

- Deformation due to thermal influences.[5]
- Deformation due to gravity.
- Play of the rotor on the auxiliary shaft (if loosened and reassembled unsystematically).

Some of these errors only occur after longer downtimes.

16.3.1.3 Scalar Error

Error whose amount can be measured or estimated but whose angle is uncertain, e.g.:

- Running errors of operating bearings, which are, however, not used for balancing.
- Play of the rotor on the operating shaft (calculated maximum value).
- Play in the universal joints of the drive joint shaft.
- Run-out errors at joints which are separated again after balancing and reassembled unsystematically.

16.3.2 Determination of Errors

Errors can be very different in the tolerance planes. They are therefore always determined for the tolerance planes separately.

Different methods can be used for this purpose:

16.3.2.1 Estimation of Errors

An estimation based on technical data—e.g. in the case of a fit—may be sufficient to check whether this error is can be neglected or not.

An estimation must base on the maximum possible value. Sometimes the angle remains undefined (scalar error), even if it would be actually a systematic error, e.g. the eccentricity of a fit.

16.3.2.2 Measuring Errors

Errors can be measured in the balancing machine, if necessary supported by a special procedure, e.g. index-balancing, see Sect. 16.6.2. In this case, the amount *and* angle of the error can be determined (systematic error).

If the clearance of a fit, which is normally a scalar error, is always pushed out in the same direction when index-balancing, this error becomes a systematic error of the rotor or the adaptor.

For measuring random errors, see Sect. 16.6.3, for scalar errors, see Sect. 16.6.4.

---

[5] Again, depending on the nature of the influences, systematic or scalar errors may arise.

16.3.2.3 Errors During Measurement

The known limitations of the measurement on the balancing machine are used as a basis, in particular the minimum achievable residual unbalance U_{mar} and the unbalance reduction ratio R_{UR}, see Chap. 13.

For a given balancing task U_{mar} is a fixed value, whereas R_{UR} must be referenced to the current measurement signal. Thus, if a bias or similar is used, this value suppressed in the measuring device (invisible) must of course also be taken into account.

16.3.3 Treatment of Errors

Errors must be reduced if they reach a magnitude which makes balancing inadmissibly difficult or even impossible.

Of course, the total number of errors also plays a role:

- If the number is small—let's say 3—they can probably be accepted if they are no greater than 10% of the balance tolerance each.
- If the number is high—let's say 10—they probably shouldn't exceed 3% each.

However, a final assessment is only possible once the combined error has been determined, see 16.3.4.

Errors must be treated differently depending on their nature, see Sects. 16.3.3.1 to 16.3.3.3.

16.3.3.1 Calculation of Systematic Errors

If two or more systematic errors of approximately the same magnitude occur, the vector sum should be formed and the magnitude of the sum vector should be used as the single error ΔU_i.

In all other cases, the amount of each error is used directly as ΔU_i.

16.3.3.2 Calculation of Random Errors

Random errors can be determined in a balancing task for different constellations, e.g.:

- Scatter from run to run for a *given* assembly.
- Scatter with *repetition of* the assembly.

The amount of the radius of the scattering range is used as the error ΔU_{ran}, see Sect. 16.6.3.

16.3.3.3 Calculation of Scalar Errors

Scalar errors basically consist from amounts, see Sect. 16.6.4 These amounts are therefore used directly as ΔU_i.

16.3.4 Determination of the Combined Error

The combined error is determined for each tolerance plane—for one plane for single-plane balancing, for two plane for two-plane balancing and for more planes for flexible rotor balancing.·

Let $\left|\Delta\vec{U}_{iA}\right|$ be the amount of a single error i in the tolerance plane A. Then the combined error ΔU_A can be calculated to:

$$\Delta U_A = \sqrt{\sum_{1}^{i} \left|\Delta\vec{U}_{iA}\right|^2} \quad g \cdot mm. \tag{16.1}$$

Thereby is:

ΔU_A is the amount of combined error in tolerance plane A, in g•mm;

$\left|\Delta\vec{U}_{iA}\right|$ is the amount of error i in tolerance plane A, in g•mm.

An analogous formula is used for tolerance plane B.

Note 1

The amounts of combined error and ΔU_A are ΔU_B usually different for different balancing machines. But even on the same balancing machine, the combined errors depend on *how* the machine is used and *what* accessories are used.

Note 2

The author recommends to apply the same system of errors also for flexible rotors, see Sect. 8.6

Note 3

There is no fixed limit to the ratio of errors to unbalance tolerance: it depends entirely on the robustness of the balancing process:

- A robust balancing process will not "derail" even with errors up to about 70% of the balancing tolerance.
- While an unstable balancing process may reach its limits with errors as low as 30% of the unbalance tolerance.

16.4 Permissible Indications for the Residual Unbalance

The permissible indications for the residual unbalances $U_{\text{reading A}}$ and $U_{\text{reading B}}$ in tolerance planes A and B during the balancing process are equal to the permissible residual unbalances $U_{per\ \text{A}}$, $U_{per\ \text{B}}$, reduced by the combined errors ΔU_{A} and ΔU_{B}:

$$U_{\text{reading A}} = U_{per\ \text{A}} - \Delta U_{\text{A}} \quad \text{g} \cdot \text{mm} \tag{16.2}$$

$$U_{\text{reading B}} = U_{per\ \text{B}} - \Delta U_{\text{B}} \quad \text{g} \cdot \text{mm} \tag{16.3}$$

From this it follows that ΔU_{A} and ΔU_{B} must be limited in size: The balancing process and equipment must be chosen appropriately, otherwise the expressions $U_{per\ \text{A}} - \Delta U_{\text{A}}$ and $U_{per\ \text{B}} - \Delta U_{\text{B}}$ can become very small, or even negative.

However, if ΔU_{A} is less than 10% of $U_{per\ \text{A}}$ or ΔU_{B} is less than 10% of $U_{per\ \text{B}}$, these errors can be neglected.

16.5 Acceptance Criteria

16.5.1 General

At the end of the balancing process, or as an agreed part of the acceptance, it is checked whether the rotor complies with the unbalance tolerance. This check is based on the readings of the residual unbalance $U_{\text{reading A}}$ in tolerance plane A and $U_{\text{reading B}}$ in tolerance plane B.

Whether or not the unbalance state of the rotor is accepted or not on the basis of this test must be clarified between the manufacturer and the customer—this is not the content of the ISO 21940-14 standard.

Note

ISO 21940-14 (2012) still contains specifications for acceptance that have become obsolete since ISO 21940-11 (2016) was published. An amendment to Part 14 is in preparation.

The details set out in part 11 are explained here (to replace the superseded details in Part 14: 2012).

16.5.2 Unbalance Readings in Tolerance

The unbalance is clearly ***within tolerance***, i.e. does not exceed the specified tolerance U_{per}, if for the unbalance indications $U_{\text{reading A}}$ <u>and</u> $U_{\text{reading B}}$ both equations Eqs. 16.4 and 16.5 are fulfilled:

$$U_{\text{reading A}} \leq U_{per\text{ A}} - \Delta U_{\text{A}} \quad \text{g} \cdot \text{mm} \tag{16.4}$$

$$U_{\text{reading B}} \leq U_{per\text{ B}} - \Delta U_{\text{B}} \quad \text{g} \cdot \text{mm} \tag{16.5}$$

16.5.3 Unbalance Readings Outside the Tolerance

The unbalance is clearly ***out of tolerance***, i.e. exceeds the specified tolerance U_{per}, if for the unbalance readings $U_{\text{reading A}}$ and $U_{\text{reading B}}$ at least one of the two equations Eq. 16.6 or Eq. 16.7 is fulfilled:

$$U_{\text{reading A}} > U_{per\text{ A}} + \Delta U_{\text{A}} \quad \text{g} \cdot \text{mm} \tag{16.6}$$

$$U_{\text{reading B}} > U_{per\text{ B}} + \Delta U_{\text{B}} \quad \text{g} \cdot \text{mm} \tag{16.7}$$

16.5.4 Region of Uncertainty

The area between *within tolerance* and *outside tolerance* is the ***range of uncertainty***. The range of uncertainty is given by the two equations Eqs. 16.8 and 16.9:

$$U_{per\text{ A}} - \Delta U_{\text{A}} < U_{reading\text{ A}} \leq U_{per\text{ A}} + \Delta U_{\text{A}} \quad \text{g} \cdot \text{mm} \tag{16.8}$$

$$U_{per\text{ B}} - \Delta U_{\text{B}} < U_{reading\text{ B}} \leq U_{per\text{ B}} + \Delta U_{\text{B}} \quad \text{g} \cdot \text{mm} \tag{16.9}$$

Note

In order to minimize the range of uncertainty, the combined errors ΔU_{A} and ΔU_{B} must be kept within narrow limits.

This requires a suitable balancing process, a balancing machine in perfect condition (see Chap. 13), reliable tooling and qualified personnel.

16.6 Special Methods for Measuring Errors

16.6.1 General

Many errors are not readily apparent: A single, short measurement run may not reveal the slightest abnormality. Sometimes problems only arise in the course of the balancing process, e.g. when the correction does not work properly: it seems to be wrong in size and/or angle. Sometimes the problems only appear as the unbalance gets smaller and closer to tolerance.

Other errors, e.g. the systematic errors, are not noticeable during the normal balancing process. They only appear when a special procedure is applied (see Sect. 16.6.2).

In order to be able to assess the different types of errors correctly and reduce them if necessary, they must be separated.

Special procedures are required to identify the various discrepancies and determine their respective significance. Three procedures for the three types of errors are described below.

16.6.2 Measuring Systematic Errors

Most systematic errors can be detected and quantified by index-balancing.

Assumption

The rotor is connected to an auxiliary part that causes an error, to be determined. The angular reference is connected to the auxiliary part.

Index-balancing can be reliably performed with the following steps:

1. The rotor is connected to the auxiliary part at any angular position—now called 0°—and the unbalance is measured: Fig. 16.2, reading A.
2. The rotor is rotated 180° relative to the auxiliary part—this angular position is called 180°—and the unbalance is measured: Fig. 16.2, reading B.
3. The rotor is measured again in the 0° position: Only if the two values for the 0° position sufficiently match, continue.
4. The points A and B are connected by a straight line and the centre X of the distance AB is formed.

Result

The vector $\vec{U}_a$ describes the unbalance caused by the auxiliary part, the vector $\vec{U}_r$ describes the unbalance of the rotor in the 0° position, and the vector $-\vec{U}_r$ the unbalance of the rotor in the 180° position.

Note 1

If the measured values from steps 1 and 3 do not agree sufficiently, the reason must be searched for. If random or scalar errors are possible, use the corresponding procedures to help Sect. 16.6.3 or 16.6.4.

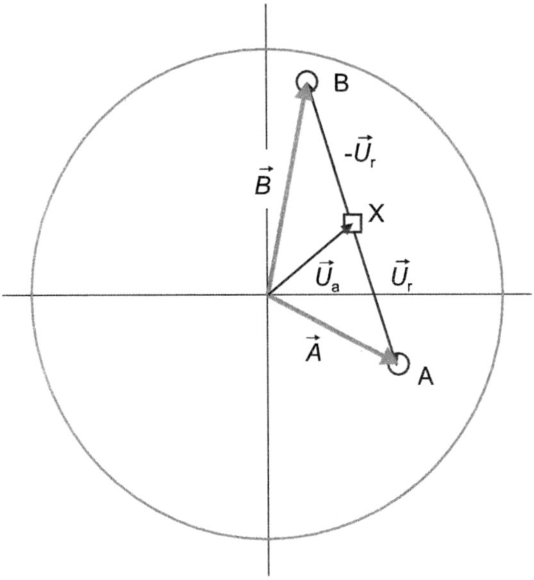

Fig. 16.2 Polar representation of index-balancing: evaluation of the measured unbalances $\vec{A}$ and $\vec{B}$ to distinguish between the unbalance of the rotor $\vec{U}_r$ or $-\vec{U}_r$ and the systematic error of the auxiliary part $\vec{U}_a$

Note 2

In this case, it was assumed that the rotor was rotated relative to the angular reference. However, if the angular reference is connected to the rotor, the vector from the centre to X describes the rotor unbalance and the vectors from X to A and X to B describe the systematic error of the auxiliary part in the positions 0° and 180°.

Note 3

Sometimes the 180° turn is not possible, e.g. because of a 120° division (or any other angle). In this case the evaluation of the measurements is more difficult (X is not on the straight line between A and B, but is the corner point of an equal-sided triangle), but follows the same principle.

Note 4

In some cases, the play of the fit of rotor and auxiliary part cannot be neglected: typically a random error. But if the play is pushed out radially—in a fixed angular relationship to the auxiliary part—in both measurements, it becomes part of the systematic error of the auxiliary part.

16.6.3 Measuring Random Errors

If it is suspected that non-negligible random errors occur, several measurement runs must *always* be made in order to detect the scatter and to determine its size.

Note

If random errors are due to different influences, further differentiation may be needed by varying or keeping boundary conditions unchanged, e.g. angular position at start, acceleration, or measuring speed.

Example

The amount of a random error can be calculated by statistical means. However, ISO 21940-14 recommends a simplified graphical method as an *approximation*, see Fig. 16.3:

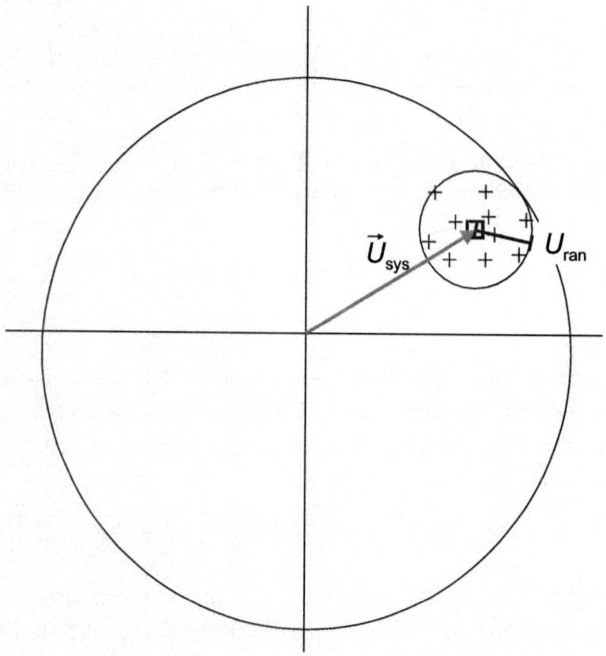

Fig. 16.3 Polar plot of the unbalance vectors of 10 runs: The mean value represents $\vec{U}_{sys}$ the systematic part, the radius of the circle U_{ran} the random error (only the magnitude of U_{ran} is relevant)

1. All unbalance vectors are entered in a vector diagram and the mean value $\vec{U}_{sys}$ over all runs is determined.
2. Then the smallest circle is drawn around the end point of the mean value, which includes all measured values. ◀

Result

The vector $\vec{U}_{sys}$ describes the systematic part of the measurements, the magnitude of the radius U_{ran} the random error.

Note 1

As the number of runs increases, the spread generally increases, but the uncertainty of the result generally decreases.

Note 2

If one of the measuring points is far outside, the cause must be investigated more closely and, if necessary, this behaviour must be avoided.

16.6.4 Measuring Scalar Errors

Scalar errors can only be detected and quantified by a longer observation time, or with the help of several runs.

Example

A part is attached to the rotor by a frictional connection—e.g. the inner ring of a anti-friction bearing. The connection is however insufficient, so that the part moves over time, i.e. its angular position changes. The angular reference is firmly connected to the rotor. ◀

Measurement

Since the rate of migration can vary greatly, a new reading is not taken until a significant change has occurred.

Result

The series of measured values forms (ideally) an arc. Its centre represents the systematic part of the measurement $\vec{U}_{sys}$, the radius of the circle the scalar error U_{sca} (Fig. 16.4).

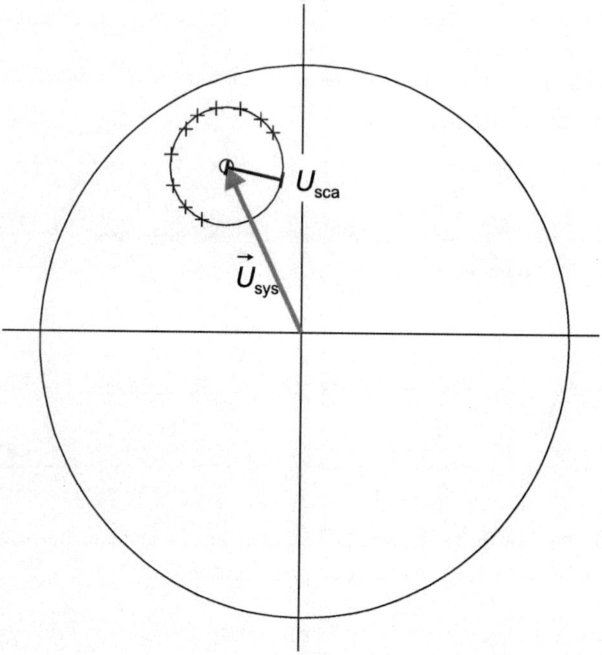

Fig. 16.4 Polar plot of the measured values of 10 runs: The centre of the circle $\vec{U}_{sys}$ represents the systematic component, the radius of the circle U_{sca} the scalar error (only the magnitude of U_{sca} is relevant)

Note 1

A similar change in the measured values also occurs when an unbalanced part of the rotor, or an auxiliary part (e.g. a pulley), twists from run to run.

Note 2

The measured values do not always form a clear circle: if, for example, random errors are superimposed, a ring of different width is created. The radius to the centre of this ring then represents the scalar error, the width of the ring is an indication for the random error.

16.7 Examples of Errors and Their Handling

16.7.1 General

ISO 21940-14 presents tabulates and describes in detail various errors and the methods to reducing them.

16.7.2 Examples

Figures 16.5 and 16.6, and Table 16.1 and the following sections explain these facts in accordance with this standard.

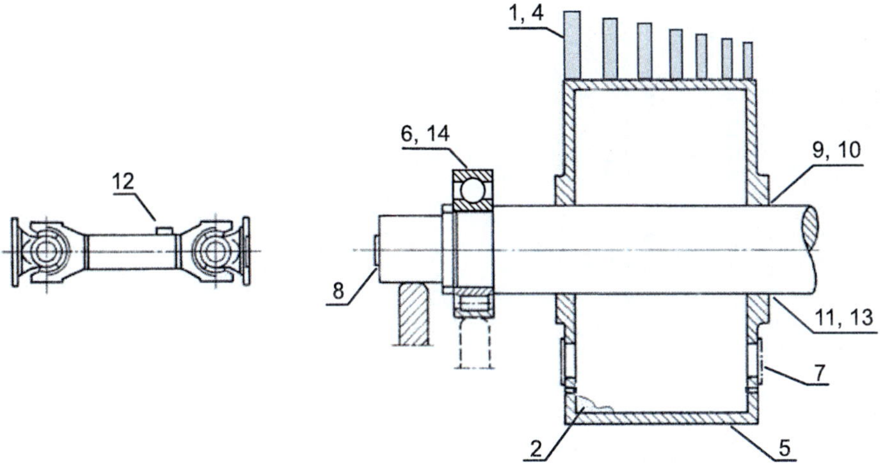

Fig. 16.5 Balance errors: Shown on a rotor with its own shaft and driven by cardan shaft. The numbers indicate the last number X of Sect. 16.7.2 X

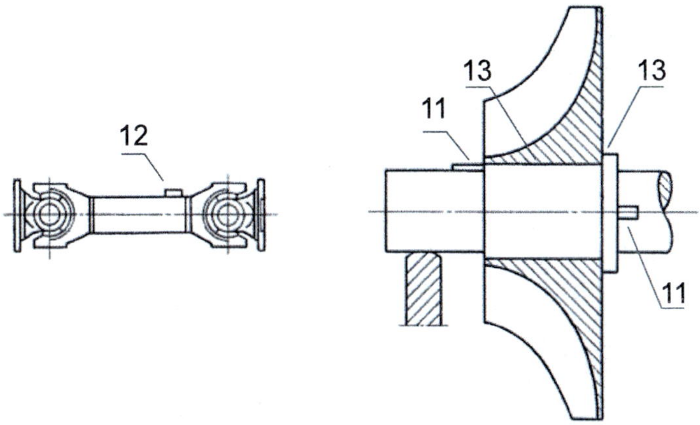

Fig. 16.6 Balance errors: Shown on a rotor with auxiliary shaft and driven by cardan shaft. The numbers refer to the last number of Sect. 16.7.2 X

Table 16.1 Examples for balance errors and their treatment: Error origin, error source, measurement, calculation, and procedures for reduction (based on ISO 21940-14, modified).

Error origin	Error source	Measuring the error	Calculation	Reduction of errors during balancing	16.7.1. X
Rotor	Movable parts, e.g. compressor blades	Measurements from several measuring runs with different starting angles		Correct systematic errors, keep random errors within limits	1
	Liquids or solids in cavities	Measurements from several measuring runs with different starting angles [a]. In case of liquids, wait until they have settled sufficiently		Remove or fix liquids resp. solids If it does not interfere later in operation, use an average over runs	2
	Thermal influences and effects due to gravity	Measurements from several measuring runs with different starting angles after significant idle times [a]		Turn rotor slowly until a stationary state is reached [b]	3
	Windage effects	Enclose rotor, or cover intake openings, or run backwards	Calculation if asymmetry occurs	Enclose rotor or cover intake openings, or run backwards	4
	Magnetized rotor	Compare readings with an additional blocked pick-up		Demagnetize rotor, shield pick-up magnetically, magnetic shield around rotor	5
	Tilted service ball bearings	Compare readings at two different speeds	Calculate the effect	Straighten out races on shaft; remachine shaft shoulders, improve balancing machine pedestals	6
	Incomplete assembly	Measurements with and without these elements	Calculate the effect	Complete rotor, quantify effects of missing parts, e.g. key, see ISO 21940-32	7
	Coupling face on rotor	Measure run-out	Calculate the effect from data	Bias run-out effects, or reduce them	8

[a] The effects of entrapped liquids or solids can be similar to those of thermal effects and may create confusion.
[b] A low speed (approx. 10 min^{-1}) is usually sufficient; the time required can be extremely different, up to 1 hour.

Table 16.1 (continued)

Error origin	Error source	Measuring the error	Calculation	Reduction of errors during balancing	16.7.1. X
Assembly	Fit clearance after balancing due to disassembly/reassembly	Disassemble/reassemble several times and measure the changes	Calculate influences from data	Disassemble/reassemble, check fit, assembly instructions	9
	Relative rotation of mounted parts during disassembly/reassembly after balancing	Measure parts with indexing procedure	Calculate influences from data	Secure parts against relative rotation, assembly instructions for reassembly	
	Parts undergo a relative rotation under service	Measurements from several runs, or observation during a longer run	Calculate influences from data	Secure parts against relative rotation	10
Accessories	Unbalance of the adaptor	Check the adaptor alone, check it with an indexing procedure		Balancing the adaptor, reducing its mass	11
	Unbalance of the drive shaft	Check with an indexing procedure	Use stored data	Balancing	12
	Run-out of adaptor	Check with an indexing procedure	Calculate influences from data	Correction of the adaptor, compensation	13
	Eccentricity of balancing bearings	Check with an indexing procedure	Calculate influences from data	Use service bearings	14
Balancing machine	Systematic and random errors of the measuring chain	Measurements from several runs, measurements of U_{mar} and URR	Calculate influences from tests	Check operation, check calibration, repair machine, make tests	15

16.7.2.1 Movable Parts

Movable parts, e.g. compressor blades of jet engines, may take a slightly different position during each run.:

These are random errors. The scatter is usually greater when the rotor is started from different angular positions, but only then a realistic picture may be obtained.

The average value from several measurement runs with different starting angles—e.g. 4 (all 90°), 6 (all 60°) or 8 (all 45°)—reduces the error compared to the individual value so significantly that the tolerance usually can be achieved.

Of course, the control value must then also be formed according to the same scheme.

16.7.2.2 Liquids or Solids in Cavities

Liquids or solids located in cavities of the rotor can vagabond. They usually result in random errors. Under special circumstances, however, they can also occur as systematic or scalar errors.

Due to standstill at different angular positions (in the case of liquids for a longer period of time), the material accumulates at the lowest place.

By different angles, different waiting times and by different accelerations, the *random* errors become apparent. By waiting for the steady state they can change from random to *scalar* errors.

Systematic errors—the material always accumulates at the same place during rotation—can often only be detected by observing the change in the measured value over time.

The errors due to liquids or solids in cavities can look very similar to that of thermal influences or effects of gravity. Only soundly planned and executed tests, as well as careful analysis, can help here.

If the effect is not negligible compared to the tolerance, the liquids or solids normally must be removed.

16.7.2.3 Thermal Influences and Effects due to Gravity

Heating

Long, slim rotors can bend due to one-sided heating and accordingly show an unbalance. Causes can be e.g.:

- Unbalance correction by welding.
- Sun exposure.

If measurement is performed very precisely, even operator's hand lying on a long shaft—during the correction process—can lead to a significant distortion.

Settlings

Rotors with long shafts also show signs of settling under the gravity of the earth if they remain in one position for a longer time.

These deformations of the rotors are temporary phenomena, but as long as the errors are too large, balancing must not be carried out. Such errors:

- Can be avoided if the angular position of the rotor is changed systematically or if it turns slowly (e.g. with 10 min^{-1}).
- Disappear, if the rotor is turned slowly until the deformations have evened out (for 10 min to 1 h).

16.7.2.4 Windage Effects

Bladed rotors tend to develop strong turbulences because of the missing stator, whereby broadband vibrations are excited.

With soft-bearing balancing machines, the excitation of the natural frequency of the rotor support can lead to random or systematic errors or even make measurement impossible. This type of machine is therefore hardly ever used for bladed rotors today.

Force-measuring balancing machines are less sensitive to such excitations, but errors can also occur here:

- Random errors become directly visible with several runs.
- Systematic errors *cannot be* detected by changing the speed, since air forces and centrifugal forces caused by unbalance primarily have the same speed dependence.
- By changing the direction of rotation, the flow is usually changed effectively, so that systematic errors can also be detected. However, they are not yet directly determinable.

The errors can generally be reduced by reducing the air flow by:

- A cylindrical housing around the rotor (resp. the bladed part), within which the air rotates with the rotor.
- Covering the suction side, e.g. by a rotating or stationary cover.

Sometimes the comparison with another medium (compared to air) is also possible to determine the error. This method is used for satellites, e.g., to determine the systematic error caused by the parabolic antennas, which does not occur during service (vacuum in space).

The symmetry of the common bladed rotors is normally good enough that they do not interfere with balancing—assuming correct moment unbalance tolerances for a narrow plane distance.

However, there are exceptions: Single-channel impellers are used e.g. for pumping liquids together with solid content. Here, the hydraulic reaction forces do not always cancel each other out completely and possibly lead to a systematic error in service, when balanced under air.

However, the aim here is not to avoid the error caused by the air flow, but to correctly detect the subsequent operating state, i.e. to replace the remaining hydraulic forces by unbalances (centrifugal forces), e.g. with an appropriate bias.

16.7.2.5 Magnetized Rotor

Some rotors exhibit a magnetic field due to their function, others have a residual magnetism, e.g. after a crack test. If this magnetic field is bipolar, it can generate a signal—synchronously with the rotor speed—in the transducers of the balancing machine, forming a systematic error.

These errors cannot be detected by changing the speed, since magnetically induced measuring signals and centrifugal forces caused by unbalance primarily have the same speed dependence.

This error can be determined by comparison with an additional transducer which is mounted in such a way that it is excited by the magnetic field but not by the unbalance.

The error can be eliminated by demagnetizing the rotor (only if not function-related), or by shielding the rotor or the transducer.

16.7.2.6 Tilted Service Ball Bearings

Due to a runout of ball bearings, constraining forces can occur rotating with the rotor speed and acting in the unbalance reading like a moment unbalance.

At service speed, these errors are normally negligible, but at balancing speed, which is usually much lower, this constraining force is measured as a much larger moment unbalance and can be disturbing.

The error (at speed 1) can be determined by measuring the unbalance at two different speeds n_1 and n_2 and using Eq. 16.10.

$$\Delta\vec{U}_1 = \frac{1}{1-(n_1/n_2)^2}(\vec{U}_1 - \vec{U}_2) \tag{16.10}$$

In case of two-plane balancing, each tolerance plane is calculated separately. In case of measurement of resulting unbalance and moment unbalance, each type of unbalance is calculated separately.

There are special types of bearings of balancing machines which are fault-tolerant, i.e. they reflect an inclined position only to a very reduced extent in the unbalance reading.

16.7.2.7 Incomplete Assembly

Parts which are symmetrical and cause sufficiently equal unbalances need not be fitted if they are not necessary for the integrity of the rotor, e.g. identical, opposing bolts on a rim.

For other parts, e.g. unassembled fans, or blades, the errors caused by the later assembly must of course be taken into account.

ISO 21940-32 comments on the problem of keyway/keyway. In this book, see Sect. 9.4 for details and possible errors.

16.7.2.8 Coupling Face on Rotor

The drive means (drive shaft, belt pulley) are coupled for balancing to the *rotor* coupling face. Run-out errors at this position lead to systematic errors which appear to be firmly connected to the rotor. They can therefore *not* be separated from the rotor by index-balancing. However, it can be determined by:

- Measuring the run-out errors and weighting them with the masses and mass moments of inertia of the coupled parts.
- Using coupling parts with different mass properties and deducing the error from the unbalance values measured this way.

The errors determined this way may be used as bias during balancing.

16.7.2.9 Fitting Errors

When disassembling and reassembling an assembled rotor, errors may occur, e.g. due to:

- Radial play.
- Inadequate face contact, e.g. due to too tight radial fit.
- Obstruction of centring due to screws.

The error is determined by repeated disassembly and assembly. If precise assembly instructions exist, work according to these instructions, if not, select and compare different procedures.

This error is reduced by improving design details, paying attention to particularly critical points in production, and establishing precise assembly instructions.

16.7.2.10 Relative Rotation of Mounted Parts

Force-locked parts are affected. Sometimes parts that define the shaft axis, e.g. bearing rings or bushings, move.[6] This relative rotation can be very slow, so that it is not detected by the usual measurement of unbalances. A determination of the error is done by observation over a longer period of time (up to 10 min).

Other parts will only show a relative rotation with a certain angular acceleration, if this overcomes the frictional force. This error can be determined by several measurement runs, applying always a clear difference in angular acceleration between starting and braking.

In both cases, the only remedy is to secure the parts better against relative rotation.

[6] It is possible that relative rotation only occurs at higher speeds – e.g. service speed – in which case it cannot be observed at the low speed on the balancing machine.

16.7.2.11 Adapter Unbalance

The unbalance of a mounting—auxiliary shaft, adapter—is included in the unbalance measurement of the rotor. The error can be determined by:

- Unbalance measurement of the adapter itself.
- Index balancing (together with run-out errors).

The error is reduced by balancing the adapter itself, or by index balancing the rotor (see Sect. 16.6.2). Thus the adapter can also be corrected for *one* type of rotor (possibly for a certain range of rotors).

Sometimes, the masses are also unnecessarily heavy and thus cause greater errors than are actually necessary. Unbalances in planes far away from the rotor also have a particularly strong effect due to the transfer unbalances (see Sect. 10.4.5.1).

16.7.2.12 Unbalance of the Drive Shaft

This error of the drive element can be separated from the unbalance of the rotor by index balancing (see Sect. 16.6.2).

16.7.2.13 Adapter Run-out

Run-out errors show up as systematic errors, but if there is also play, random errors will also occur.

This error can be separated from the unbalance of the rotor by index balancing. The influence of the play can also be reduced (see Sect. 16.6.2).

16.7.2.14 Eccentricity of Balancing Bearings

Balancing bearings are used to protect the service bearings or to take advantage of other characteristics e.g. reduced clearance. They are not identical to the service bearings, so the errors can add up in the worst case.

The error of the balancing bearing can be separated from the unbalance of the rotor by index balancing (see Sect. 16.6.2). However, the errors of the service bearings cannot be avoided; therefore balancing bearings are not applicable in case of very close unbalance tolerances.

16.7.2.15 Systematic and Random Errors of the Measurement Chain

Balancing machines can show errors to an unacceptable degree. Unfortunately, this cannot always be detectable during normal machine operation. If errors occur, one should:

- Check the mode of operation.
- Repeat calibration and compare the values.
- Repair balancing machine.

Statistical tools can be used for the verification (see Sect. 13.2), or tests recommended by ISO 21940-21 (see Sects. 13.4 to 13.7) may help.

The results of the tests according to ISO 21940-21 can be collected and summarized to an overview, so that the errors of the measuring chain for different rotor types and different speeds are directly available.

16.7.3 Specialties When Measuring Unbalances

Is there a special situation with rotor balancing compared to other technical machining processes such as turning or milling?

Unfortunately, the various standards do not yet provide any information on this. But at least some specialties of rotor balancing can be mentioned:

1. There is no generally accepted *standard* for the unbalance. With the product of mass and length, the mass can be determined quite accurately, the radius is already less accurate. However, only a limited precision is required for measuring the unbalance, see Sect. 14.4.
2. However, the zero point of the measurement—and this is what the unbalance tolerance and every measured value refers to—cannot be determined exactly.
3. Most permissible residual unbalances *could* be measured on a balancing machine with errors of 5 to 10%.[7] For economic reasons, however, this possibility is often not used, e.g. by applying lower than optimal balancing speeds.
4. In addition, there are the errors due to the boundary conditions during balancing (see Sect. 4.8), which are usually considerably larger.
5. In almost all balancing machines, measuring and machining coincide, i.e. the unbalance is corrected on the basis of the measured values, often in several steps (an iterative procedure).
6. The measurement of the residual unbalance is usually carried out after balancing in the same balancing machine, sometimes later—for checking—in another one.

16.7.3.1 Errors When Measuring on a Balancing Machine

ISO 21940-21 specifies two characteristic values of the balancing machine the smallest achievable residual unbalance U_{mar} and the unbalance reduction ratio R_{UR} (see Sects. 13.4 and 13.5).

[7] These percentages seem generous when compared with those of a weight balance, for example. However, it must be borne in mind that only the tolerance zone is to be measured here at the end of the balancing process, and that the next quality level is a factor of 2.5 away, and still a factor of 1.6 for the finest gradation.

The resulting errors are already recorded as individual errors, see Sect. 16.7.2.15. If these individual errors are small enough, the balancing machine is suitable for measuring the other errors, as well as the final state of the rotor after balancing.

16.7.3.2 Errors when Measuring Without a Balancing Machine

The residual unbalance can also be determined without a balancing machine, e.g. at the installation site with aids and procedures such as those used for in-situ balancing (see Chap. 18).

This eliminates the errors which occur during the balancing process due to other boundary conditions. However, the service condition can also deviate from the delivery condition: these influences must be estimated.

The errors caused by the measuring process are also not easy to assess. In addition to the errors of the measuring device in the specific task, they include the individual errors when setting the test unbalances and the influence of the scatter from run to run.

Protective Measures for Balancing

17

Contents

H. Schneider, *Rotor Balancing*, https://doi.org/10.1007/978-3-662-66049-2_17

17.1 Preliminary Note

When designing and using balancing machines, it must be ensured that hazards from the operation of the machine are kept to a minimum.

Dangers can arise in the vicinity of the machine e.g. by touching rotating parts, but also at a distance from the machine by parts which detach from the rotor during rotation.

The International Standard ISO 7475[1] for protection during balancing was completely revised and expanded in 2002.

This standard was further developed to become Part 23 of the ISO 21940 series of standards, published in 2012.[2]

17.2 General

The manufacturer (or the seller) of the balancing machine will be able to offer and, if necessary, supply appropriate protection, if the user responsibly fixes the risks of his balancing process.

To analyse the risks from the process different areas must be considered, they can be conditioned by the:

- Rotors.
- Correction method.
- Operating sequences.
- Installation conditions.

This information may also serve the seller to advise the user to such an extent that the user can establish the necessary degree of protection on his own.

For balancing machines with integrated correction, all these data are usually known or defined during the design of the balancing system, e.g. by the manufacturer. Machines of this type are therefore usually supplied including all protective devices.

The standard ISO 21940–23 therefore refers to the measuring station of a balancing machine. The standard is very extensive and goes into great detail – only the background and the essential principles can be explained here.

[1] ISO 7475:2002: Mechanical vibration – Balancing machines – Enclosures and other protective measures for the measuring station.

[2] The current edition is ISO 21940-23 (2012): Mechanical vibration – Rotor balancing – Part 23: Enclosures and other protective measures for the measuring station of balancing machines.

17.3 Dangers Due to the Rotor

What are the hazards directly associated with the rotating rotor? In over 99% of all balancing processes, the workpiece rotates with speeds between 100 and 3000 min^{-1}, depending on the rotor and the balancing machine, but much higher speeds can also occur.

In general, the rotational energies – opposed to translational energies –- are typically underestimated in practice.

Examples

A typical electric motor armature of 1 t contains at 500 min^{-1} a rotational energy E_{rot} = 50,000 N·m.

- With this energy, the rotor could "run up the wall" 5 m high, or accelerate an entangled person to a speed of about 14 m/s (50 km/h).
- The slowness of the human eye does not allow to detect fast movements, the consequence: protruding parts on a rotating body (e.g. screw heads, blades) cannot be seen in front of the well visible base body, i.e. the risk of accidental contact is high.
- Even at balancing speed, which is generally lower than the service speed, parts can detach from the rotor, e.g. balancing weights that are not yet finally fastened.
- For the rotor example above, a balancing mass detached from a radius of 250 mm will have a speed of 13 m/s.
- A person at a distance of 1 m would be hit after 0.07 s, at a distance of 10 m after 0.7 s. Even if this person would observe that the weight detaches from the rotor, he would have no chance to get out of the way. ◀

17.4 Protection Classes

It would be wrong to issue uniform regulations based on such considerations. For a light rotor with a smooth surface and material removal by grinding the balancing process is absolutely safe. A large rotor, on the other hand, which could even burst at balancing speed, requires very extensive protective measures.

ISO 21940–23 defines four different protection classes (Table 17.1), based on two criteria:

- Protection against contact,
- Protection against detached particles or rotor parts.

Sometimes it is advisable to combine the protection requirements of classes A and B, e.g. when touching a rotor is dangerous *and* particles with limited energy may detach.

Table 17.1 Protection classes A to D, defined by the need for protection against contact and the resistance of the protection against detached parts (following ISO 21940–23)

Necessity for guards	from guards to burst-proof construction			
	no	yes		
Necessary protection measure, resp. Hazards-criteria	Spectacles, goggles or face-shield needed	Minimum protection required to avoid physical contact with work piece, and no particles shall be released	Area-specific energy up to ≈340 mN·m/mm² Absolute energy up to ≈2000 N·m Impulse up to ≈200 kg·m/s	Above values of • specific energy, • absolute energy and • impulse in class C
Protection class	A	B	C	D

17.4.1 Protection Against Contact

If contact with the rotor—or parts of the machine, e.g. the drive—is dangerous, this contact must be prevented either by a suitable protection or by shutting down the entire balancing machine.

In most cases, the protection must be opened for the balancing process in order to bring the rotor in and out, or to perform an unbalance correction. During this opening the drive must be reliably locked.

17.4.2 Protection Against Particles or Parts

In this case, particles or parts can fly off the rotor at balancing speed, but should be caught by a guard.

Three criteria are important here:

1. area-specific energy,
2. absolute energy,
3. impulse.

17.4.2.1 Area-specific Energy

The area-specific energy E_{spec} is a measure of the penetration potential of a particle flying off the rotor. It can either be calculated, or determined from the diagram in Fig. 17.1.

Here, the absolute energy E_{abs} of a particle is related to the smallest area the particle could penetrate the protection. The highest values are generally achieved here by small, fast particles with an elongated shape (relatively small cross-section), e.g. blades.

The smallest cross-sectional area is specified in ISO 21940-23 for a number of different bodies, as shown in Fig. 17.2.

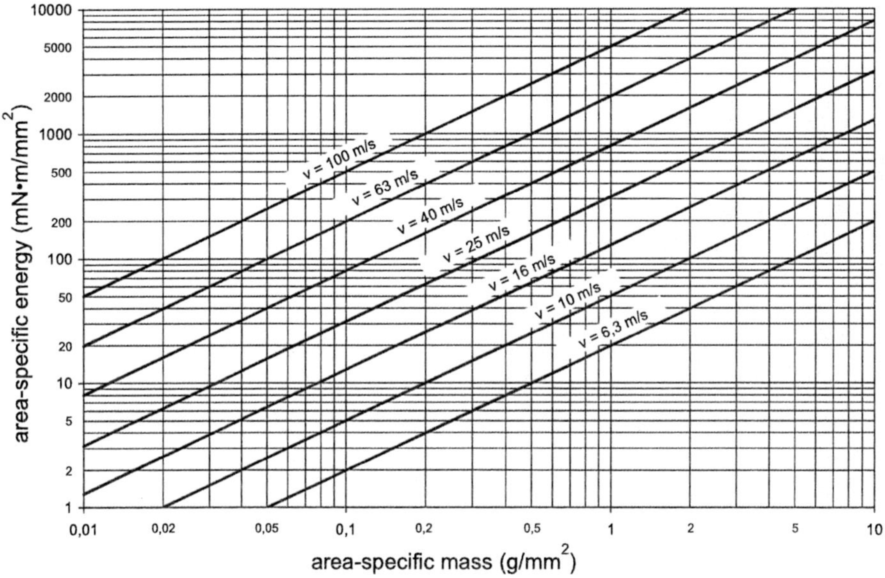

Fig. 17.1 Area-specific energy as a function of the area- specific mass and the velocity of the particle or part (according to ISO 21940-23)

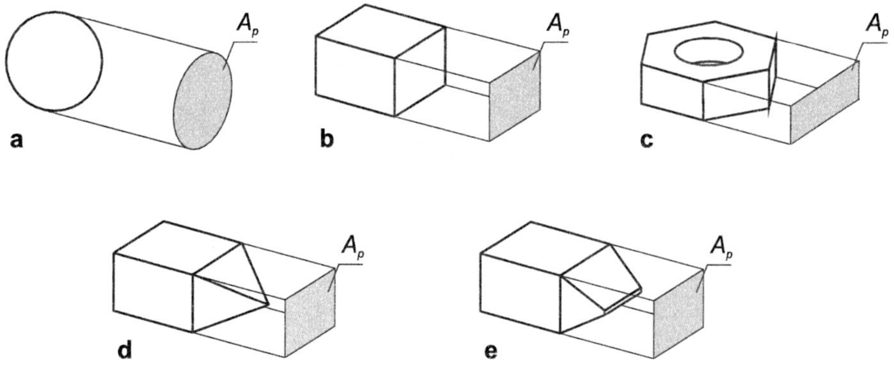

Fig. 17.2 Smallest cross-sectional area A_p of various bodies shapes, according to ISO 21940-23: ball (**a**), cuboid (**b**), nut (**c**), cuboid with cone (**d**), cuboid with chisel (**e**)

17.4.2.2 Absolute energy

The absolute energy E_{abs} of a particle or part loads the protection as a whole and individual components, e.g. windows and sheet metal, as well as all connections. It can either be calculated, or determined from the diagram in Fig. 17.3.

In case of an impact, no openings shall occur, through which particles or parts can escape.

17.4.2.3 Impulse

During the impact of a rotor part, a free-standing protection can be displaced or even be knocked over due to the impulse exchange. Here, the most critical case is a large mass with relatively low speed.

Displacement must be kept within limits and knock over must be prevented, e.g. by reliably guiding or securing the protection.

The impulse only plays a role in special cases and is therefore not taken into account in the marking of the protections.

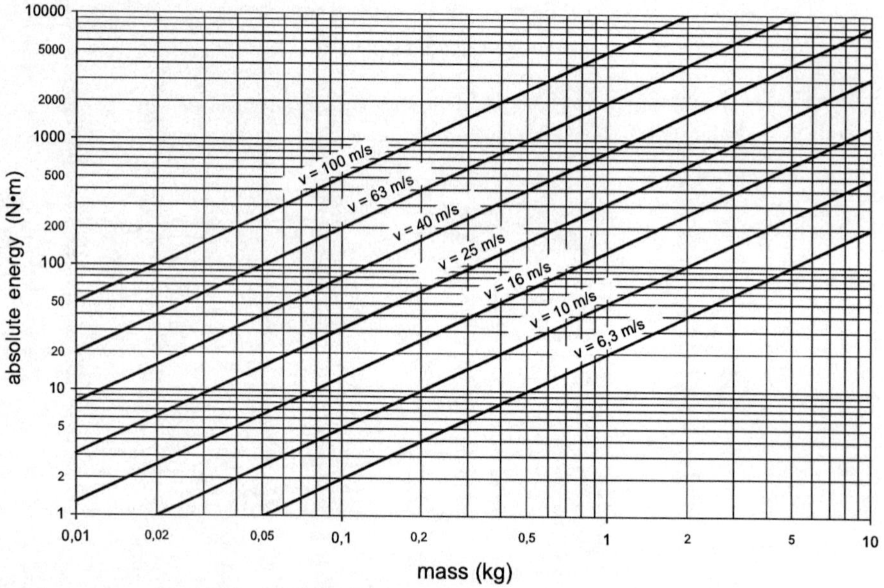

Fig. 17.3 Absolute energy as a function of the mass of the particle or part and the velocity (according to ISO 21940-23)

17.5 Choice of Enclusures

The standard describes in detail how the penetration potential of a given particle can be determined by calculations and tests—e.g. with a series of standard projectiles.

From this, *with a safety factor*, a limit for the area-related energy $P_{res,qual}$ is obtained, which the enclosure must fulfil.

A similar procedure also sets the limit on the absolute energy $E_{abs,qual}$ that the enclosure must satisfy.

17.6 Examples of Protection Classes

17.6.1 Class 0[3]

To control the unbalance of the rollers of an aircraft engine bearings (Fig. 17.4), the rollers are accommodated in a static air bearing and driven from above by a tangential belt drive.

This is one of the very rare cases where there is no danger whatsoever from the rotor: the surface is smooth, the mass is small, so is the balancing speed and therefore the kinetic energy, and there is no correction, so there is neither material removed nor added. Therefore, no measures are required: class 0 is sufficient.

Fig. 17.4 Example for protection class 0. Checking the unbalance of rollers of a roller bearing: no measures required (but see FN 3)

[3] Unfortunately, protection class 0 is no longer listed today in ISO 21940–23, although it has a certain significance in practice, see example.

17.6.2 Class A

For electric armatures that are corrected by axial drilling, class A (Fig. 17.5) may be correct. Even if there is no danger from the rotor itself, precautions must be taken against the possibility that drilling chips fly off.

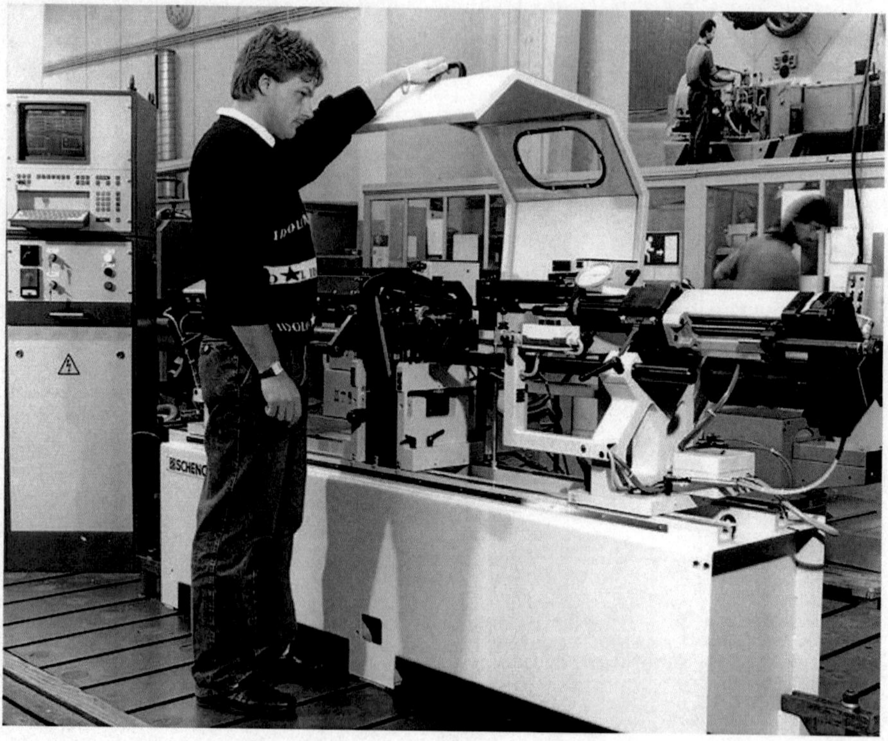

Fig. 17.5 Example for protection class A. Protective shield when correcting electric armatures by drilling

17.6.3 Class B

If, in the case of larger rotors, the detachment of particles or parts is reliably prevented, but if contact is dangerous, protection class B can be implemented (Fig. 17.6).

Fig. 17.6 Example for protection class B. Protective fencing of the machine with interlocking the access door or drive

17.6.4 Class C

Very often, protection class C is correct (Fig. 17.7), especially for universal balancing machines with varying balancing program.

Fig. 17.7 Example for protection class C: Pentration-proof enclosure of the entire machine against detached parts, with interlocking of access or drive

17.6.5 Class D

For high-speed balancing and spin testing (strength test) of a high-speed rotor, e.g. of a compressor rotor, the highest protection class D (Fig. 17.8) shall be used.

While protection classes A and (sometimes) B can be established without any problems by the users of balancing machines on their own, for C and even more so for D a considerable amount of calculation and sometimes testing is required.

Fig. 17.8 Example of protection class D: Longitudinally movable protection against possible burst of compressor rotors, with interlocking of the protection or the drive

17.7 Protection Class C for Universal Balancing Machines

For all balancing machines intended for a specific rotor program, appropriate protection must be provided from the beginning. Often, the supplier equips the balancing machine accordingly, but the user can also take responsibility for the appropriate protection.

For universal balancing machines used in maintenance and repair, sometimes the rotors are not known in advance. For this case, in which the hazard cannot be defined in advance, ISO 21940–23 helps with recommended standard values, graded according to machine size Table 17.2.

17.7.1 Design of the Protection

If the penetration resistance of a protection cannot be calculated with sufficient certainty, tests must be carried out with the materials used and, if necessary, with assemblies or components. Standardised test bodies are used for these tests.

The permissible load of the protection is smaller than the calculated or tested limit load by a safety factor of 2.

17.7.2 Marking of the Protection

All protective devices (A to D) must be clearly marked so that they can be checked during daily use whether the existing protection is also correctly adapted to the respective—possibly new—balancing task.

Class C protection is marked with the two values for the energy. C 600/230, for example, means that it can withstand a maximum absolute energy of 600 N·m and an area-specific energy of 230 mN·m/mm^2.

Table 17.2 Universal balancing machines—recommendation for enclosures of protection class C, if the possible hazards cannot be stated in advance (according to ISO 21940-23)

	Balancing machine capacity	kg	1,5	8	50	250	1.500	8.000	50.000
criterion	Enclosure qualified for absolute energy $E_{abs, qual}$	N·m	2	6	20	60	200	600	2000
criterion	Enclosure qualified for area-specific energy $P_{res,qual}$	mN·m/mm^2	30	50	70	110	160	230	340
	Designation of enclosure		C 2/30	C 6/50	C 20/70	C 60/110	C 200/160	C 600/230	C 2000/340

17.8 Hazards and Safety Requirements

In line with national and international safety standards, in recent years even this balancing standard has also put greater emphasis on describing general risks and their avoidance – starting with the control system, relating to noise emission and extending to requirements for the marking of key parameters of the balancing machine for transport and installation, as well as for the operating instructions.

Here, reference is made only to the possible problems in the balancing process itself: The systematics of the tables contained in the standard are shown here in excerpts, e.g. on "Hazards and sources of hazards", Table 17.3:

The users of ISO 21940-23, i.e. the manufacturer or supplier, or the user of a balancing machine must carry out a risk assessment which corresponds to the planned current use of the balancing machine.

Protective measures are to be taken to avoid any hazards that may occur or at least to reduce them to an acceptable level.

For this purpose, ISO 21940–23 provides a very detailed table (Table 17.4) with the protection requirements and protective measures matched to the sources of danger, as well as notes on verification, with the following structure:

Table 17.3 List of significant specific hazards and examples of hazard sources for the measuring station of balancing machines (extract from ISO 21940-23; the range between 1.1 and 5 has been omitted)

Item	Specific hazard	Examples for hazard source	Associated activity	Related danger zone
1	Mechanical			
1.1	Crushing	Workpiece moving	Loading the workpiece	Between rotor and pedestal
↕	↕	↕	↕	↕
5	Human errors	Inadvertent operation of controls, misuse of guard-controls	Measuring unbalance during set-up	Around rotor

Table 17.4 List of safety requirements and/or protective measures and their verification procedures (extract from ISO 21940-23; the range between 1.1 and 6 has been omitted).

Item	Hazard source	Safety requirements and protective measures	Verification
1	Mechanical		
1.1	Disengagement or failure of the end-drive coupling	An enclosure around the universal joint shaft shall prevent the whipping around of the shaft if not coupled to a rotor. …	By visual inspection or measurement
↕	↕	↕	↕
6	Missing and/or incorrectly positioned safety-related measures/means	Equipment and accessories for adjusting and maintaining of the machine which are not readily available to the user shall be provided.	By visual examination

Note

The manufacturer or the supplier of a balancing machine often *designs* and *supplies* Class C protection. For Class D they at least *dimension* the protection.

However, it is the user's responsibility to maintain the safety of his equipment, e.g. to adjust the protection of the balancing machine, or have it adapted, if the ***balancing task changes***.

In-Situ Balancing

18

Contents

18.1 Preliminary Note

In-situ balancing is balancing machines or entire machine trains running under service conditions either at the service site or in the test field, if the vibration level is not satisfactory and therefore should be improved (Fig. 18.1).

To make this decision of unsatisfactory vibration level, the vibrational condition must be measured and compared with permissible limit values (tolerances).

H. Schneider, *Rotor Balancing*, https://doi.org/10.1007/978-3-662-66049-2_18

Fig. 18.1 In-situ balancing of a centrifugal fan

18.2 Vibration Limits

For the assessment of the vibrations of machines under service condition two series of ISO Standards are used:

- ISO 10816: Mechanical vibration—Evaluation of machine vibration by measurements on non-rotating parts,
- ISO 7919: Mechanical vibration—Evaluation of machine vibration by measurements on rotating shafts.

The entire standards work[1] is very extensive—over ten parts—so that only a few key words can be given here.

The ***vibrations*** of the machines are usually examined in a broadband manner, i.e. scales are formed which describe this mixture (and not the individual frequency components).

[1] In 2016, the basic parts ISO 7919–1 and ISO 10816–1 were revised and published together in a new standard ISO 20816–1.

ISO 10816 uses the effective value of the vibration velocity v_{eff} (Eq. 18.1):

$$v_{eff} = \sqrt{\frac{1}{T}\int_0^T v^2(t)dt} \quad \text{mm/s} \tag{18.1}$$

In ISO standards this expression is called v_{rms} (root-mean-square value).

ISO 7919 is interested in the ***shaft movement*** in the bearing and uses the peak-to-peak value of the vibration displacement here—because of the risk of rubbing.

Both series of standards for the evaluation of the vibration condition are based on four zones (A to D): ISO Standards

Zone A	Vibrations of newly commissioned machines
Zone B	Machines can run continuously without restrictions
Zone C	Machines can only be operated for a limited time
Zone D	Vibrations are dangerous and can cause damage to the machine

Both series of standards specify limit values for different machine groups and graduated according to size, which are then varied again according to the type of support—rigid or elastic installation.

18.3 Task

Although the unbalance component of the vibration condition is only *one* cause of vibration among many, it can contribute significantly to the vibration level. The unbalance of a rotor or a machine cannot be determined directly, but only by measuring and evaluating the unbalance effect.

Different areas and different methods can be used for in-situ balancing:

- The bearing vibrations (on the housing).
- Absolute shaft vibrations.
- Relative shaft vibrations (in the bearing housing).

As a general rule, the method that is most informative should be selected; this depends on the design type and the bearing arrangement.

Machines whose vibrations are monitored, or which are regularly checked or balanced, have permanently installed vibration transducers. In these cases, all types of measurement can occur.

In all other cases, transducers must be fitted for in-situ balancing; here, measurement of bearing vibrations is most common.

Since the unbalance mainly affects the rotational frequency vibrations, this component must be extracted with high accuracy from the vibration mixture—measured, displayed in amount and angle and then evaluated.

18.4 Theory of In-situ Balancing

In-situ balancing only promises long-lasting success if the following conditions are met:

- The unbalance accounts for a significant portion of the measured vibration value.
- The machine is not too sensitive to unbalances (see Sect. 8.8).
- The machine is not too susceptible to unbalances (see Sect. 8.8).

If the unbalance only accounts for an marginal portion of the measured vibration condition, balancing, even if performed really good, would make no sense. In this case, the main causes for the unsatisfactory vibration condition must be sought.

A good standard for acceptable unbalances is ISO 21940-11 (see Sect. 7.4). Even if it would be possible by in-situ-balancing to reach a condition far below these tolerances, this path should not be followed: even small unbalance changes (see Sect. 8.8) would cause the vibrations to rise again above the desired level.

Often a situation too near to a resonance is the problem—sometimes also with rotors with rigid behaviour—if the setting-up is not sufficient.

18.4.1 Causes for Unbalances

If a balanced rotor does not run smoothly does not run smoothly in service, this can have various causes (see Sect. 18.4).

Some examples, that can be contributed to unbalances, are mentioned here (details will be given in the other sections of the book):

- Incorrect or insufficient balancing.
- Elastic or plastic deformation of the rotor (see Chap. 4).
- Manufacturing or assembly errors of subassemblies or assembled rotors (see Sect. 9.3).
- Thermal deformations (see Sect. 16.3.2.3).
- High wear or deposits during service (see Sect. 8.8).

18.4.2 Difficulties

Balancing is made considerably more difficult by the service condition and by the rotor housing:

Contrary to a balancing machine, the service bearing generally does not avoid the resonance area as clearly, and the damping is also often no longer negligible. In addition, both bearing supports of the machine can have quite different characteristics.

Therefore, all problems related to the determination of the unbalance angle, the calibration of the amount and the plane separation are much more complex than on a balancing machine (see Sect. 12.3.11).

18.4.3 Methodology

First of all, it must be clarified how many correction planes and measurement planes are required. One can assume that each rotor system has at least two bearing planes.

If balancing in one plane is sufficient (see Sect. 4.6.3), at least one measuring plane and one correction plane are required.

If two-plane balancing is needed (see Sect. 4.6.4), two measuring planes are also required. These measuring planes must be selected in a way that they receive signals from the two correction planes as different as possible.

A two-bearing rotor with more than two correction planes, e.g. a rotor with shaft-elastic behaviour (see Sect. 4.6.5), also requires only two measuring planes, unless additional requirements are added—e.g. concentricity in certain planes outside the bearing positions.

The more bearing planes there are, the more measuring planes are required. Unfortunately, a generally applicable, simple rule on how to define the number of measuring planes and correction planes cannot be established, as there are too many different influences.

The procedure used for in-situ balancing is easy to understand: The *vibration* vectors measured at the beginning (initial unbalance) are calibrated and adjusted by test runs with arbitrarily applied unbalances in such a way that they can be read as *unbalance* vectors.

18.4.3.1 Correction in a Single Plane

Only one measurement plane is required (Fig. 18.2).

The steps for single-plane balancing are:

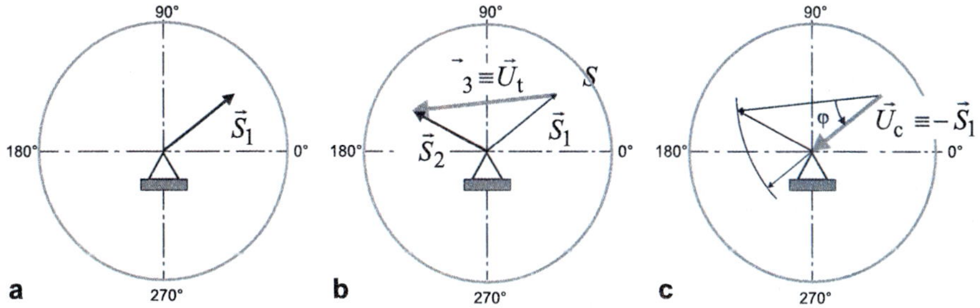

Fig. 18.2 Vector diagram for polar correction in a single plane. Indication of the measuring device for the different steps

1st run

The vibration vector $\vec{S}_1$ (on the basis of the arbitrarily attached angular reference system) is measured, Fig. 18.2a.

2nd run

By applying an arbitrary but defined test unbalance in the correction plane, the vibration state is typically changed. The vibration vector $\vec{S}_2$ is measured, Fig. 18.2b.

Evaluation

The difference between $\vec{S}_2$ and $\vec{S}_1$ is the vibration vector $\vec{S}_3$; ; caused by the test unbalance (for formation of $\vec{S}_3$ see Sect. 2.3.1).

Balancing now intends, to change the vibration vector $\vec{S}_1$ *not* so that vector $\vec{S}_2$ results, but in a manner that zero (or the desired tolerance) is achieved. To achive this, the vector $\vec{S}_3$– for *polar* balancing—must be rotated by a certain angle in the right direction and changed in size accordingly, Fig. 18.2c.

Example 1

In Fig. 18.2 the lengths of the vibration vectors (drawn with an arbitrarily chosen scale) are $S_1 = 27$ mm and $S_2 = 25$ mm. The test unbalance $U_t = 1000$ g·mm, is added at 0° of the reference system on the rotor. ◀

Task

How to define the necessary unbalance correction?

Solution

The *vector* of the test unbalance results in $S_3 = 43$ mm, the angle in $\varphi = 32°$.

The required correction unbalance is calculated from:

$$U_c = U_t \frac{S_1}{S_3} = 1000 \frac{27}{43} = 630\,\mathrm{g} \cdot \mathrm{mm},$$

The *angular position*: turned by 32° relative to 0° on the rotor, in the direction of increasing angle (corresponding to the angular markings 0° to 270°).

If the correction is to be performed in *components* rather than *polar*, the evaluation of the same vibration vectors $\vec{S}_1$, $\vec{S}_2$ and $\vec{S}_3$ (as Fig. 18.2) looks somewhat different.

Correction is intended in 120° components; the three correction positions 1, 2, 3 are labelled 0°, 120°, 240° (the angles on the rotor), Fig. 18.3.

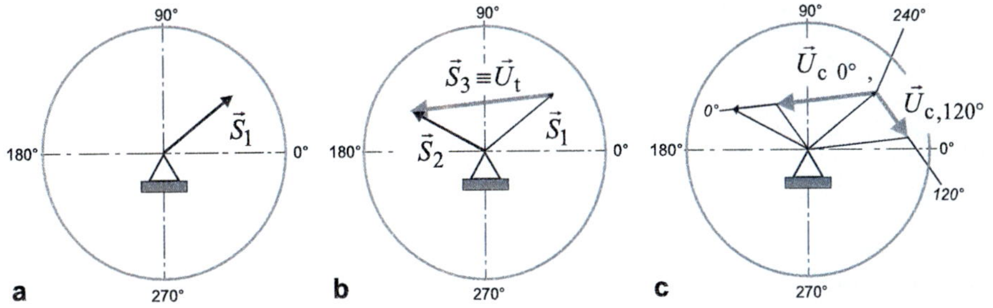

Fig. 18.3 Vector diagram for correction in 120° components in a single plane. Same initial situation as Fig. 18.2. The effective direction of the test unbalance at 0° on the rotor (position 1) and the counting direction of the display determine the position of the components 120° (position 2) and 240° (position 3)

Example 2

The initial situation is the same as in example 1, but correction is done in 3 components. The test unbalance $U_t = 1000$ g·mm was placed in position 1 (0°). ◀

Task

How large are the unbalance corrections in the components?

Solution

The lengths for the correction vectors $U_{c,0°} = 26$ mm and $U_{c,120°} = 17$ mm result from the vector diagram.

The calculation is carried out with formulas analog to the polar correction. Corrections are required in position 1: $U_{c,0°} = 605$ g·mm and in position 2: $U_{c,120°} = 395$ g·mm.

If there are several possible combinations of correction positions to reach zero, the one with the least masses is usually chosen; these are always positions adjacent to the unbalance vector.

The evaluation of the measurement results, i.e. the determination of the unbalance correction (or masses), can be carried out graphically without any problems in the case of single-plane balancing—see Figs. 18.2 and 18.3 –, but modern measuring devices with computers simplify this task considerably.

18.4.3.2 Correction in Two Planes

To solve this problem, two measuring planes (e.g. bearing planes A and B) and two correction planes I, II are necessary, Fig. 18.4.

The steps for two-plane balancing are:

1st run

The vibration vectors $\vec{S}_1$ in both measurement planes (A, B) with respect to the arbitrarily attached angular reference system are measured, Fig. 18.4.

2nd run

By applying an arbitrary but defined test unbalance T_I in the correction plane I, the vibration state is typically changed. The vibration vectors are $\vec{S}_2$ measured in both measuring planes. The differences between $\vec{S}_2$ and $\vec{S}_1$ are the vibration vectors caused by the test unbalance (for formation of $\vec{S}_3$ see Section 2.3.1):

The vector $\vec{S}_3$ in measurement plane A is the intentional change due to the test unbalance T_I, the vector $\vec{S}_3$ in measurement plane B is the unavoidable influence of this test unbalance.

3rd run

The test unbalance in correction plane I is removed and a test unbalance T_{II} is placed in correction plane II, thus the vibration state is typically changed. The vibration vectors $\vec{S}_4$ are measured in both measuring planes. The differences between $\vec{S}_4$ and $\vec{S}_1$ are the vibration vectors $\vec{S}_5$ caused by the test unbalance T_{II} (see Sect. 2.3.1).

The vector $\vec{S}_5$ in measurement plane B is the intentional change due to the test unbalance, the vector $\vec{S}_5$ in measurement plane A is the unavoidable influence of this test unbalance.

Evaluation of Fig. 18.4

Two different situations can be distinguished.

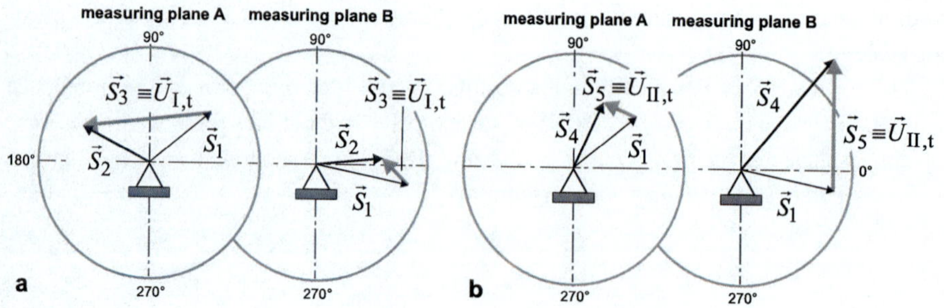

Fig. 18.4 Vector diagrams for the measured values for two-plane balancing, with test unbalance T_I in correction plane I (**a**) and test unbalance T_{II} in correction plane II (**b**)

1. The Influences Are Small

If the unavoidable influences are small enough: each balancing plane can be treated as single-plane balancing.

The unbalance in one correction plane is corrected first, then the other. Unfortunately, this also changes (worsens) the condition achieved in *one* measurement plane when the *other correction* plane is corrected.

Because of this (small) plane influence, the machine can only be balanced in several steps (iteratively). If it is difficult to apply the correction masses, these steps can of course also be carried out on paper first, before the summed-up correction takes place on the rotor.

2. The Influences Are Big

If the influence is bigger—in borderline cases so big that the desired tolerance cannot be achieved even with a very large number of steps in both measuring planes—other graphical or numerical methods can be used to evaluate the measured values.

In both cases, it is much easier to use a computer with a special balancing program, or to use a measuring device with an integrated computer, display see Fig. 18.5.

18.4.3.3 Correction in More Than Two Planes

A distinction must be made whether it is:

1. A rotor with two bearing planes, which behaves in a shaft-elastic manner and therefore has more than two correction planes, or whether it is
2. A multiple-bearing rotor with a corresponding number of correction planes and measuring planes.

1. Rotor with Shaft-Elastic Behaviour

In case of a rotor with *shaft-elastic behaviour* (see Sect. 6.4), the ***low-speed*** balancing is performed like a two-plane balancing. The mass sets for the individual flexural modes are determined the same way as on a balancing machine (see Sect. 6.4.6).

For balancing at ***several speeds,*** a separate test run is required for each flexural mode in order to find the correct angular position and size of the correction mass sets.

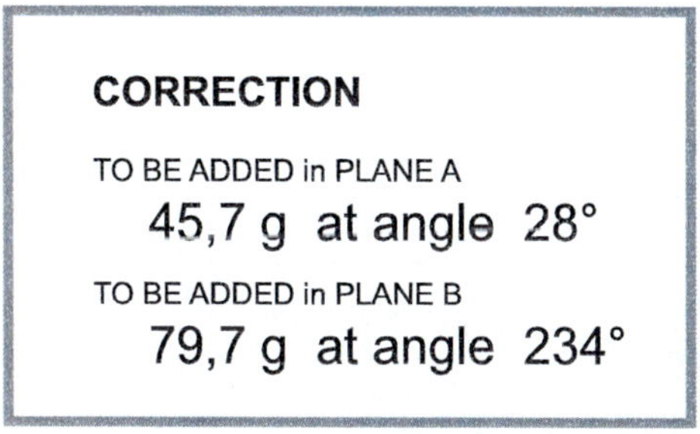

Fig. 18.5 Instruction for corrections in two planes on a measuring device with computer; with a good correction plane separation

Difficulties are often encountered when, for example, the correction planes are not sufficiently accessible.

It may also be that for in-situ balancing the speed range is not entirely available. In these cases, a compromise must be achieved with less information in fewer planes.

2. The Multiple Bearing Rotor

The *multiple-bearing rotor* (not shaft-elastic) causes difficulties in principle:

The measured values are not only dependent on the unbalances—two planes would still be sufficient for this—but also on the alignment of the bearing journals and the bearing conditions.

If, for example, the bearing journals are not in line, circumferential forces or displacements will inevitably occur at the bearings. Since these bearing reactions, in contrast to the unbalance reactions, occur independently of the speed, in principle it is possible to detect them by measuring at different speeds (see Sect. 16.3.2.6).

In general, there is no advantage in balancing the force reactions of the bearings at service speed as well, it is better to eliminate the cause. In this case, an adjustment of the rotor and its bearing or a mechanical rework will usually be necessary.

This category includes not only rigidly coupled rotors with a total of more than two bearing planes, but also elastically coupled rotors with two bearings each, where a constraint is exerted from one rotor to the other via the coupling element.

18.5 Practice of In-situ Balancing

In many cases, only single-plane balancing is required when in-situ balancing is necessary. Even typical rotors with two correction planes, which have already been balanced on a balancing machine, but subsequently now require correction due to wear, repair or similar, can often be sufficiently corrected in one plane only.

Balancing rotors with shaft-elastic behaviour is nowadays increasingly carried out in production on special high-speed balancing machines, so that in this case, too, the entire procedure (see Sect. 10.4.7) is no longer necessary in the in-situ condition, and smaller retouches are sufficient after repair and for maintenance.

18.5.1 Measuring Techniques

During balancing on a balancing machine forces or vibrations on the bearings are used for measurement. The task basically is similar for in-situ balancing: the bearing reaction is also measured with magnitude and angular position. It is therefore obvious to use the same measuring principles as a basis, but the measuring devices are adapted to the specific task.

Fig. 18.6 Balancing a fan in in-situ condition: portable measuring device, vibration transducer, and reference transducer (contactless scanning at the pulley)

In contrast to the balancing machine, where the transducers and often also the angular position encoder are permanently installed, these parts usually have to be reattached for in-situ balancing (Fig. 18.6).

18.5.2 Measuring Planes

If measurement at the bearings is chosen, the measuring planes should correspond as closely as possible to the bearing planes.

It is important that there are no elastic or plastic links, e.g. covers, in between.

Basically, measurements should be taken horizontally and vertically in each measuring plane, Fig. 18.7.

In critical cases, e.g. with multiple bearing rotors, an additional axial measurement may be helpful.

18.5.3 Boundary Conditions

To make in-situ balancing successful, many boundary conditions must be all right.

Some of them have to be checked before starting the measurements, others have to be observed during the measurements.

The following boundary conditions should be ensured before the first measurement is carried out or at least the first correction is applied:

1. The support and coupling conditions must be checked carefully to avoid constraints with all the resulting consequences.
2. Once selected, the same speed and the same system state must always be used as a basis for a series of measurement runs (which are evaluated together).

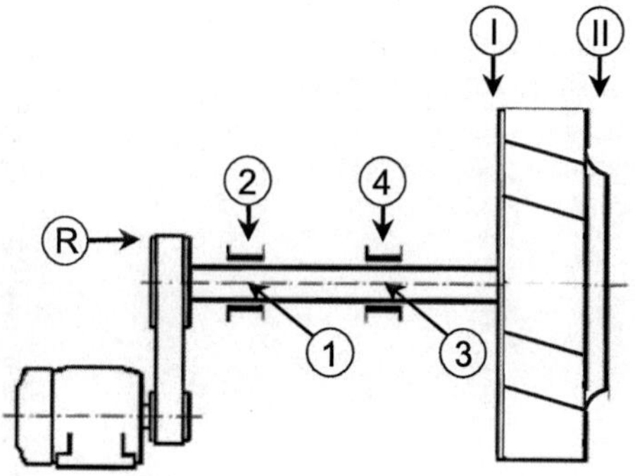

Fig. 18.7 In-situ balancing. Principle arrangement of vibration transducers (horizontally: 1 and 3, vertically: 2 and 4), reference transducer (R) and correction planes I and II

3. The measured values must be reproducible, i.e. each run with the same unbalance condition and the same boundary conditions must result in the same vibration condition and thus the same measured values.
4. A systematic evaluation of the measurements is only possible if the vibration system behaves in a linear and phase-true manner: doubling the test unbalance must cause a change in the vibration twice as large; if the test unbalance is rotated by the angle φ, the corresponding vibration vector must also rotate by the same angle φ.

If a machine has already been balanced once in the service condition and the evaluation is still available, then the necessary correction can be concluded from the first measuring run without further test runs. Provision: the same condition for the machine, the same position of the transducers and reference transducers, and the same measuring instrumentation.

18.6 ISO 21940-13[2]

This standard provides important information for medium and large machines for the practical execution of in-situ balancing. The focus is on safety during the process (machine and human) and the requirements for adequate logging and archiving means.

[2]ISO 21940: Mechanical vibration – Rotor balancing – Part 13 (2012): Criteria and safeguards for the in-situ balancing of medium and large rotors.

Symbols, Vocabulary and Definitions

19

Contents

19.1 Preliminary Note

Some important information for reading this book and understanding rotor balancing is summarised here:

1. The symbols used in the book see Sect. 19.2.
2. Vocabulary and definitions of rotor balancing; based on ISO 21940, Part 2 see Sect. 19.3.[1]

[1] In 1974, ISO Standard 1925 "Balancing – Vocabulary" became an essential aid in the field of rotor balancing. The most important terms of rotor balancing were specified – in English, French

H. Schneider, *Rotor Balancing*, https://doi.org/10.1007/978-3-662-66049-2_19

19.2 Symbols

Unless expressly stated otherwise, in this book symbols are used with the following meaning:

Majuscules	
$-\vec{C}_I; \vec{C}_{II}$	Couple unbalance in planes I and II (g · mm)
$-\vec{C}_{rI}; \vec{C}_{rII}$	Resultant couple unbalance in planes I and II (g · mm)
D	Damping factor
$\vec{F}$	Force (N)
$\vec{F}_{A,B}$	Centrifugal bearing forces (N), related to the bearing planes A or B
$\vec{G}$	Gravitational force of a mass on earth (N)
$\vec{G}_{A,B}$	Bearing forces due to the gravitational force of a rotor (N), related to bearing planes A or B
J	Moment of inertia (kg · m^2)
L	Distance between bearing planes or length of rotor (mm)
$\vec{M}$	Torque (N · m)
$\vec{M}_u$	Unbalance moment = moment of centrifugal force (N · m)
P	Power (N · m/s)
$\vec{P}$	Moment unbalance (g · mm^2)
$\vec{P}_r$	Resultant moment unbalance (g · mm^2)
$\vec{P}_{per}$	Permissible moment unbalance (g · mm^2)
T	Time period (s)
$\vec{U}$	Unbalance (g · mm)
$-\vec{U}_I; \vec{U}_{II}$	Couple unbalance (g · mm); forms the moment unbalance with the distance b between the planes
$\vec{U}_{1,2}$	Components of unbalance in directions 1 or 2 (g · mm)
$\vec{U}_{I,II}$	Unbalances in planes I or II; for a rotor with rigid behaviour: complementary unbalances (g · mm)
$\vec{U}_{A,B}$	Unbalances in bearing planes A or B (g · mm)
$\vec{U}_c$	Correction unbalance (g · mm)

and German – and defined. The current standard for balancing technology terms is ISO 21940, Part 2. It has been greatly reduced in scope and explanations (compared to the last ISO 1925). I continue to use a very detailed representation here to promote understanding.

$\vec{U}_{ne}$	Equivalent modal unbalance in the nth flexural mode (g · mm)
$\vec{U}_r$	Resultant unbalance (g · mm)
$\vec{U}_s$	Static unbalance (g · mm)
$\vec{U}_t$	Test unbalance (g · mm)
$\vec{U}_i$	Initial unbalance (g · mm)
$\vec{U}_{per}$	Permissible residual unbalance for the rotor (g · mm)
$\vec{U}_{per\ I,II}$	Permissible residual unbalances for correction planes I or II (g · mm)
$\vec{U}_{pers}$	Permissible residual unbalance as a static unbalance (g · mm)
W	Work, energy (N · m)
Minuscules	
$\vec{a}$	Acceleration (m/s²)
a	Distance of correction plane I to bearing plane A (mm)
b	Distance between correction planes I and II (mm)
b	Arc length (m)
c	Distance of correction plane II to bearing plane B (mm)
d	Distance of mass centre to a bearing plane (mm)
$\vec{e}$	Specific unbalance; eccentricity of mass centre (μm)
$\vec{e}_{per}$	Permissible specific residual unbalance (μm)
h_I	Distance of correction plane I to mass centre (mm)
h_{II}	Distance of correction plane II to mass centre (mm)
l	Lever arm for creating of a moment unbalance (mm)
m	Mass; rotor mass (kg)
n	Speed (min^{-1})
$\vec{r}$	Radius (mm)
$\vec{r}_c$	Correction radius (mm)
r_i	Inertia radius (m)
$\vec{s}$	Displacement (m)
t	Time (s)
u	Unbalance mass (g)
u_c	Correction mass (g)
$u_{c\ I,II}$	Correction masses in planes I or II (g)
$\vec{v}$	Velocity (m/s)
x	Instantaneous value of vibration displacement (m)

$\hat{x}$	Amplitude of vibration displacement (m)
z	Coordinate in longitudinal direction of rotor
Greek letters	
$\vec{\alpha}$	Angular acceleration (rad/s^2)
$\vec{\varphi}$	Plane angle (rad)
φ	Phase angle (rad)
φ_0	Initial phase angle (rad)
$\phi(z)$	Mode shape function
ω	Angular frequency (rad/s)
ω_e, ω_n	Resonant angular frequency (rad/s)
Ω	Angular frequency of service speed (rad/s)
Indices	
1, 2, 3	Components in directions 1, 2, 3
I to V	Related to correction planes I to V
A, B	Related to bearing planes A, B
c	Related to correction
e	Related to equivalent modal unbalances
k	Counting index for unbalance planes, k = 1 to K
m	Related to moment unbalance
n	Counting index for modal expressions, n = 1, 2, 3…
per	Permissible
r	Resultant
s	Related to static unbalance
t	For tests
u	Related to unbalance
x, y, z	Related to rectangular coordinates

19.3 Vocabulary and Definitions

For a good common understanding in a specialist field like rotor balancing technology, it is important to use the correct terms with the standardised meaning – both nationally and internationally.

The following vocabulary and definitions have been compiled on the basis of ISO 21940, Part 2. It has been greatly reduced in the scope and in the declarations. For didactical reasons, I continue to use a very detailed presentation here, e.g. by using also the older ISO 1925.

If the old definitions no longer fit, they have been omitted, and then there are only *Author's notes*. New terms are marked with *new*.

The terms are split into the following sections according to ISO 1925. Within in each section they are listed in *alphabetical order*.

- **Mechanics**: Sect. 19.3.1
- **Rotor systems**: Sect. 19.3.2
- **Unbalance**: Sect. 19.3.3
- **Balancing**: Sect. 19.3.4
- **Balancing machines**: Sect. 19.3.5

No longer included as separate sections today in standard ISO 21940, Part 2, but specified here:

- **Flexible Rotors:** Sect. 19.3.6
- **Rotating rigid free-bodies:** Sect. 19.3.7
- **Balancing machine tooling**: Sect. 19.3.8

The German and French terms corresponding to the English keyword (from ISO 1925) are given in the left-hand column.

Sometimes alternatives are common for one term, in which case all alternatives are listed with a reference to the detailed definition.

19.3.1 Mechanics—Mechanik—Mécaniques

Drehachse axe de rotation	**axis of rotation:** Instantaneous line about which a body rotates. AUTHOR'S NOTES: If the bearings are not isotropic, the axis of rotation is generally not stationary. For rigid bearings the axis of rotation coincides with the shaft axis, for non-rigid bearings not necessarily. For shaft-elastic rotors there is no satisfactory definition so far
Massenmittelpunkt centre de masse	**centre of mass:** Today not listed in ISO 21940-2. AUTHOR'S NOTE: For balancing, the concept of the centre of mass, i.e. the point where the mass of a body can be thought of as being concentrated, is sufficient.
	flexural mode: see Sect. 19.3.6
Biegeresonanzdrehzahl	**flexural resonance speed:** see Sect. 19.3.6
Hauptträgheitsachsen axes principaux d'inertie	**principal axis of inertia**: One of three mutually perpendicular axes intersecting each other at a given point about which the products of inertia of a solid body are zero. Note 1: In balancing, the term principal axis of inertia is used to designate the one central principal axis most nearly coincident with the shaft axis of the rotor
Resonanzdrehzahl vitesse de résonance	**resonance speed:** Characteristic speed at which a resonance of a system is excited Note 1: In the context of balancing, a resonance speed is related to the once-per-revolution component of the vibration only.
Starrkörper-Schwingungsform	**rigid-body mode:** Mode shape of a rotor corresponding to a rigid-body-mode resonance speed for a given support system.
Resonanzdrehzahl der Starrkörper-Schwingungsform	**rigid-body-mode resonance speed:** Resonance speed of a rotor at which flexure of the rotor can be neglected.
Drehzahl vitesse	**speed:** Angular velocity of a rotor. NOTE 1: Speed is measured in revolutions per unit time or in angles (radians) per unit time. NOTE 2: The quantities most frequently used for specifying speed are: n rotational speed measured in revolutions per minute; f rotational frequency measured in revolutions per second; ω angular velocity measured in radians per second.

19.3.2 Rotors—Rotoren—Rotors

Auswuchtdrehzahl vitesse d'équilibrage	**balancing speed:** The speed at which a rotor is balanced
Lagerständer support du palier	**bearing support:** The component or group of components that transmits the load from the bearing to the foundation

elektrische Rundlaufabweichung excentricité électrique	**electrical runout:** Error that can be included in the runout measurement when non-contact transducers are used. Note: These deviations may be due to residual magnetism or electrical inhomogeneity of the measured component or to effects affecting the calibration of the transducer
Anbauteil accessoire	**fitment:** Component of a rotor which has to be mounted on a shaft, a balancing mandrel or a balancing adaptor so that its unbalance can be measured. Example: Couplings, pulleys, pump impellers, blower fans and grinding wheels.
nachgiebiges Rotorverhalten rotor à comportement flexible	**flexible rotor behaviour:** Rotor where the flexure caused by its unbalance distribution *cannot* be neglected with respect to the agreed *unbalance tolerance* at any speed up to the maximum service speed. Note 1: Flexible behaviour includes *shaft-elastic behaviour*, *settling behaviour* (i.e. unbalance indication irreversibly changes with the first run-up) and *component-elastic behaviour* (i.e. unbalance indication reversibly changes with speed due to displacement of rotor components other than the shaft)
nachgiebiger Rotor rotor flexible	**flexible rotor:** colloquial term for a rotor with flexible behaviour AUTHOR'S NOTE: This term – a classification of the rotor – should only be used in connection with the rotor concept, see Sect. 4.2.2
Fundament assise	**foundation:** The structure that supports the mechanical system. Note: In the context of balancing and vibration of rotating machines, the term foundation is usually applied to the heavy base structure on which the entire machine is mounted
halbe Passfeder demi-clavette	**half key:** key which is used during balancing and has the unbalance of that part of the (full) key which, after assembly, fills either the groove in the shaft or the groove in the mounted part. Note 1: The unbalance of the half key for a given shaft may be different from the unbalance of the half key required for the associated attachment, assuming equal length of half keys, because of differences in distance from the centreline of the shaft, keyway depth and clearance. Note 2: To calculate the unbalance of the two half keys, divide the full key along the curved line of separation between the shaft and the attachment. Half of the height clearance of the key to the keyway must be considered for each of the two half keys
Innenbord-Rotor rotor entre paliers	**inboard rotor:** A rotor that has its centre of mass between the journals. Note: In order to describe the rotor accurately, it may be necessary to specify the position of the centre of mass and the correction planes also
isotrope Lagerung support de palier isotrope	**isotropic bearing support:** bearing arrangement with the same dynamic characteristics in each radial direction

Lagerzapfenachse axe du tourillon	**journal axis:** The mean straight line joining the centroids of the cross-sectional contours of a journal
Lagerzapfenmittelpunkt centre du tourillon	**journal centre:** The point of intersection of the journal axis with that radial plane of the journal in which the resultant transverse bearing force acts
Lagerzapfen tourillon	**journal:** The part of a rotor that is radially supported or guided by a bearing in which it rotates
	key: see *half-key*
örtliche Massen-Exzentrizität excentricité locale de masse	**local mass eccentricity:** for axially thin rotor elements cut out perpendicular to the shaft axis, the distance of the *centre of mass* of each individual rotor element from the *shaft axis*
Massen-Exzentrizität excentricité de masse	**mass eccentricity:** Radial distance between a centre of mass and the shaft axis. Note 1: A similar expression: specific unbalance.
Außenbord-Rotor rotor en porte-à-faux	**outboard rotor:** Rotor that has its centre of mass located other than between the journals. Note: In order to describe the rotor accurately, it may be necessary to specify the positions of the centre of mass and the correction planes also
fliegend porte-à-faux	**overhung:** A position outside the bearing distance; e.g. overhung mass, overhung correction plane. AUTHOR'S NOTE: In order to describe the rotor accurately, it may be necessary to specify the positions of the centre of mass and the balancing planes also
vollkommen ausgewuchteter Rotor rotor parfaitement équilibré	**perfectly balanced rotor:** an ideal rotor without unbalance
Zentrierung pion de centrage goujon	**pilot, spigot:** Design of the interface between rotor components to ensure their concentricity. AUTHOR'S NOTE: The centring has to keep play, radial and axial runout within reasonable limits compared to the specified unbalance tolerance
starres Rotorverhalten rotor à comportement rigide	**rigid rotor behaviour:** The *flexure* of a rotor caused by its unbalance distribution *can* be neglected with respect to the agreed unbalance tolerance at any speed up to the maximum service speed
starrer Rotor rotor rigide	**rigid rotor:** colloquial term for a rotor with rigid behaviour AUTHOR'S NOTE: This term – a classification of the rotor – should only be used in connection with the rotor concept, see Sect. 4.2.2.

Rotor rotor	**rotor:** A body, capable of rotating. Note: The term rotor is sometimes applied to bodies which do not have journals, for example disc-shaped bodies such as a flywheel. In the sense of the definition, during balancing such a disc-shaped body only becomes a rotor when it is placed on a shaft with journals. AUTHOR'S NOTE: Or with the aid of an *adaptor* placed on the spindle of a *balancing machine*
Rotorverhalten comportement d'un rotor	**rotor behaviour:** The rotor behaviour is defined by: • the behaviour of the unbalances over the speed range, • the types of unbalance to be corrected or checked, • the ability of the rotor to maintain the position of its mass elements and their centres of mass relative to each other over the speed range, or to change it in different ways
Betriebsdrehzahl vitesse de service	**service speed:** The speed at which a rotor rotates under service conditions
Schaftachse axe de l'arbre	**shaft axis:** The line joining the journal centres which follows the deflected shape of the rotor due to gravity or any other constant force
Rundlaufabweichung bei niedriger Drehzahl excentricité à basse vitesse	**slow speed runout:** Real or apparent runout measured at low speed on the rotor surface where no vibration caused by unbalance occurs. NOTE: Depending on the transducer used, the runout at low speed may contain mechanical, magnetic or electrical components. AUTHOR'S NOTE: Often the runout at low speed is measured at rotational frequency so that it can be subtracted vectorially from later measurements of the same surface at high speed. This way, the unbalance-related vibration can be measured on its own
Gesamte angezeigte Rundlaufabweichung excentricité totale indiquée	**total indicated runout:** The difference between the maximum and minimum values of the radius of a cross-sectional contour. This is measured in a cross-sectional plane perpendicular to a fixed axis of rotation

19.3.3 Unbalance—Unwucht—Balourd

The definitions in this section mainly refer to the unbalance condition in rotors with rigid behaviour. However, some terms and definitions can also be applied to rotors with flexible behaviour. For specific terms and definitions in this area, see Sect. 19.3.6

Unwuchtbetrag module du balourd	**amount of unbalance:** The product of the *unbalance mass* and the *radial distance* of its centre of mass from the shaft axis. AUTHOR'S NOTE: A common unit of the unbalance amount is g · mm
Unwuchtwinkel angle du balourd	**angle of unbalance:** Polar angle at which an *unbalance mass* is located with reference to a given rotating coordinate system, fixed in a plane perpendicular to the *shaft axis* and rotating with the rotor

Auswuchtgütestufe qualité d'équilibrage	**balance quality grade:** Rotor unbalance classification which is the product of the *specific unbalance* and the *maximum service speed*
Auswuchtgüte qualité d'équilibrage	**balance quality:** An evaluation of the *achieved unbalance condition*; the product of the specific unbalance *e* with the maximum service angular velocity of the rotor in mm/s. Author's Note: This term is outdated: In ISO 21940-2, only the *balance quality grade* is defined
Zentrum der Unwucht centre de balourd	**centre of unbalance:** The position on the *shaft axis* of that reference plane of the *resultant unbalance* for which the magnitude of the *resultant moment unbalance* reaches a minimum
begrenzte Urunwucht balourd initial réduit	**controlled initial unbalance:** An *initial unbalance* which has been limited by balancing the individual parts or careful design, manufacture and assembly of the rotor
Unwuchtpaar couple de balourds	**couple unbalance** – $\vec{U}_{CI}$; $\vec{U}_{CII}$**:** Two *unbalance vectors* of the same amounts and opposing directions, acting in two different planes perpendicular to the shaft axis. Note 1: A couple unbalance is an alternative representation of a moment unbalance with the amount of each of the two unbalance vectors calculated by dividing the amount of the moment unbalance by the distance between the two planes of the couple unbalance
dynamische Unwucht déséquilibre dynamique	**dynamic unbalance:** The state of unbalance that corresponds to the central principal axis of inertia having any inclined and offset position relative to the shaft axis. Note 1: This condition can be produced by adding a *static unbalance* und a *couple unbalance* to a perfectly balanced rotor. Note 2: This condition can be produced equivalently by adding *two unbalance vectors* acting in two different planes perpendicular to the shaft axis to a perfectly balanced rotor. Author's note: This definition is sufficient for rotors with rigid behaviour. For a complete description of the unbalance state of the rotor in the general case (also rotors with shaft-elastic behaviour) the modal unbalances may be added (see Sect. 4.2.1)
Urunwucht balourd initial	**initial unbalance:** *unbalance* of any kind that exists in the rotor before *balancing* Author's note: See also *start unbalance*
Momentenunwucht moment de balourds	**moment unbalance:** Component of the unbalance that corresponds to an inclined *central principal axis of inertia*, intersecting the shaft axis at the *centre of mass*. Note 1: The dimension of the moment unbalance is mass times length squared. Note 2: This condition can be produced by *two unbalance vectors* with equal amounts and opposite directions, acting in two different planes perpendicular to the shaft axis, which thereby constitute a *couple unbalance*.

Restunwucht balourd résiduel	**residual unbalance:** Unbalance of any kind that remains after balancing
resultant moment unbalance résultant moment balourd	**resultant moment unbalance**: Vector sum of the moments of all unbalance vectors distributed along the rotor with respect to an arbitrarily chosen plane perpendicular to the shaft axis. AUTHOR'S NOTE. This plane must be the plane of the *resultant unbalance*, i.e. only the plane of the *resultant unbalance* can be *arbitrarily* selected
resultant unbalance résultant balourd	**resultant unbalance:** Vector sum of all unbalance vectors distributed along the rotor (see Sect. 5.3.2). AUTHOR'S NOTE: The *resultant unbalance* is the same for each radial plane, but magnitude and angular position of the *resultant moment unbalance* depend on the axial position chosen for the *resultant unbalance*
bezogene Unwucht balourd spécifique	**specific unbalance:** Unbalance amount divided by the mass of the rotor. Note 1: the specific unbalance calculated with the amount of static unbalance is numerically equivalent to the mass eccentricity. AUTHOR'S NOTE: In the case of a general rotor (with rigid behaviour) with both types of unbalance, *e* is a calculated quantity containing the effect of the *resultant unbalance* and the *resultant moment unbalance*, i.e. the total permissible unbalance U_{per}
Anfangsunwucht	**start unbalance:** *new* AUTHOR'S NOTE: If the balancing process takes place in several individual steps – e.g. balancing of the individual parts, then of the entire rotor – the unbalance condition at the beginning of the first step is the *initial unbalance*, and for any further step it is the *start unbalance* of this step
statische Unwucht balourd statique	**static unbalance $\vec{U}_s$:** Component of the unbalance that corresponds to a parallel misalignment of the *central principal axis of inertia* of the rotor with respect to the *shaft axis*. NOTE 1: The amount of the static unbalance is equal to the product of the total mass of the rotor and the radial distance of its *centre of mass* from the *shaft axis*
	unbalance condition: see *unbalance*
unbalance vector vecteur balourd	**unbalance vector:** A vector whose magnitude is the *amount of unbalance* and whose direction is given by the *unbalance angle*
Unwucht, Unwuchtzustand déséquilibre, balourd	**unbalance:** Condition that exists in a rotor when vibratory forces or motions are imparted to it and its bearings from centrifugal forces of mass eccentricities. AUTHOR'S NOTE: The unbalance must be defined independently, e.g. as the product of the unbalance mass and the radius of its centre of mass from the shaft axis

19.3.4 Balancing—Auswuchten—Équilibrage

Unwuchttoleranz tolérance déséquilibre	**balance tolerance:** Defined maximum unbalance amount below which the unbalance condition is considered to be permissible. AUTHOR'S NOTE: This term may comprise *modal unbalance tolerances* if applied to rotors with *flexible behaviour*
schrittweises Auswuchten équilibrage progressif	**balancing in stages during assembly:** Stepwise procedure of adding one or two components of a group of rotor components to a balanced shaft and balancing this assembly; and repeating this process until the assembly is completed
	balancing in-situ: see *in-situ balancing*
Auswuchten équilibrage	**balancing:** The process by which the mass distribution of a rotor is examined and, if necessary, adjusted to ensure that *residual unbalances* are within specified limits
Vorhalteunwucht	**bias unbalance:** Unbalance applied to an auxiliary shaft or drive adapter to create an intended unbalance condition
Kalibriermasse masse d'étalonnage	**calibration mass:** A known unbalance mass used for calibration
Komponentenausgleich correction du composant	**correction in components:** Unbalance correction in a correction plane by adding or removing mass in two or more locations (at predefined angular positions)
Ausgleichsmasse masse de correction	**correction mass:** A known unbalance mass attached to a rotor in a given correction plane at a known radial distance from the shaft axis for the purpose of correction. NOTE: The same correction can be effected by removing mass from the opposite side of the rotor or by moving masses to different angular positions
Ausgleichsebene plan de correction, plan d'équilibrage	**correction plane:** A plane perpendicular to the shaft axis of a rotor in which correction for unbalance is applied
Ausgleich méthode de correction	**correction:** Part of the balancing procedure by which the mass distribution of a rotor is adjusted to reduce unbalance. NOTE 1: Corrections are usually made by adding material to the rotor, removing material from the rotor or by shifting material to a different angular position on the rotor
Differenz-Testmassen masses d'essais différentielles	**differential test masses:** Two masses, representing different amounts of unbalance, added to a rotor in the same transverse plane at diametrically opposite positions. NOTE: Differential test masses are used in cases where a single test mass is impractical

Differenz-Unwucht balourd différentiel	**differential unbalance:** The difference in unbalance between the two differential test masses
	dynamic balancing: Obsolete term, see *two-plane balancing*
hochtouriges Auswuchten équilibrage à haute vitesse	**high speed balancing:** Procedure for *balancing* at a balancing speed at which the rotor exhibits *flexible behaviour*. AUTHOR'S NOTE: Several speeds may be required here
Auswuchten auf Umschlag équilibrage par indexage	**index balancing:** Procedure whereby the rotor is indexed relative to the balancing tooling to enable the evaluation and, if necessary, the correction of the rotor's unbalance without the influence of the tooling. NOTE 1: More generally on multipart rotors, where parts can be rotated with respect to each other, it is possible to use index balancing to evaluate and, if necessary, correct the unbalance contribution from each part separately
Umschlag-Unwucht balourd d'indexage	**indexing unbalance:** The change in unbalance when two components of an unbalanced rotor assembly are indexed against each other, see *index balancing*
Eindrehen indexage	**indexing:** Incrementally rotating a rotor or a component of a rotor assembly for the purpose of bringing it into a desired angular position AUTHOR'S NOTE: The task is to bring the rotor – e.g. for a correction – into a certain angular position. Mentioning "component of a rotor assembly" can lead to confusion with *index balancing*.
Betriebsauswuchten équilibrage in situ	**in-situ balancing:** The process of *balancing* a rotor in its operational situation (with operational drive, bearings and support structure) instead of in a *balancing machine*
niedertouriges Auswuchten équilibrage à basse vitesse	**low speed balancing:** Procedure for rotor balancing at a *balancing speed* where the rotor exhibits *rigid behaviour*
Massenzentrieren centrage de masse	**mass centring:** The process of determination the *principal axis of inertia* of the rotor and, in a subsequent machining process, machining rotor elements (journals, centrings, or reference surfaces) to bring the *axis of rotation* determined by these elements into close proximity to the principal axis of inertia.
Messebene plan de mesure	**measuring plane:** A plane perpendicular to the shaft axis in which measurements are performed to determine an unbalance vector

zulässige Restunwucht balourd résiduel maximal admis	**permissible residual unbalance:** Numerical value of the unbalance tolerance. Note: The permissible residual unbalance is either an overall value for a rotor derived from the balance quality grade or a portion of that value allocated to each of a number of planes perpendicular to the shaft axis; e.g. correction plane, measuring plane, or reference plane
Ebenen-Verlagerung transposition des plans	**plane transposition:** A procedure to determine the unbalance for planes other than those in which it was originally measured, see *transfer unbalances*
polarer Ausgleich correction polaire	**polar correction:** Correction of the unbalance of a correction plane in the angular position of the measured unbalance (by removing material) or opposite to it (by adding material)
	progressive balancing: see *balancing in stages during assembly*
	proving mass, see *test mass*
	proving plane, see *test plane*
Viertelpunkte points de quadrature	**quarter points:** An expression describing the position of the optimum balancing planes of a rotor with *shaft-elastic behaviour*, according to ISO 21940-12 (see section 10.3.5)
Bezugsebene plan de référence	**reference plane:** Any plane perpendicular to the shaft axis to which an amount of unbalance is referred. AUTHOR'S NOTE: More correct would be *"an unbalance"* because the vector character can also be relevant
Ein-Ebenen-Auswuchten équilibrage dans un seul plan	**single-plane balancing:** The process by which the mass distribution of a rotor is examined and, if necessary, corrected, usually in a single correction plane, to ensure that the residual *resultant unbalance* is within a specified tolerance
	static balancing: Obsolete term, see *single-plane balancing*
Anfälligkeit gegen Unwuchten susceptibilité au déséquilibre	**susceptibility to unbalance:** Indication of the likelihood of a rotor having a change of unbalance over a certain period of operation
Testmasse masse d'essai	**test mass:** Precisely defined mass to be attached to a rotor at a precisely defined position to provide a precisely defined change of the state of unbalance, frequently used in combination with other test masses to establish a test mass set for the evaluation of a balancing machine
Testebene plan d'essai	**test plane:** Plane perpendicular to the shaft axis of a rotor in which a test mass may be attached

Toleranzebene plan de tolerance	**tolerance plane:** Reference plane for which an unbalance tolerance is specified
Transferunwuchten	**transfer unbalances:** Unbalances measured for planes other than those in which the *permissible residual unbalance* has been reached, see section 10.4.5.1
Probemasse	**trial mass:** Unbalance mass chosen arbitrarily (or by prior experience with similar rotors) and attached to a rotor to determine the rotor response
Nachwuchten équilibrage de compensation	**trim balancing:** Correction of small residual unbalances, often carried out in-situ. AUTHOR'S NOTE: It should be called final rebalancing
Zwei-Ebenen-Auswuchten équilibrage dans deux plans	**two-plane balancing:** The process by which the mass distribution of a rotor is examined and, if necessary, corrected, usually in two correction planes, to ensure that *residual unbalances* are within specified limits. AUTHOR'S NOTE: This is sufficient only for rotors with *rigid behaviour*
Ebene des Schwingungsaufnehmers plan du transducteur de vibration	**vibration transducer plane:** Plane perpendicular to the shaft axis in which a vibration transducer is located

19.3.5 Balancing Machines—Auswuchtmaschinen—Machines à équilibrer

Betragsanzeige indicateur de balourd	**amount indicator:** The analogue or digital display of a balancing machine for the amount of unbalance or its effect
Winkelanzeige indicateur d'angle	**angle indicator:** A device used to indicate the angle of unbalance
Winkellagengeber générateur de référence d'angle	**angle reference generator**: A device in balancing which generates a reference signal by which the angular position of the rotor is defined
Winkelbezugsmarke marques de référence d'angle	**angle reference mark:** A mark placed on the rotor to denote an angle reference system fixed on the rotor; it may be optical, magnetic, mechanical or radioactive. AUTHOR'S NOTE: In series production, a fixed angular reference mark is sometimes dispensed: By scanning a regular pitch on the rotor, a speed-frequency signal is obtained considering its number of divisions

Auswuchtlager palier d'équilibrage	**balancing bearings:** Special anti-friction bearings – often with reduced clearance – to balance a rotor in a low-speed balancing machine. AUTHOR'S NOTE: Balancing bearings, sometimes called auxiliary bearings, are used primarily in the balancing of jet engine rotors because the large clearance of cold engine bearings can cause significant balance errors
Auswuchtmaschine machine à équilibrer	**balancing machine:** A machine or facility that enables the unbalance in a rotor to be measured and corrected. NOTE 1: Single-plane balancing machines can only be used for single-plane balancing. NOTE 2: Two-plane balancing machines can be used to carry out single-plane balancing and two-plane balancing
Auswuchtlauf cycle d'équilibrage	**balancing run:** A procedure consisting of a measuring run and the associated correction process (see Sect. 12.3.16)
Ersatzmasse masse de compensation	**bob weight:** A mass applied e.g. to a crank-bracket of a crankshaft during balancing to simulate rotating and oscillating masses of the piston/connecting rod assembly
Einstellvorgang étalonnage	**calibration:** The process of adjusting a balancing machine so that the unbalance indicators read in terms of selected correction units in specified correction planes for a given rotor and other essentially identical rotors. Note: This may include the adjustment of the angular position, if necessary
Fliehkraft nutzende Auswuchtmaschine machine à équilibrer centrifuge	**centrifugal balancing machine:** A balancing machine which provides for the support and rotation of the rotor and has a means of measuring the once-per-revolution vibratory forces or motions due to the unbalance in the rotor
angegebene kleinste erreichbare Restunwucht $U_{\text{ker'ang}}$ balourd résiduel réalisable déclaré $U_{\text{mar,dec}}$	**claimed minimum achievable residual unbalance $U_{\text{mar,cl}}$:** Amount of the minimum achievable residual unbalance stated by the balancing machine manufacturer. Note: This value normally depends on the balancing speed
Kompensationseinrichtung compensateur	**compensator:** Facility built into a balancing machine to adjust the unbalance reading. Example: Adjustment of the unbalance reading can be necessary to compensate for the effect of the initial unbalance. AUTHOR'S NOTE: Modern measuring devices usually compensate by calculation. Compensators can simplify processes, e.g. *index-balancing* or biasing of keys/keyways

Ausgleichsebenen-Einflussverhältnis taux d'interaction entre les plans de correction	**correction plane interference ratio I_{AB}, I_{BA}:** Ratio of the change in unbalance indication for one correction plane and the change in unbalance indication for the other correction plane of a proving rotor inducing the change in indication. NOTE 1: The interference ratio of correction plane A is calculated by $I_{AB} = U_{AB}/U_{BB}$, Where: U_{AB} is the change in unbalance indication for correction plane A induced by a change in unbalance in correction plane B; U_{BB} is the associated change in unbalance indication for correction plane B NOTE 2: For the correction plane interference ratio for correction plane B, I_{BA}, exchange A and B in Note 1. AUTHOR'S NOTES: I_{BA} is therefore calculated $I_{BA} = U_{BA}/U_{AA}$ The correction plane influence ratio should be minimal in a balancing machine where *plane separation* has been carefully performed. The ratio is usually expressed as a percentage
Ausgleichsebenen-Einfluss interaction entre les plans de correction	**correction plane interference:** The change in balancing machine indication for one correction plane A of a rotor observed for a certain change in unbalance in the other correction plane B (and vice versa). Author's note: See also *correction plane interference ratio*
Gegengewicht contrepoids	**counterweight:** A weight added to a rotor to reduce a calculated unbalance at a desired location. NOTE: Such counterweights can be used to bring an asymmetric body into a state of balance or to reduce bending moments within a body (e.g. in crankshafts)
Unwuchtpaar-Einflussverhältnis I_{SC} taux d'interaction du couple de balourds I_{SC}	**couple unbalance interference ratio I_{SC}:** Ratio of the change in static unbalance and the change in couple unbalance inducing the change in indication: $I_{SC} = U_S/U_C$ Where: U_S is the change in the static unbalance indication induced by a change in the couple unbalance; U_C is the change in couple unbalance. AUTHOR'S NOTE 1: It would be better to relate the *static unbalance* to the *moment unbalance*. For the couple unbalance always the *plane distance* has to be mentioned. AUTHOR'S NOTE 2: This interference ratio is commonly used when testing *single plane balancing machines*
Zyklenrate capacité de cycles	**cycle rate:** The number of starts and stops that a balancing machine for a given rotor (having a specified moment of inertia) and for a given balancing speed can perform per hour without damage of the machine

direkt anzeigende Auswuchtmaschine machine à équilibrer à lecture directe	**direct reading balancing machine**: A balancing machine that can be set to any two measuring planes and can indicate the unbalance in angular position and mass units, e.g. in g, without significant *correction plane interference*; and this without requiring individual calibration for the first rotor of a kind
Doppelkompensator compensateur double	**double compensator:** device in a balancing machine by which the influence of systematic error of the balancing tooling on the unbalance indication is eliminated (see index-balancing). AUTHOR'S NOTE: In the case of computer-based measuring instruments, compensation may be performed by calculation
Boden-Boden-Zeit durée d'équilibrage	**floor-to-floor time:** Duration of balancing a rotor, including all balancing and measuring runs together with the times for loading and unloading. NOTE: The time is normally expressed in minutes. AUTHOR'S NOTE 1: If there is more than one rotor in a balancing machine at a time (e.g. in a transfer machine), the time taken between two consecutive rotors leaving the machine is used. AUTHOR'S NOTE 2: If the floor-to-floor time varies from rotor to rotor, the arithmetic mean may be used. AUTHOR'S NOTE 3: When calculating for a longer period of time, e.g. one year, the time required for machine maintenance and others may be included. In this case it is referred to as gross floor-to-floor time
Schwerpunktwaage machine à équilibrer par gravité	**gravitational balancing machine:** A balancing machine that provides for the support of a *rigid rotor* under non-rotating conditions and provides information on the amount *and* angle of the static unbalance
kraftmessende Auswuchtmaschine machine à équilibrer à palier rigide	**hard-bearing balancing machine:** A balancing machine having a balancing speed range below the natural frequency of the combined rotor and support system
Betriebsauswuchtgerät matériel d'équilibrage in situ	**in-situ balancing equipment:** A measuring device that provides information for the balancing of an assembled machine unit that is not mounted in a balancing machine
Messlauf essai pour mesurage	**measuring run:** Procedure of rotating a rotor to be balanced at balancing speed to obtain information on unbalance and correction. AUTHOR'S NOTE: A measuring run normally consists of several steps (see Sect. 12.3.16)
Einrichten der Maschinenmechanik mis aux point mécanique	**mechanical adjustment:** Mechanical preparation of a balancing machine to balance a rotor
kleinste erreichbare bezogene Restunwucht $\mathbf{e}_{ker}$ balourd résiduel réalisable specifique	**minimum achievable residual specific unbalance e_{mar}:** The smallest amount of residual specific unbalance that a balancing machine is capable of achieving under given conditions

kleinste erreichbare Restunwucht U_{ker} balourd résiduel réalisable déclaré	**minimum achievable residual unbalance U_{mar}:** the smallest amount of residual unbalance that a balancing machine is capable of achieving under given conditions. AUTHOR'S NOTE: For consideration concerning balance errors, this value is required for different speeds and possibly for different rotor configurations (see Sect. 12.3.8)
tote Masse masse parasite	**parasitic mass:** On a *balancing machine*, any mass, other than that of the rotor being balanced, that is moved by the *unbalance* force(s) developed in the rotor
kalibrierte Einstellung étalonnage permanent	**permanent calibration:** The feature of a *hard-bearing balancing machine* that permits it to be calibrated once and for all, so that it remains calibrated for any rotor within the capacity and speed range of the balancing machine. NOTE: The balancing machine should be set to different rotor dimensions. AUTHOR'S NOTE: The permanent calibration is a prerequisite for a *direct reading balancing machine*.
Anzeige einer Phantom-Unwucht indication de balourd fictif	**phantom unbalance indication:** Faulty unbalance indication caused by once-per-revolution signals that are not related to unbalance. NOTE: Such indications can be caused by a insufficient angular flexibility with respect to the vertical axis of the pedestal when balancing in plain bearings or roller bearings, by a nearly immovable joint of a cardan shaft, by alternating forces due to a bent rotor journal, by a rotating magnetic field and other similar influences. AUTHOR'S NOTE: The phantom unbalance indication can be separated from the unbalance indication by various means (see clause 16.7)
Ebenentrennung séparation des plans	**plane separation:** capability of a balancing machine to minimize the correction plane interference ratio. AUTHOR'S NOTE: This term is also used to designate the corresponding work step. In the case of computer-based measuring instruments the separation is usually done by calculation
praktische Ausgleichseinheit unité pratique de correction	**practical correction unit:** A unit corresponding to a unit value of the unbalance amount indicated on a balancing machine. For convenience it is associated with a specific radius and correction plane and is commonly expressed as units of an reasonably chosen quantity, such as drill depth of a given diameter, weight, length of wire solder, plugs and wedges
Wirtschaftlicheit rendement	**production efficiency:** The ability of a machine to assist the operator in balancing a rotor so that the rotor is brought to a specified residual unbalance within the shortest possible time. AUTHOR'S NOTE: For more information on measuring run, balancing run, floor-to-floor time, production rate and cycle time, see the respective definitions and Sect. 12.3.16

Fertigungsrate taux (capacité) de production	**production rate:** Reciprocal term to *floor-to-floor time*. AUTHOR'S NOTE 1: The production rate is expressed in units of time, using longer time periods such as hour, shift, day or year. AUTHOR'S NOTE 2: In the case of *gross floor-to-floor time*, the reciprocal is the *production rate gross*.
automatische Auswuchteinrichtung dispositif à autoéquilibrage	**self-balancing device:** A device which automatically compensates for changes in the unbalance of a machine during operation
Einstellen der Messeinrichtung réglage	**setting the measuring device**: On a direct-reading balancing machine the operation to enter data on the position of the correction planes, the bearing positions, the balancing radii and (if necessary) the balancing speed into the measuring device
Ein-Ebenen-Auswuchtmaschine machine à équilibrer à un seul plan	**single-plane balancing machine:** Gravitational or centrifugal balancing machine that provides information for accomplishing single-plane balancing. NOTE: Single-plane balancing can be carried out on a pair of knife edges without rotation of the rotor but is now more usually carried out on centrifugal balancing machines. AUTHOR'S NOTE: Another possibility is offered by gravitational balancing machines
	slave bearing: see *balancing bearing*
wegmessende Auswuchtmaschine machine à équilibrer à palier souples	**soft bearing balancing machine:** Balancing machine having a balancing speed range above the natural frequency of the combined rotor and support system
nutzbarer Durchmesser diamètre utile	**swing diameter:** Maximum workpiece diameter that can be accommodated on a balancing machine
Zwei-Ebenen-Auswuchtmaschine machine à équilibrer à deux plans	**two-plane balancing machine:** A *centrifugal balancing* machine that provides information for performing two-plane balancing. NOTE: Two-plane balancing machines are sometimes used to accomplish single-plane balancing
Unwuchtreduzierverhältnis R_{UR} rapport de réduction du balourd R_{RB}	**unbalance reduction ratio** R_{UR}: Ratio of the reduction in the unbalance by a single unbalance correction step to the initial unbalance as shown by: $R_{UR} = \frac{U_1 - U_2}{U_1} = 1 - \frac{U_2}{U_1}$ Where: U_1 is the amount of the initial unbalance; U_2 is the amount of unbalance remaining after *one* correction. AUTHOR'S NOTE: For considerations concerning errors, this value is needed at different speeds and possibly for different rotor configurations (see Chap. 16)
Vektormessgerät appareil de mesurage de vecteur	**vector measuring device:** A device for measuring the amount and angle of an unbalance and for displaying them in terms of an unbalance vector, usually by means of a point or crossing lines

Winkelbeweglichkeit um die Vertikale liberté de l'axe vertical	**vertical axis freedom**: The freedom of a horizontal balancing machine bearing support or bearing housing to rotate by a few degrees about the vertical axis through the centre of the support

19.3.6 Flexible Rotors—Nachgiebige Rotoren—Rotors flexibles

	balancing according to eigenmodes: see *modal balancing*
äquivalente Unwucht in der *n*-ten Eigenform balourd modal équivalent d'ordre *n*	**equivalent n^{th} modal unbalance U_{ne}:** Minimum single unbalance equivalent to the nth modal unbalance in its effect on the nth flexural mode. Note 1: The calculated quantity *equivalent n^{th} modal unbalance* corresponds to an unbalance imagined to be fixed within the main part of the rotor in that plane perpendicular to the shaft axis, where the deflection of the corresponding n^{th} flexural mode has its maximum absolute value. Note 2: An unbalance equal in amount and position to the equivalent n^{th} modal unbalance affects some modes other than the nth flexural mode when attached to the rotor. Author's Note: Note 2 is uncomplete: Since this unbalance is a single unbalance, it influences all unbalances, i.e. the low-speed unbalances as well as all modal unbalances (as long as it is not fixed in a node of the other mode shapes). The equivalent unbalance in the nth flexural mode is therefore not intended for unbalance correction, but serves only to quantify this modal unbalance.
Biegeeigenform mode de flexion	**flexural mode:** Mode shape of a rotor corresponding to a flexural resonance speed for a given support system
Biegeresonanz-Drehzahl vitesse critique de flexion	**flexural resonance speed:** Resonance speed of a rotor at which flexure of the rotor cannot be neglected
hochtouriges Auswuchten équilibrage à haute vitesse	**high-speed balancing:** Procedure for balancing at one or more balancing speeds where the rotor exhibits flexible behaviour
Einflusskoeffizient (EFK)	**influence coefficient (IC):** The ratio (complex-valued) of the vector change of vibration in a specified *measuring plane* to the vector of the change of the *unbalance* in a specified plane in the rotor at a specified *speed*, when each of the vectors is expressed by a complex number. Note 1: The local sensitivity is equal to the magnitude of the influence coefficient

örtliche Empfindlichkeit sensibilité locale	**local sensitivity:** Ratio of the magnitude of a change of the measured vibration vector in a specified measuring plane to the magnitude of the corresponding change of the unbalance in a specified plane in the rotor at a specified speed. AUTHOR'S NOTE: If this ratio is not based on magnitudes, but on vectors, it is the *influence coefficient* (IC)
niedrigtouriges Auswuchten équilibrage à basse vitesse	**low-speed balancing:** A procedure for balancing at a balancing speed where the rotor exhibits rigid behaviour
modaler Vergrößerungsfaktor M_n facteur modal d'amplification M_n	**modal amplification factor** M_n: Ratio of the magnitude of the modal displacement vector to the magnitude of the modal eccentricity for the nth flexural mode. NOTE 1: Modal amplification factor is a nondimensional quantity. It is expressed for the nth mode as: $M_n = \frac{\left(\frac{\Omega}{\omega_n}\right)^2}{\sqrt{\left[1-\left(\frac{\Omega}{\omega_n}\right)^2\right]^2+4\zeta_n^2\left(\frac{\Omega}{\omega_n}\right)^2}}$ Where: Ω is the angular velocity expressed in radians per second; ω_n is the undamped natural frequency expressed in radians per second; ζ is the modal damping ratio. NOTE 2: The modal amplification factor at the flexural resonance speed is called *modal sensitivity*
Unwuchttoleranz einer Eigenform tolérance de l'équilibre modal	**modal unbalance tolerance:** That amount of equivalent modal unbalance that is specified as the maximum up to which the unbalance state in that mode is considered to be acceptable
modales Auswuchten équilibrage modal	**modal balancing:** A procedure for balancing of rotors with *shaft-elastic behaviour* in which corrections are made to ensure that the *residual modal unbalances* (as well as the amplitude of vibration generated by it) in the separate flexural modes are within specified limits. AUTHOR'S NOTE: Today, one should not take vibrations as a basis, but unbalance tolerances, i.e. tolerances for the *low-speed unbalances* and the *equivalent modal unbalances*
modaler Dämpfungsgrad ζ_n rapport modal d'amortissement ζ_n	**modal damping ratio ζ_n:** This dimensionless quantity is a measure of the damping effect on the nth flexural mode
	modal sensitivity: see *modal amplification factor*
Mehr-Ebenen-Auswuchten équilibrage multiplan	**multiplane balancing**: Balancing procedure that requires correction in more than two correction planes. AUTHOR'S NOTE: Details of a systematic approach and the various procedures are described in Sect. 10.4
höherharmonische Schwingung vibration à un multiple de la fréquence de rotation	**multiple-frequency vibration:** Vibration at a frequency corresponding to an integral multiple of the rotational speed. NOTE: The vibration may be caused by un-isotropy of the rotor, non-linear characteristics of the rotor/bearing system, or other causes

n-te modale Unwucht balourd modal d'ordre n	**n^{th} modal unbalance:** Unbalance distribution which only affects the nth flexural mode of a rotor and support system. AUTHOR'S NOTE: This unbalance is purely theoretical, because in practice only backward orthogonal unbalance sets can be used and even if all unbalances are corrected together (e.g. with the help of a computer programme), only the relevant flexural modes are treated, not all flexural modes occurring on the rotor
Resonanzdrehzahl des starren Rotors vitesse critique du mode-rotor-rigide	**rigid-body-mode resonant speed:** Resonant speed of a rotor at which flexure of the rotor can be neglected
Empfindlichkeit gegen Unwuchten sensibilité au déséquilibre	**sensibility to unbalance:** A measure of the change in vibration of a machine in response to a change in unbalance. NOTE: It is expressed numerically by the quotient of the amount of the change vector of the vibration and the amount of the change vector of the unbalance. AUTHOR'S NOTE: This is especially interesting for rotors with flexible behaviour
Eigenfunktion der n-ten Biegeeigenform $\phi_n(z)$	**shape function of the nth flexural mode $\phi_n(z)$:** Mathematical expression for the deflection shape of the rotor in the corresponding flexural mode, normalized so that the maximum deflection is unity. NOTE 1: Frequently it is assumed, that the modes are mutually orthogonal and the system is axially symmetric. This is not applicable in all cases
Anfälligkeit gegen Unwuchten susceptibilité au déséquilibre	**susceptibility to unbalance:** Indication of the likelihood of a rotor having a change of unbalance over a certain period of operation AUTHOR'S NOTE: ISO 21940-31 handles this topic together with *sensibility of machines to unbalance*, where flexural resonances are decisive. But susceptibility to unbalance does not only occur with flexible rotor behaviour
thermisch bedingte Unwucht balourd causé par un phénomène thermique	**thermally induced unbalance:** Change of the unbalance state exhibited by a rotor due to a change of its temperature distribution. Note: This change of the unbalance state may be permanent or temporary
Abstimmungs-Verhältnis vitesse sans dimension	**tuning ratio:** For the nth flexural mode, the ratio of the service speed to the corresponding resonance speed, i.e. Ω/ω_n. NOTE: It is sometimes also referred to as η_n

19.3.7 Rotating Rigid Free-Bodies—Rotierende starre freie Körper—Corps-libres rigides en rotation

These definitions apply to rotating rigid free-bodies. However, if such a body is placed in a balancing machine – with the appropriate tooling – it can be described as a rotor. In this case the definitions given in Sects. 19.3.1 to 19.3.5 apply.

As a typical example for rotating rigid free-bodies **satellites** may be used, rotating in orbit.

Auswuchten des freien starren Körpers équilibrage d'un corps-libre rigide	**balancing of a rigid free-body:** Procedure by which the mass distribution of a rigid free-body is checked and, if necessary, adjusted to ensure that the location of the *principal axis of inertia* is within specified limits.
konstruktive Achse axe de construction	**design axis:** Axis about which the position of parts and assemblies are designed and about which the final body is intended to be balanced. NOTE: In the ideal case, the design axis and spin axis coincide
Lage der Hauptträgheitsachse emplacement de l'axe principal	**principal axis location:** The axis location is defined by the *offset* of the centre of mass from the design axis and the *tilt angle* of the principal axis of inertia from the *design axis*
Unwucht des freien starren Körpers balour d'un corps-libre rigide	**rigid free-body unbalance**: On a balancing machine the condition that exists in any rotating rigid free-body, when an unbalance about the design axis exists
freier starrer Körper corps-libre rigide	**rigid free-body:** System of particles with rigid integral connections and no external constraints
rotierender freier starrer Körper corps-libre rigide en rotation	**rotating rigid free-body:** A rigid free-body rotating about an axis. NOTE: The rotation axis is not stationary if it is not a *central principal axis*

19.3.8 Balancing Machine Tooling—Zubehör zu Auswuchtmaschinen—Outillage de machine à équilibrer

	balancing arbor: see *mandrel*
Vorhaltmasse masse d'un balourd de compensation	**bias mass:** The mass attached to the mandrel to create a desired unbalance bias
Vorhaltunwucht	**bias unbalance:** Unbalance applied to a mandrel or drive adaptor to create an intended unbalance bias

Einstellrotor rotor d'étalonnage	**calibration rotor:** A rotor (usually the first of a series) used for the calibration of a balancing machine
Ersatzrotor, Dummy rotor équivalent	**dummy rotor:** An accessory that replaces a rotor or part of a rotor in a balancing procedure. It must have appropriate stiffness and the same dynamic properties as the replaced part: i.e. the position of the centre of mass, mass and moments of inertia must be the same (at least the difference of relevant moments of inertia)
Hilfswelle faux arbre, arbre d'équilibrage	**mandrel:** A machined shaft on which a work piece is mounted for *balancing*
Meisterrotor rotor de contrôle	**master rotor:** A *calibration rotor* with provisions for adding calibration masses at a known location and used for periodically checking the calibration of a *balancing machine*
Testrotor rotor de vérification	**test rotor:** A rigid rotor of suitable mass designed for testing balancing machines and balanced sufficiently to permit the introduction of exact unbalances by means of additional masses with high reproducibility of magnitude and angular position
Unwuchtvorhalt einer Hilfswelle balourd de compensation d'un faux arbre	**unbalance bias of a mandrel**: A known amount of unbalance added to a mandrel. NOTE: Biasing a mandrel generally serves the purpose of either • compensating for the residual unbalance that run-out of the mandrel's rotor mounting surface causes when this single mandrel is used in balancing a series of rotors of the same mass, or • introducing a specific unbalance at a specific angular position for the purpose of balancing parts which, after being removed from the mandrel, are to have a specified unbalance

Documents for Calculations

20

Contents

20.1 Preliminary Note

Information on the calculation of physical data for the balancing process, as well as for unbalance and unbalance correction, is summarised here.

20.2 Decimal Multiples and Decimal Parts

Decimal multiples and parts of units can be designated by placing certain prefixes before the unit name, Table 20.1. They base on ISO 80000.[1]

[1] SO 80,000 Quantities and units, consists of various parts.

H. Schneider, *Rotor Balancing*, https://doi.org/10.1007/978-3-662-66049-2_20

Rules

- Prefixes are placed without space in front of the unit sign.
- Never use more than one prefix to denote a decimal multiple or part of a unit, e.g.: 1/1000 μm = 1 nm (nanometre). The expression 1 mμm (millimicrometre) is therefore not permitted.
- Preference should be given to steps of thousands, i.e. multiples 10^3, 10^6, 10^9, 10^{12} and parts 10^{-3}, 10^{-6}, 10^{-9}, 10^{-12}.

Table 20.1 Overview of the powers of ten from 10^{-12} to 10^{12}

Power of ten	Number	Prefix name	Prefix symbol
10^{12}	1.000.000.000.000	tera	T
10^{9}	1.000.000.000	giga	G
10^{6}	1.000.000	mega	M
10^{3}	1.000	kilo	k
10^{2}	100	hecto	h
10^{1}	10	deca	da
10^{0}	1	–	–
10^{-1}	0,1	deci	d
10^{-2}	0,01	centi	c
10^{-3}	0,001	milli	m
10^{-6}	0,000.001	micro	μ
10^{-9}	0,000.000.001	nano	n
10^{-12}	0,000.000.000.001	pico	p

20.3 Conversion Factors for SI Units and Inch/Pound Units

Unfortunately, some of the world's major industrialised countries still use units that do not correspond to the SI system of measurement. So it can happen that inch, pound and derived units are mentioned when describing a balancing task. The following information is intended to facilitate the transition from one system of measurement to the other, Table 20.2.

Table 20.2 Conversions between SI units and inch/pound units

Length	1 m (Meter)	= 3,28 ft (foot)
		= 39,4 in (inch)
	1 ft = 12 in	= 0,305 m
	1 in	= 25,4 mm (Millimeter)
	1 mil = 1 thou = 1/1000 in	= 25,4 µm (Mikrometer)
	1 µ in (micro inch)	= 0,0254 µm
Force	1 N (Newton)	= 0,225 lb (pound force)
	1 lb = 16 oz (ounce force)	= 4,45 N
Mass	1 kg (Kilogramm)	= 2,20 lb (pound mass)
		= 35,3 oz (ounce mass)
	1 g (Gramm)	= 0,0353 oz
		= 0,564 dram
	1 lb = 16 oz	= 0,454 kg
	1 oz = 16 dram	= 28,4 g
	1 dram	= 1,77 g
Unbalance	1 kg · mm	= 0,0868 lb · in
		= 1,39 oz · in
	1 g · mm	= 0,0222 dram · in
	1 lb · in = 16 oz · in	= 11,5 kg · mm
	1 oz in = 16 dram · in	= 0,720 kg · mm
	1 dram · in	= 45,0 g · mm
	1 g · in (Gramm · inch)	= 25,4 g · mm
	1 kg · mm ^{2}	= 0,0547 oz · in ^{2}
	1 oz · in ^{2}	= 18,3 kg · mm ^{2}
Torque	1 N · m (Newton · Meter)	= 0,738 lb · ft (pound·foot)
	1 lb · ft	= 1,36 N · m
Moment of inertia	1 kg · m ^{2}	= 23,7 lb · ft ^{2}
		= 0,738 slug ft ^{2}
	1 lb · ft ^{2}	= 0,0421 kg · m ^{2}
	$1\ \text{slug} \cdot \text{ft}^2 = \frac{1\ \text{lb}}{32{,}2\ ^{\text{ft}}/_{\text{s}^2}}\ \text{ft}^2$	= 1,36 kg · m ^{2}

20.4 Nomogramms, Diagramms

20.4.1 Nomogramms

With the help of nomograms, various rough calculations can be made quickly, see. Fig. 20.4 to Fig. 20.11:

- Fig. 20.4 Allocation of the permissible residual unbalance to 2 planes
- Fig. 20.5 Moment unbalance due to a skewed rotor
- Fig. 20.6 Drive data for the selection of the drive shaft
- Fig. 20.7 Estimation of the moment of inertia of a rotor
- Fig. 20.8 Choice of the cardan shaft depending on the permissible residual unbalance
- Fig. 20.9 Balancing speeds and acceleration times of different drives
- Fig. 20.10 Drive power for fans on a balancing machine
- Fig. 20.11 Flexural resonance speeds of an idealised rotor with shaft-elastic behaviour under different support conditions

For the simplest case, the initial values are marked on two axes. The connecting straight line intersects the third axis – here the result can be read, Fig. 20.1.

Sometimes even with three axes there are different input and output options, Fig. 20.2.

Sometimes two steps are necessary, e.g. to convert the drive power during service to that during balancing, Fig. 20.3.

More complicated cases with more than three axes and evaluations with different input data are also possible. In these cases, the possibilities are marked in the nomograms.

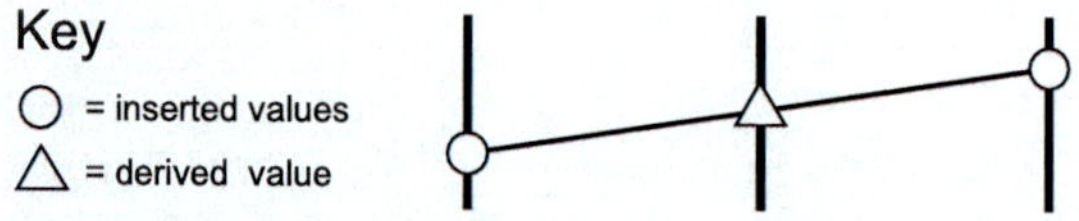

Fig. 20.1 Two axes on the outside for the input, one axis in the middle for the result

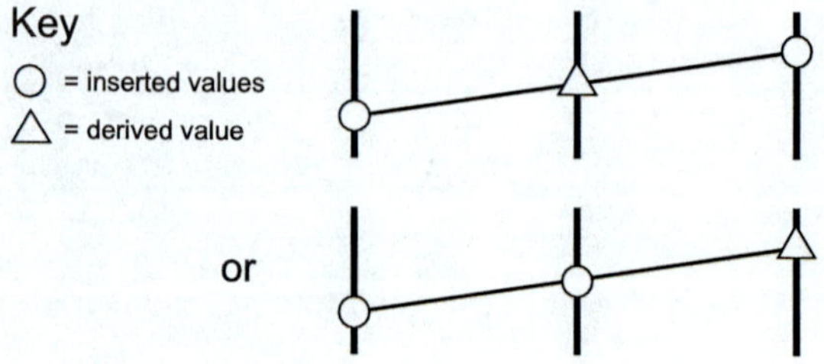

Fig. 20.2 The result axis can be in the centre or on the outside

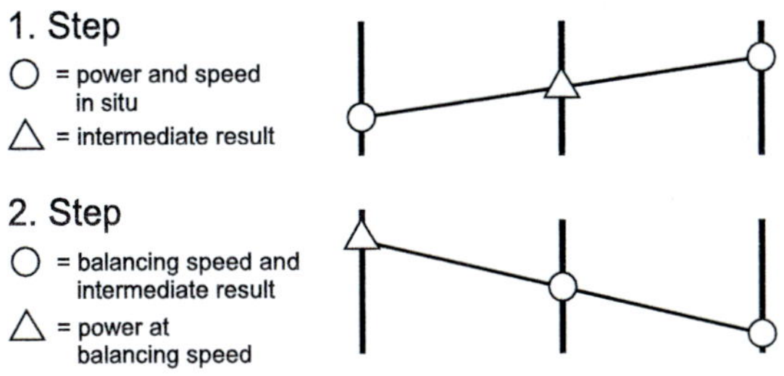

Fig. 20.3 An evaluation in two steps, example Fig. 20.10: Drive power of a fan at balancing speed

20.4.2 Diagrams

These diagrams serve to quickly determine some physical data, e.g.: to estimate masses or unbalances – for different materials, bores with/without drill point, or for very special balancing methods, see Fig. 20.12 to Fig. 20.25:

- Fig. 20.12 Unbalances due to geometric errors of disc, or tube
- Fig. 20.13 Masses of round steel from 1 to 10 mm diameter
- Fig. 20.14 Masses of round steel from 10 to 100 mm diameter
- Fig. 20.15 Masses of flat steel
- Fig. 20.16 Masses for drill cones from 0.1 to 40 mm drill depth
- Fig. 20.17 Masses for drill diameters from 0.5 to 1.5 mm, up to 2.6 mm depth
- Fig. 20.18 Masses for drill diameters from 1 to 5 mm, up to 13 mm depth
- Fig. 20.19 Masses for drill diameters from 5.5 to 10 mm, up to 26 mm depth
- Fig. 20.20 Masses for drill diameters from 5 to 50 mm, up to 400 mm depth
- Fig. 20.21 Conversion of correction masses from steel to other metals
- Fig. 20.22 Correction mass for several full-depth holes and one partial borehole
- Fig. 20.23 Effective unbalance and effective radius for radial drilling
- Fig. 20.24 Effective unbalance with two angular-spread unbalances
- Fig. 20.25 Effective unbalance and effective radius with correction over a larger circumference

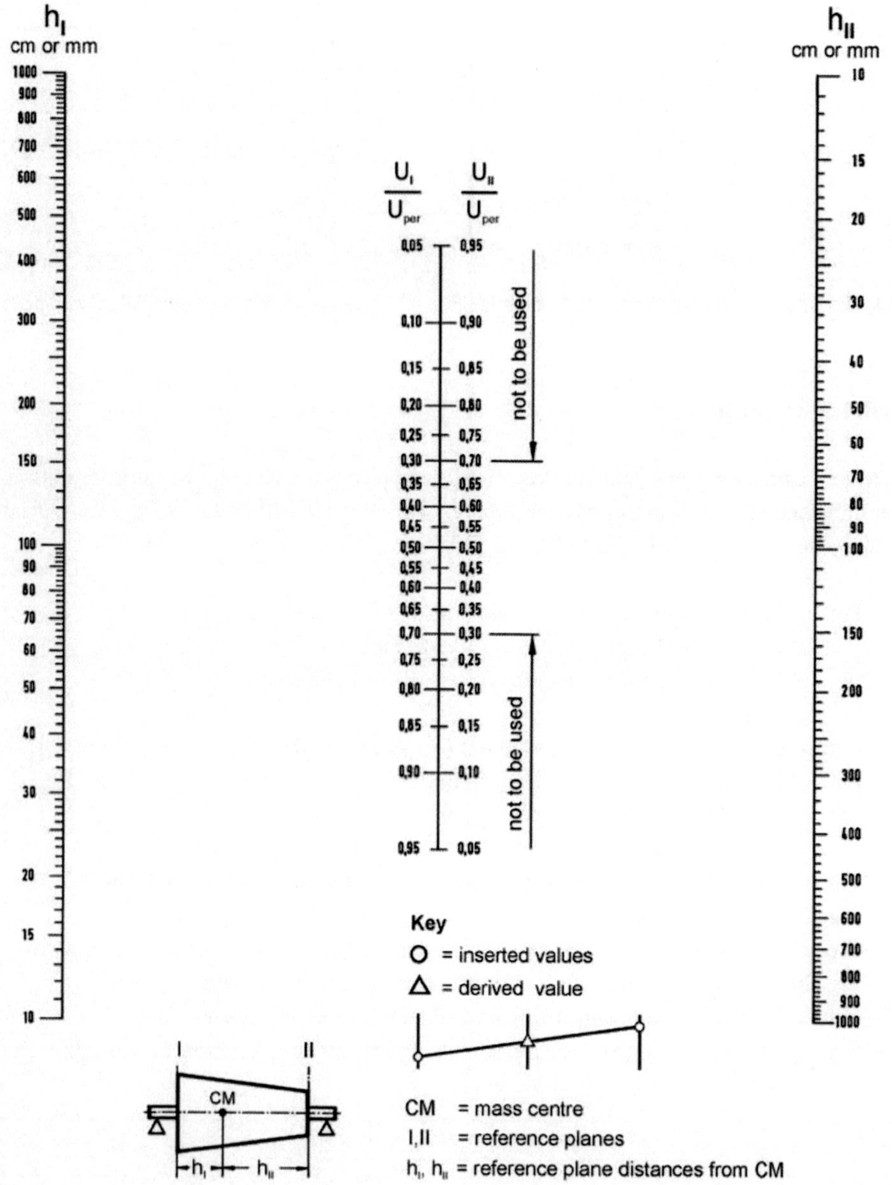

Fig. 20.4 Nomogram for the distribution of the permissible residual unbalance U_{per} (in mass centre CM) to the reference planes I and II. (see Sect. 7.5.2)

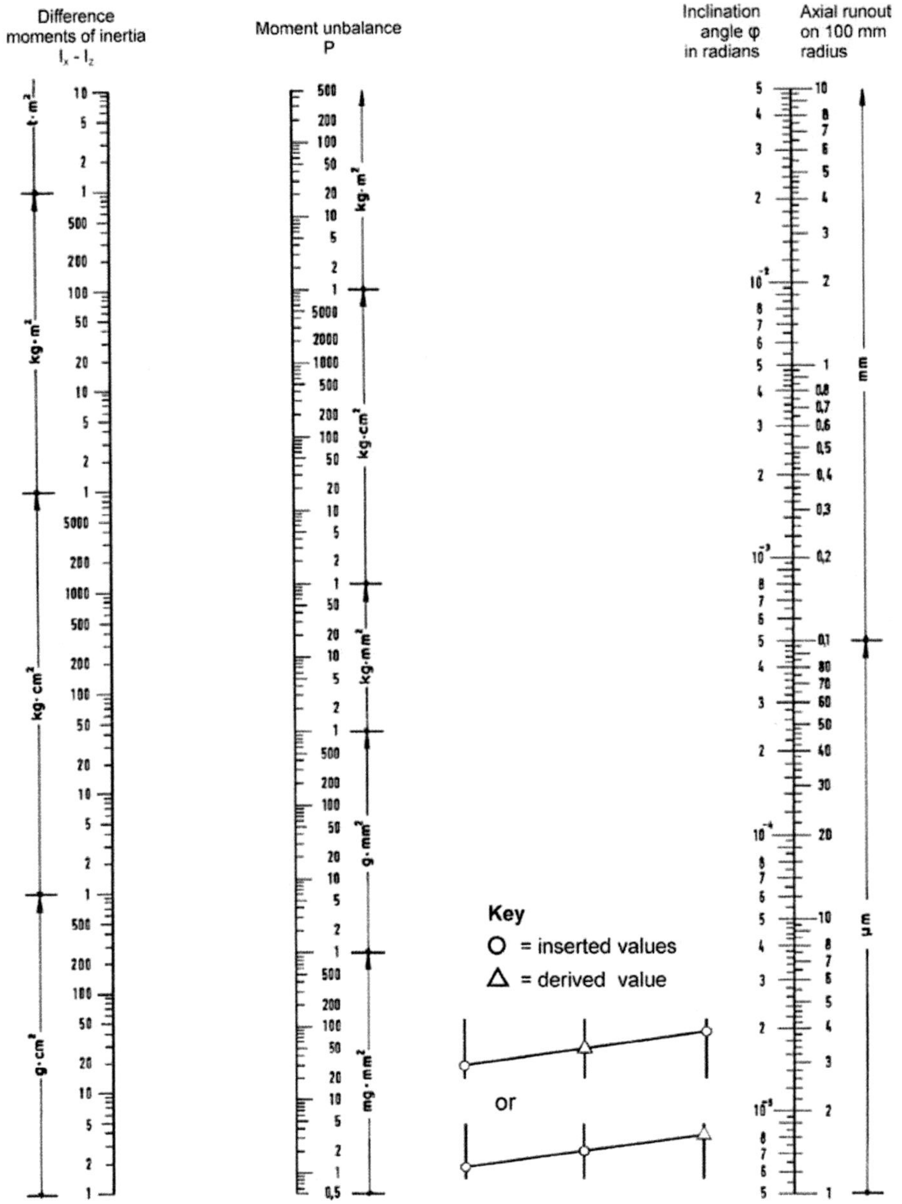

Fig. 20.5 Nomogram for the relationship between the moments of inertia J_x and J_z, the skew (axial runout) and the moment unbalance P

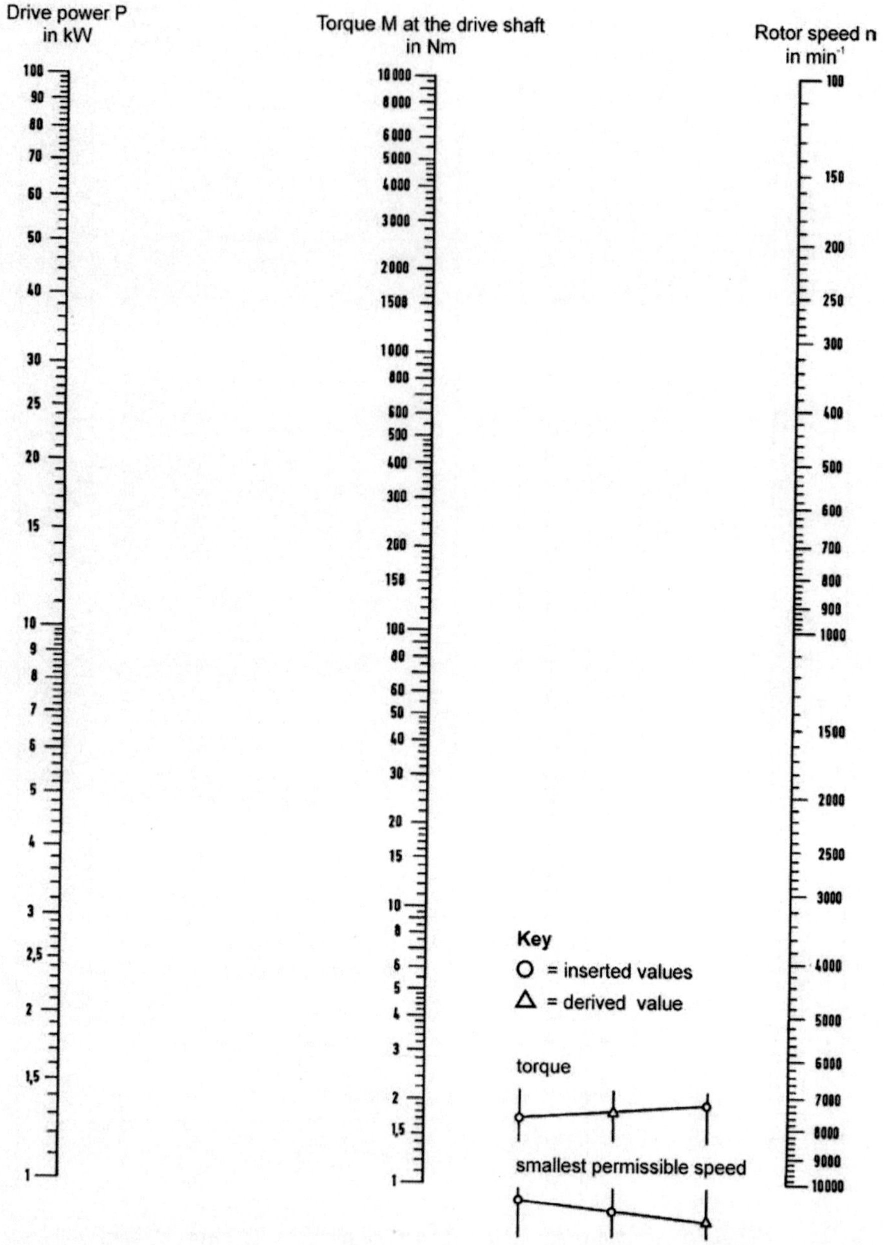

Fig. 20.6 Nomogram to determine the occurring torque from drive power and balancing speed, or the smallest permissible rotor speed for a given drive shaft (assumption: max. occurring torque = nominal torque)

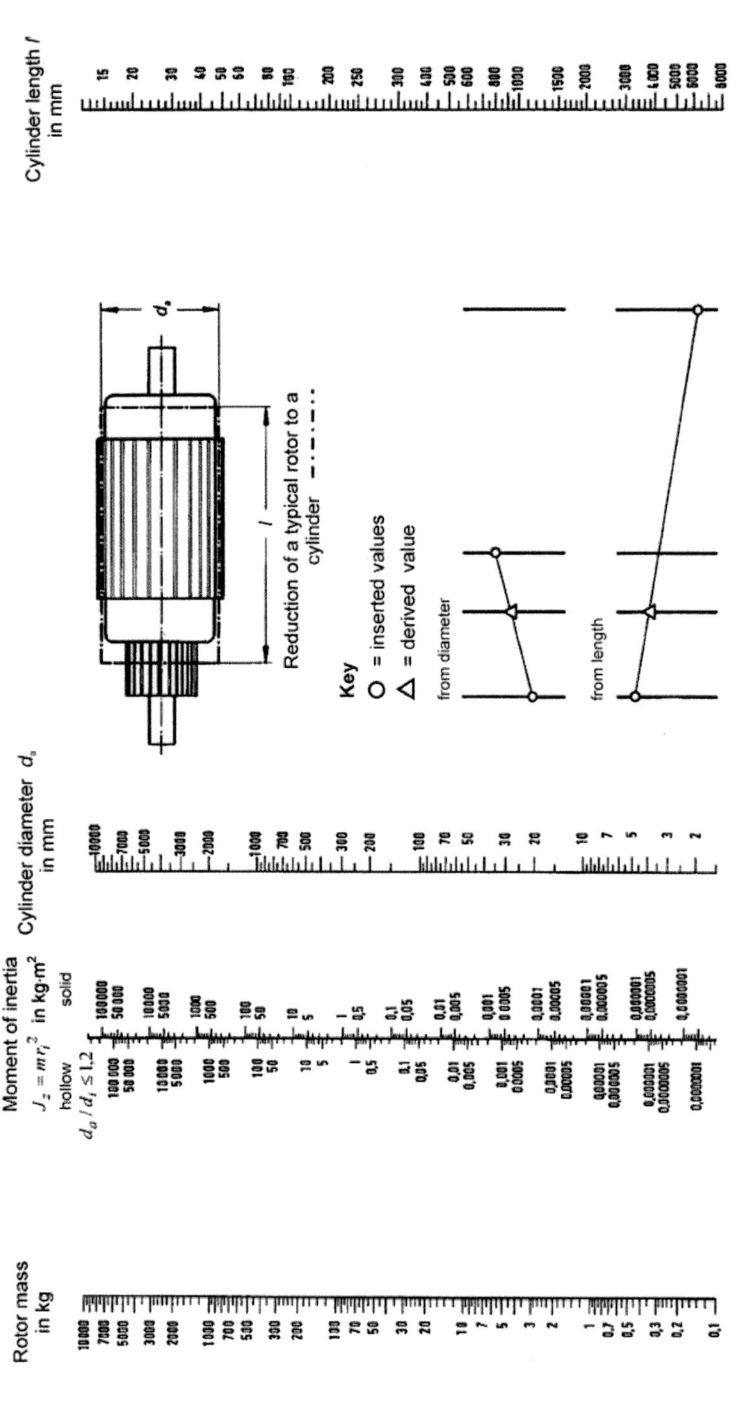

Fig. 20.7 Nomogram to determine the moment of inertia from the rotor data (for steel with density $\rho = 7{,}85$ g/cm³)

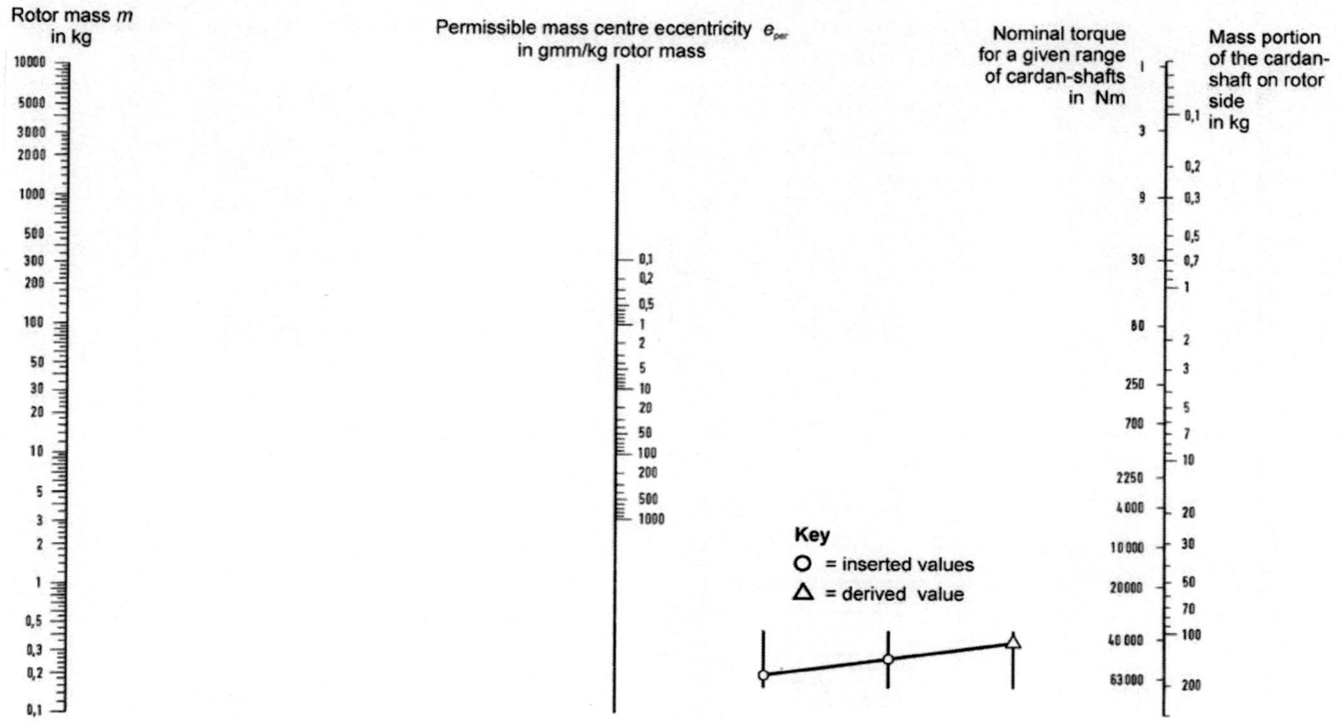

Fig. 20.8 Nomogram to determine the maximum permissible drive shaft as a function of the rotor mass and the intended balancing quality, given as the permissible centre of mass eccentricity

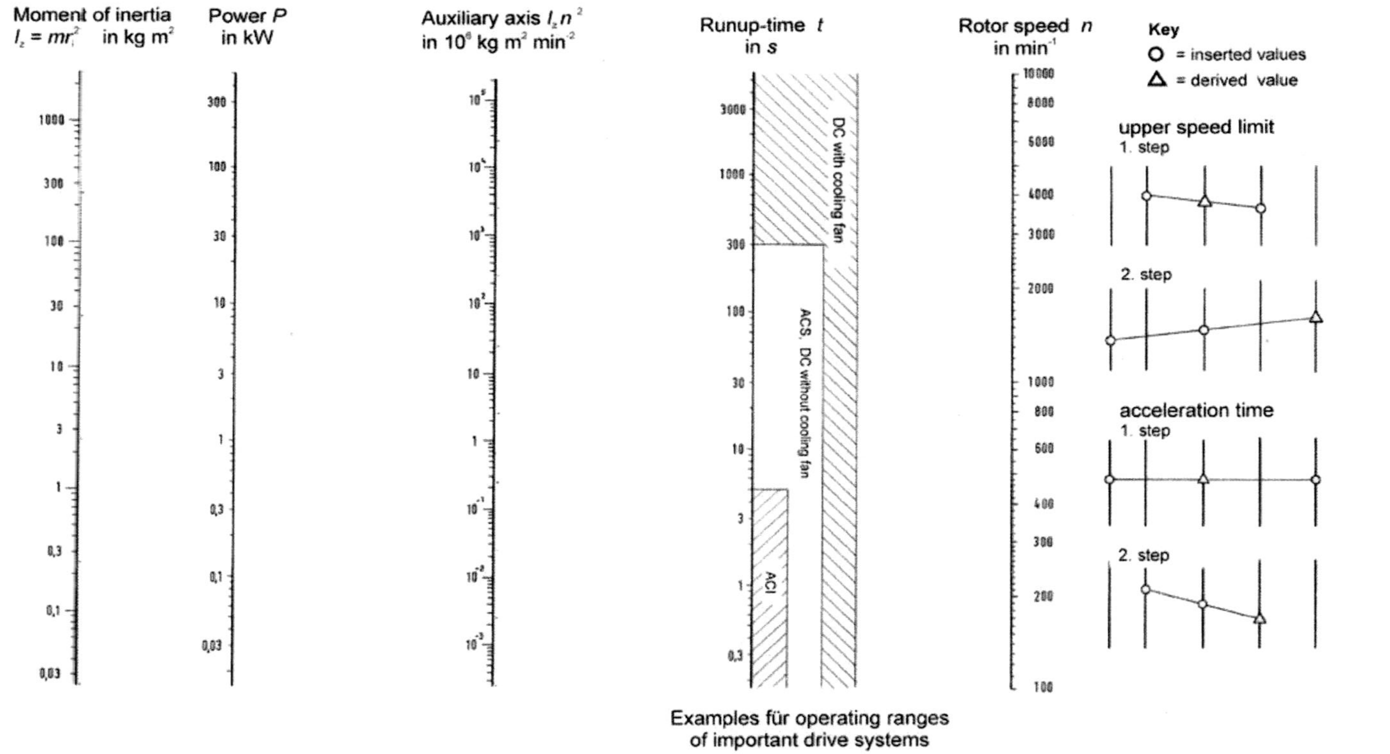

DC: Direct current motor ACI: Alternative current induction motor ACS: Alternative current synchronous motor

Fig. 20.9 Nomogram to determine the maximum permissible balancing speed or run-up time due to rotor moment of inertia (friction losses are not taken into account)

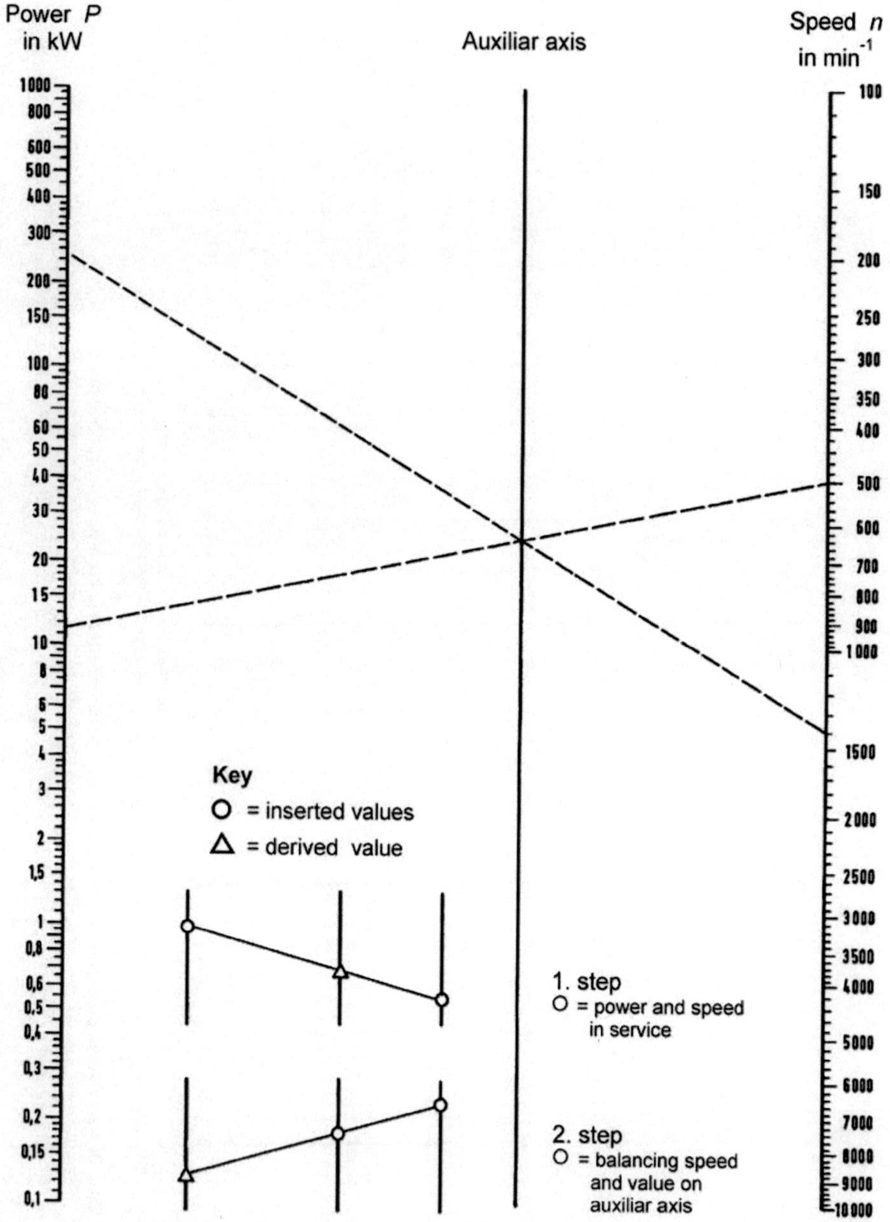

Fig. 20.10 Nomogram to determine the required drive power when balancing fans (assumption: $P \sim n^3$)

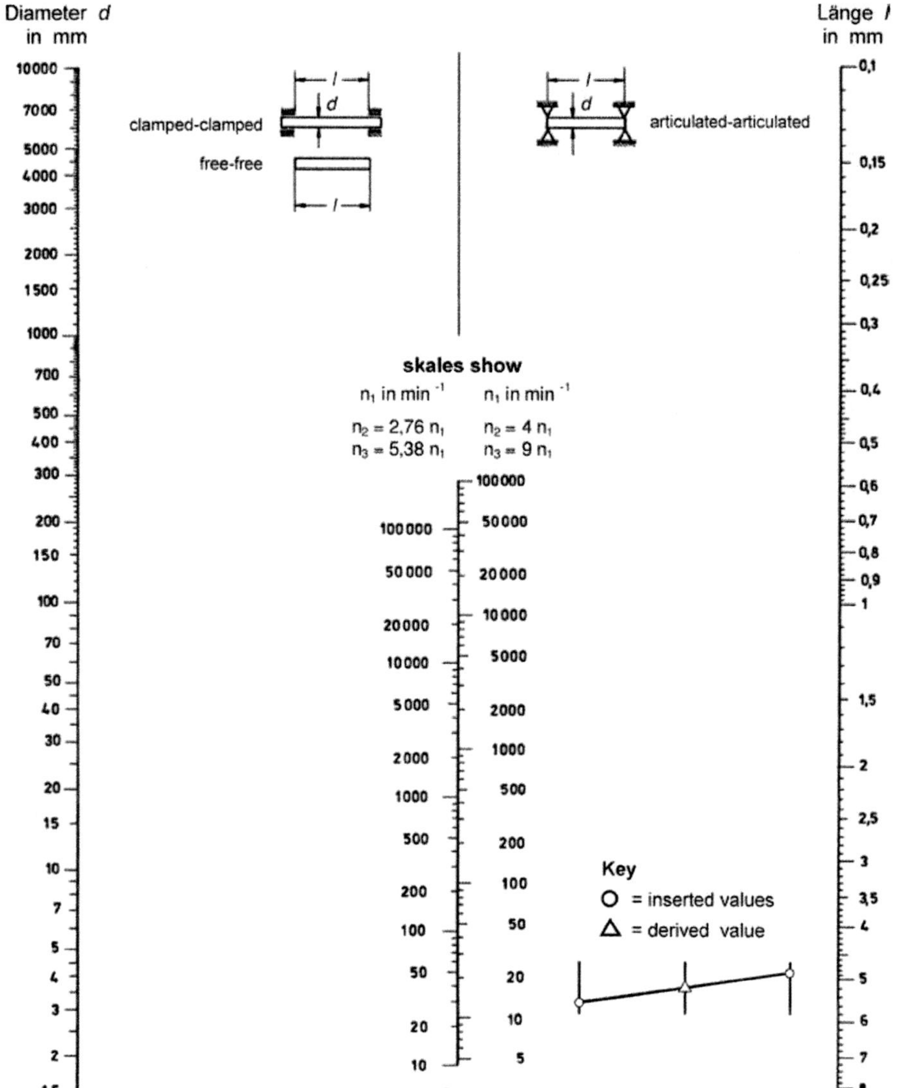

Valid for steel with density $\rho_S = 7{,}8\ \mathrm{g/cm^3}$ and elasticity module $E_S = 20{,}5 \cdot 10^6\ \mathrm{N/cm^2}$.
For other materials with data ρ_W and E_W the resulted speed can be recalculated with:

$$n_W = n_S \sqrt{\frac{E_W}{E_S}\,\frac{\rho_S}{\rho_W}}$$

All values are for solid rolls. For hollow rolls with outer diameter d_a and inner diameter d_i calculate:

$$n_{hollow} = n_{solid} \sqrt{1 + \left(\frac{d_i}{d_a}\right)^2}$$

Fig. 20.11 Nomogram to determine the flexural resonant speeds of (idealised) shafts; free – free: very soft bearings; clamped – clamped and articulated – articulated: rigid bearings

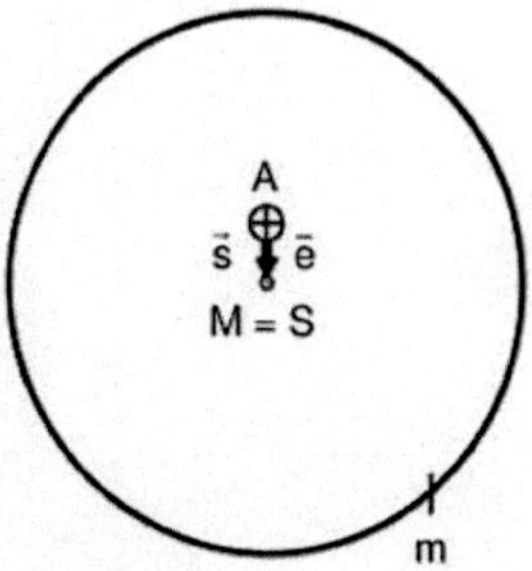

Solid disk

Position eccentric to shaft axis

A = shaft axis
M = centre point of disk
S = mass centre
s = centre point eccentricity
m = disk mass

e = mass centre eccentricity

$\vec{U} = \vec{s}m; \ \vec{e} = \vec{s}$

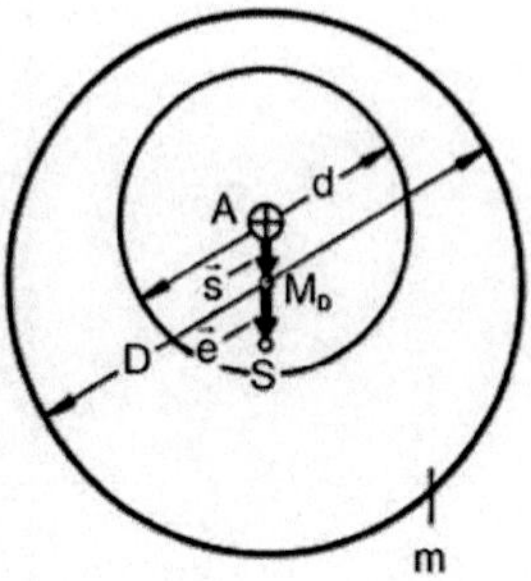

Hollow disk

Hollow concentric to shaft axis, outer diameter eccentric to shaft axis

D = outer diameter
d = inner diameter
M_D = outer diameter centre point

$$\vec{U} = \vec{s}m\frac{D^2}{D^2 - d^2}; \qquad \vec{e} = \vec{s}\frac{D^2}{D^2 - d^2}$$

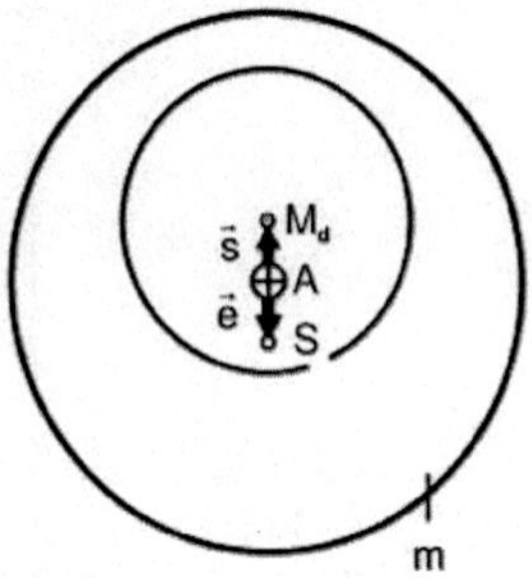

Hollow disk

Hollow eccentric to shaft axis, outer diameter concentric to shaft axis

D = outer diameter
d = inner diameter
M_d = hollow diameter centre point

$$\vec{U} = -\vec{s}m\frac{d^2}{D^2 - d^2}; \qquad \vec{e} = -\vec{s}\frac{d^2}{D^2 - d^2}$$

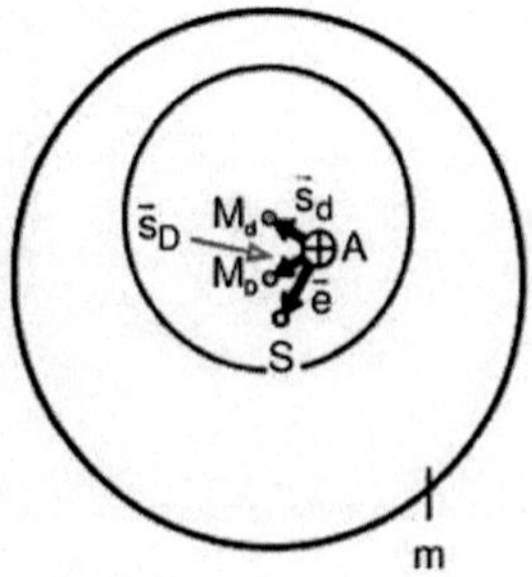

Hollow disk

Hollow eccentric to shaft axis by $\vec{s_d}$, outer diameter eccentric to shaft axis by $\vec{s_D}$

$$\vec{U} = \vec{s}_D m\frac{D^2}{D^2 - d^2} - \vec{s}_d m\frac{d^2}{D^2 - d^2} = m\frac{\vec{s}_D D^2 - \vec{s}_d d^2}{D^2 - d^2}$$

$$\vec{e} = \vec{s}_D\frac{D^2}{D^2 - d^2} - \vec{s}_d\frac{d^2}{D^2 - d^2} = \frac{\vec{s}_D D^2 - \vec{s}_d d^2}{D^2 - d^2}$$

Fig. 20.12 Some typical cases for calculating the unbalance from geometric errors: homogeneous disc with the same outer diameter, solid and with hole

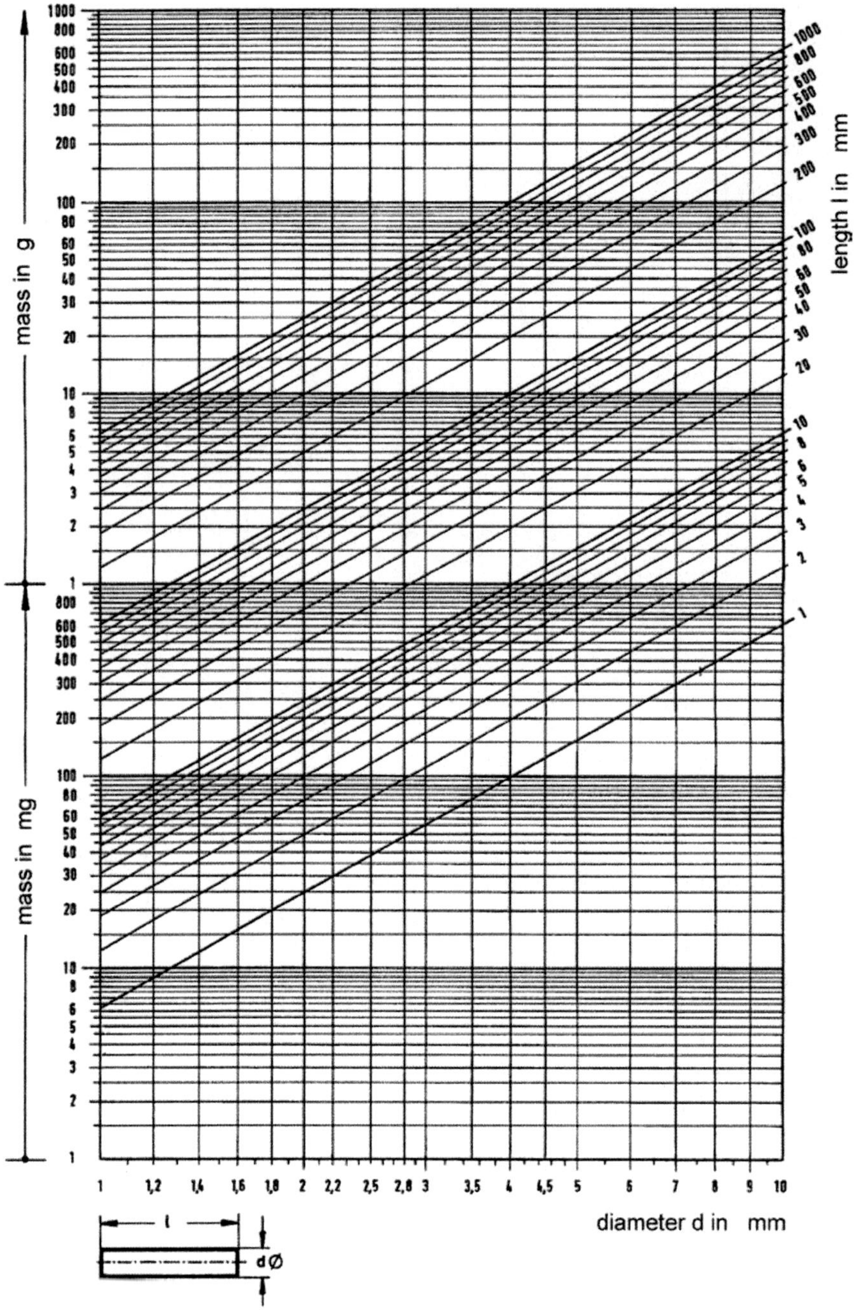

Fig. 20.13 Masses for round steel from 1 to 10 mm diameter and lengths from 1 mm to 1000 mm (density $\rho = 7{,}85\ g/cm^3$)

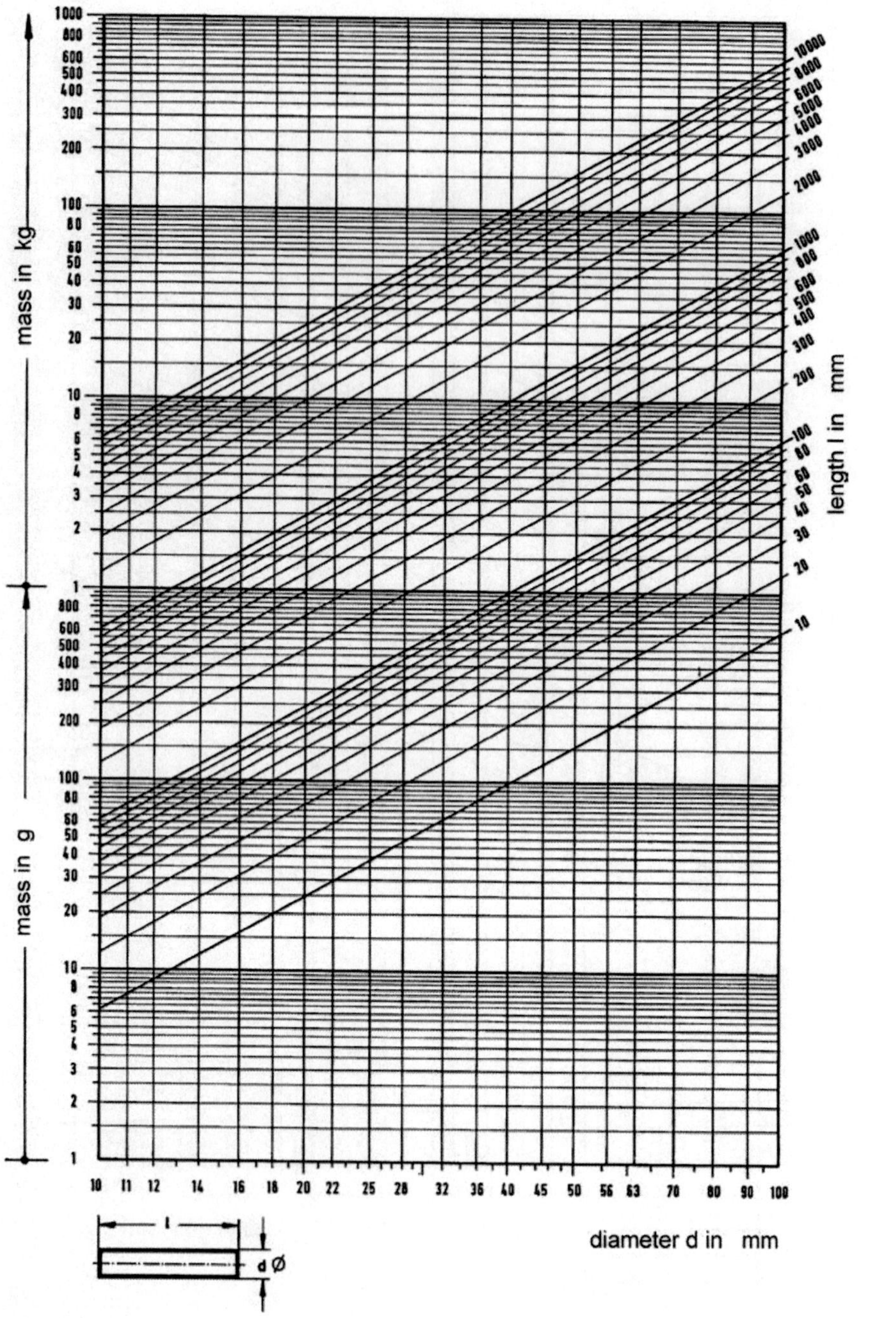

Fig. 20.14 Masses for round steel from 10 mm to100 mm diameter and lengths from 10 mm to 10 000 mm (density $\rho = 7{,}85\ g/cm^3$)

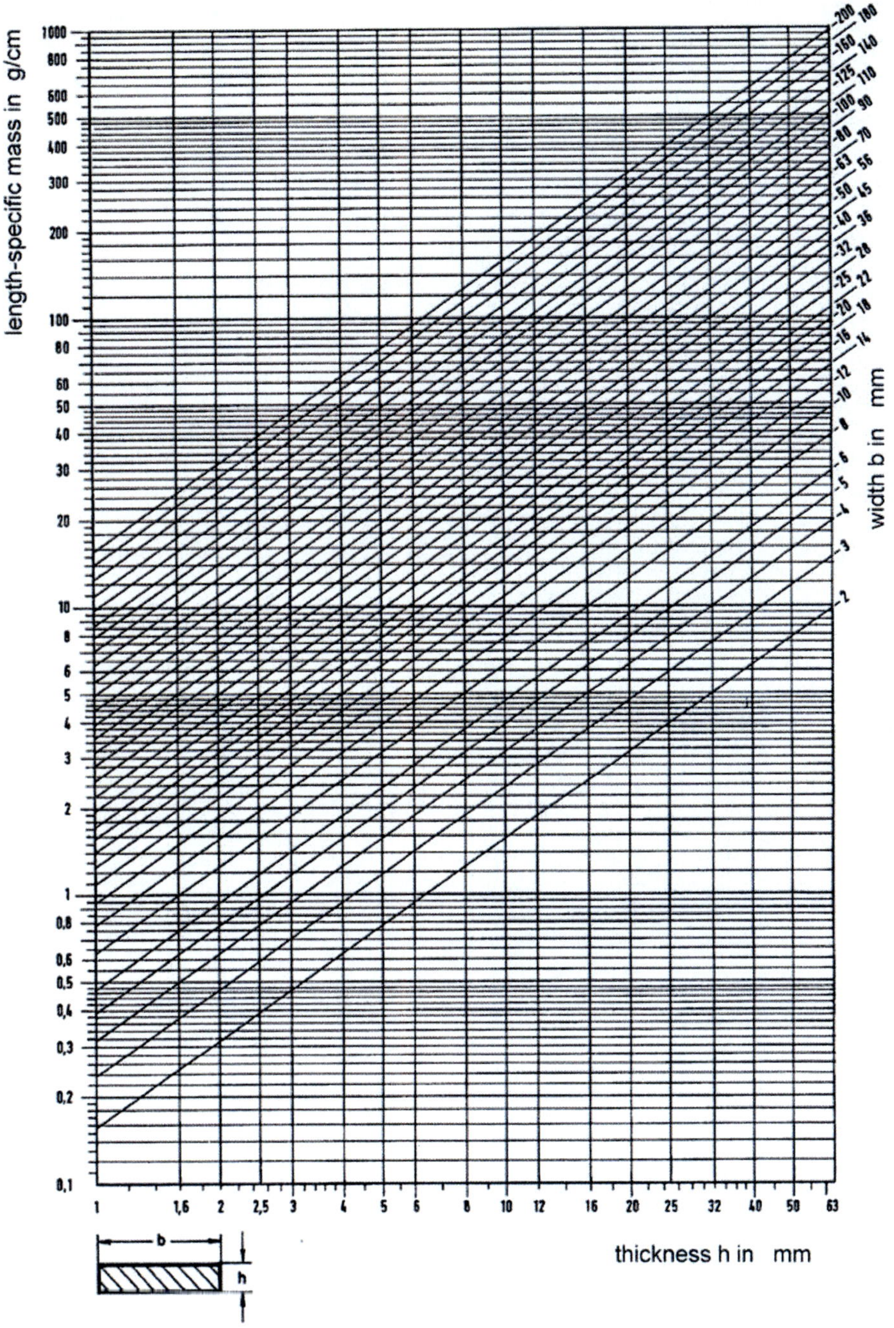

Fig. 20.15 Masses per cm length for flat steel from 1 to 63 mm thickness and widths from 2 to 200 mm (density $\rho = 7{,}85\ g/cm^3$)

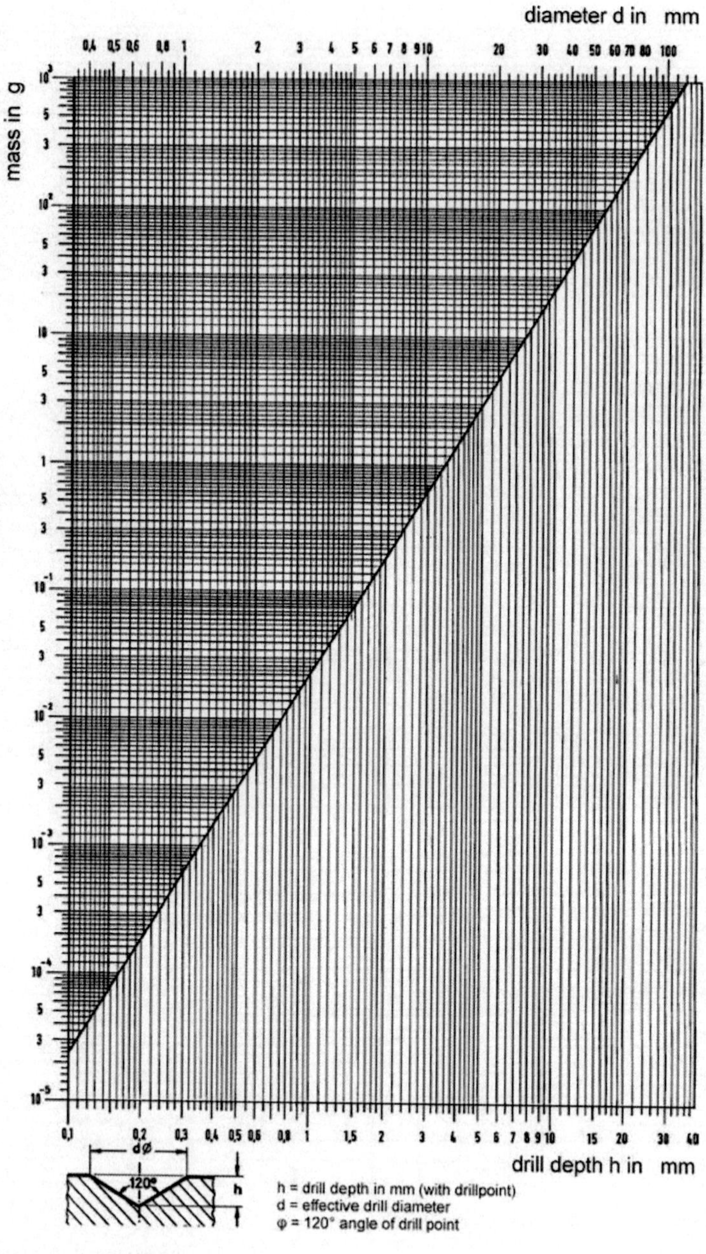

Fig. 20.16 Masses for drill cone from 0.1 mm to 40 mm drill depth (density $\rho = 7,85$ g/cm^3)

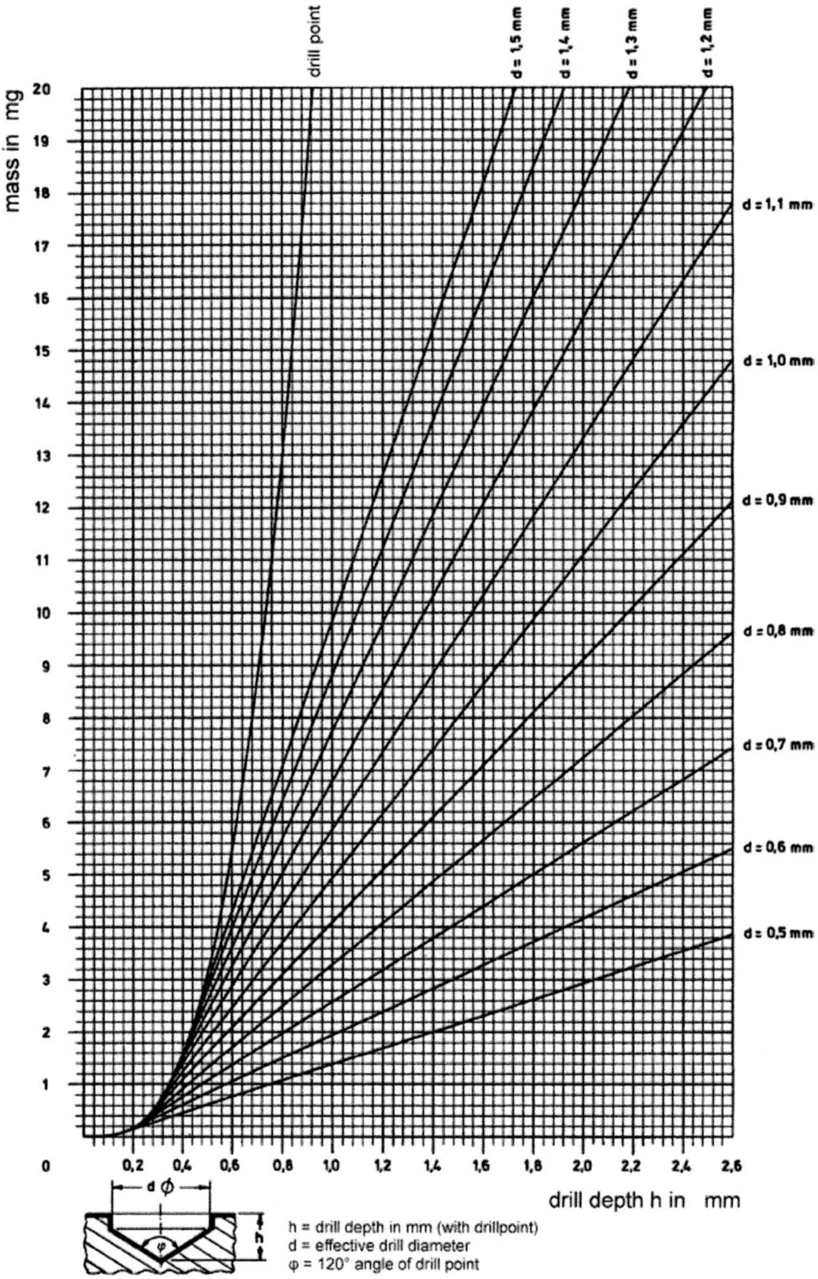

Fig. 20.17 Masses for drill diameters from 0.5 mm to 1.5 mm and of up to 2.6 mm drill depth (density $\rho = 7{,}85\ g/cm^3$)

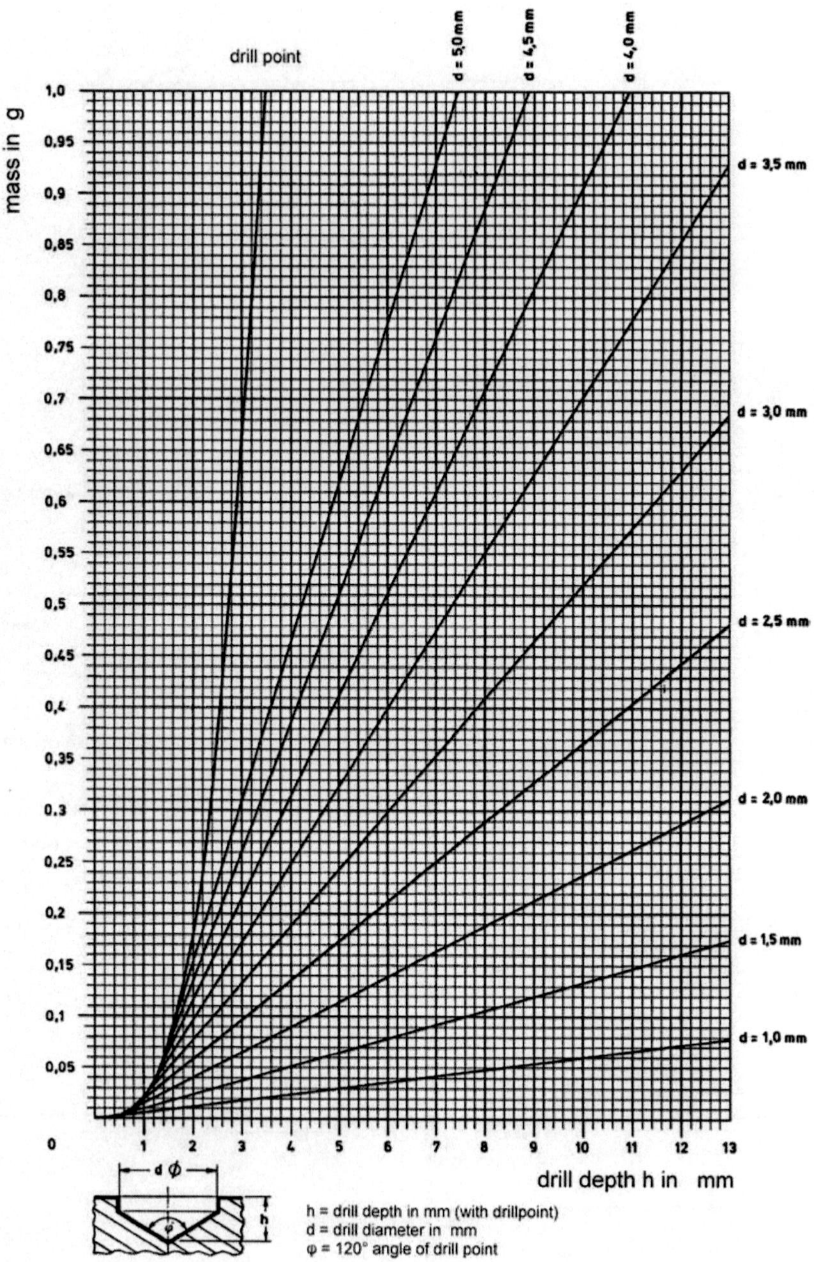

Fig. 20.18 Masses for drill diameters from 1 to 5 mm and of up to 13 mm drill depth (density $\rho = 7{,}85\ g/cm^3$)

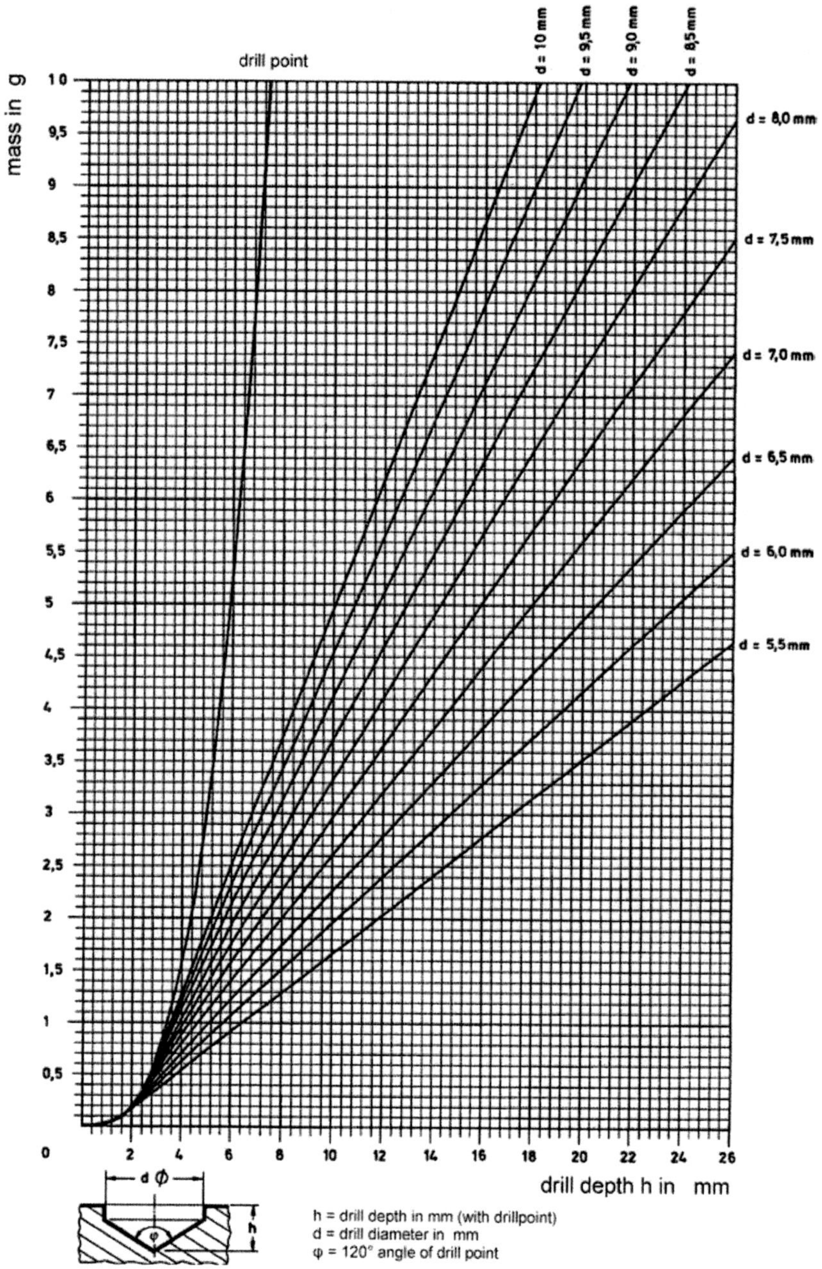

Fig. 20.19 Masses for drill diameters from 5.5 mm to 10 mm and up to 26 mm drill depth (density $\rho = 7{,}85\ g/cm^3$)

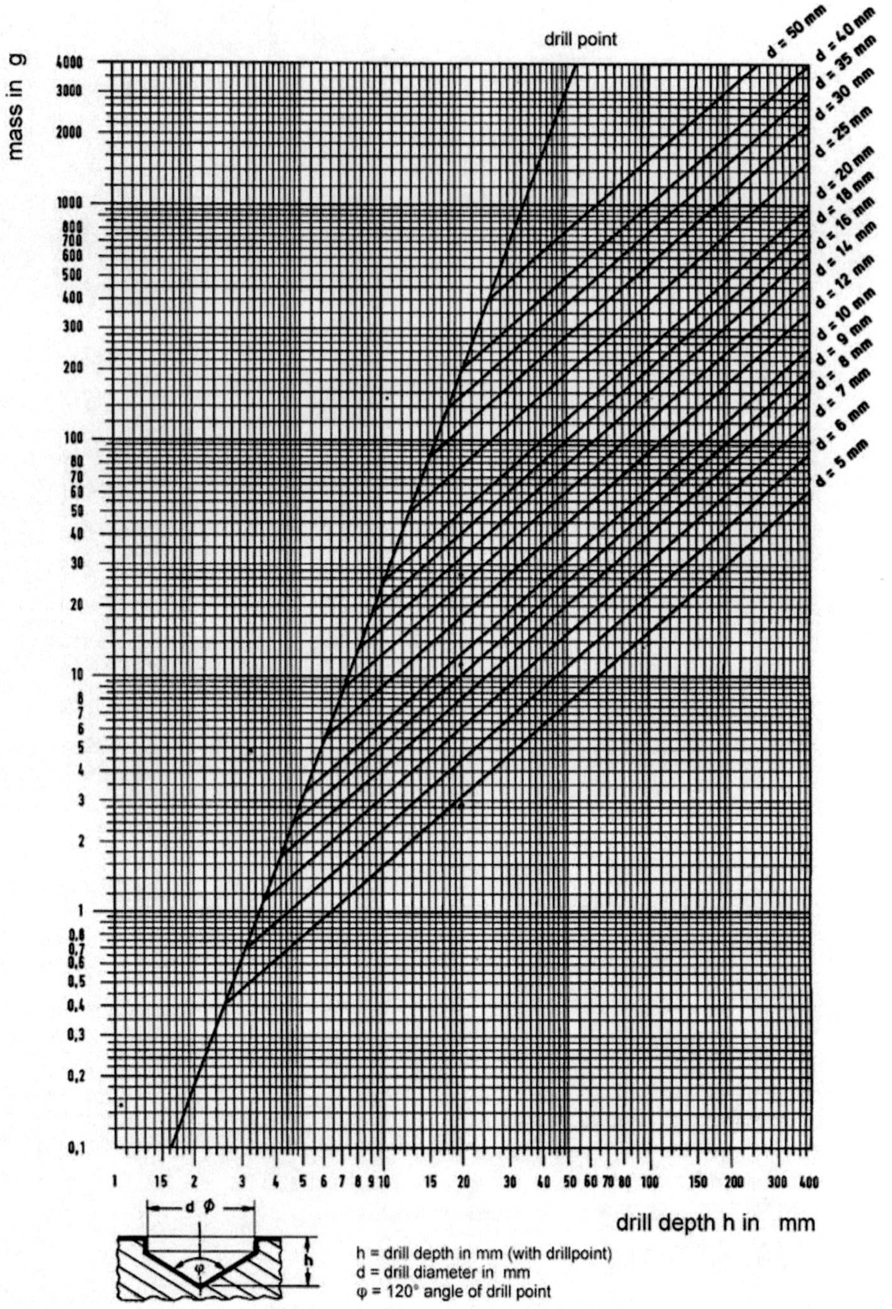

Fig. 20.20 Masses for drill diameters from 5 to 50 mm and up to 400 mm drill depth (density $\rho = 7{,}85$ g/cm^3)

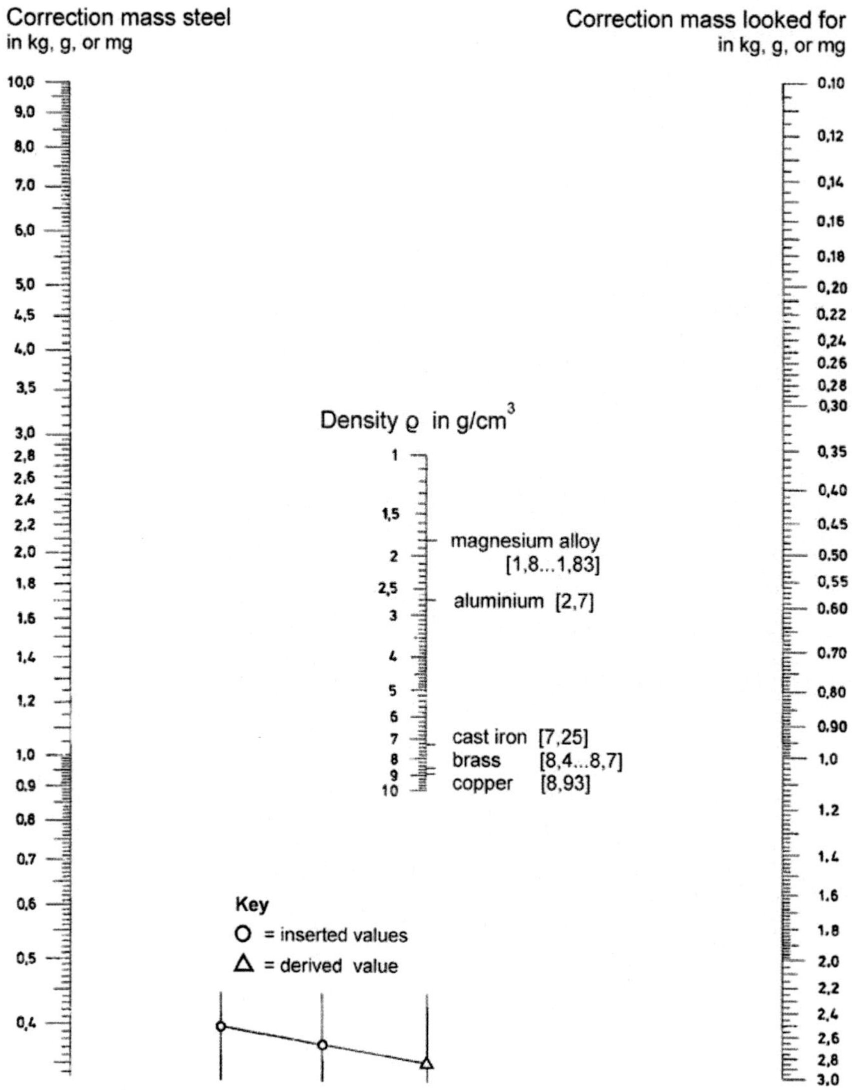

Fig. 20.21 Nomogram for converting the correction masses from steel (density $\rho = 7{,}85$ g/cm³) to other metals with the same volume

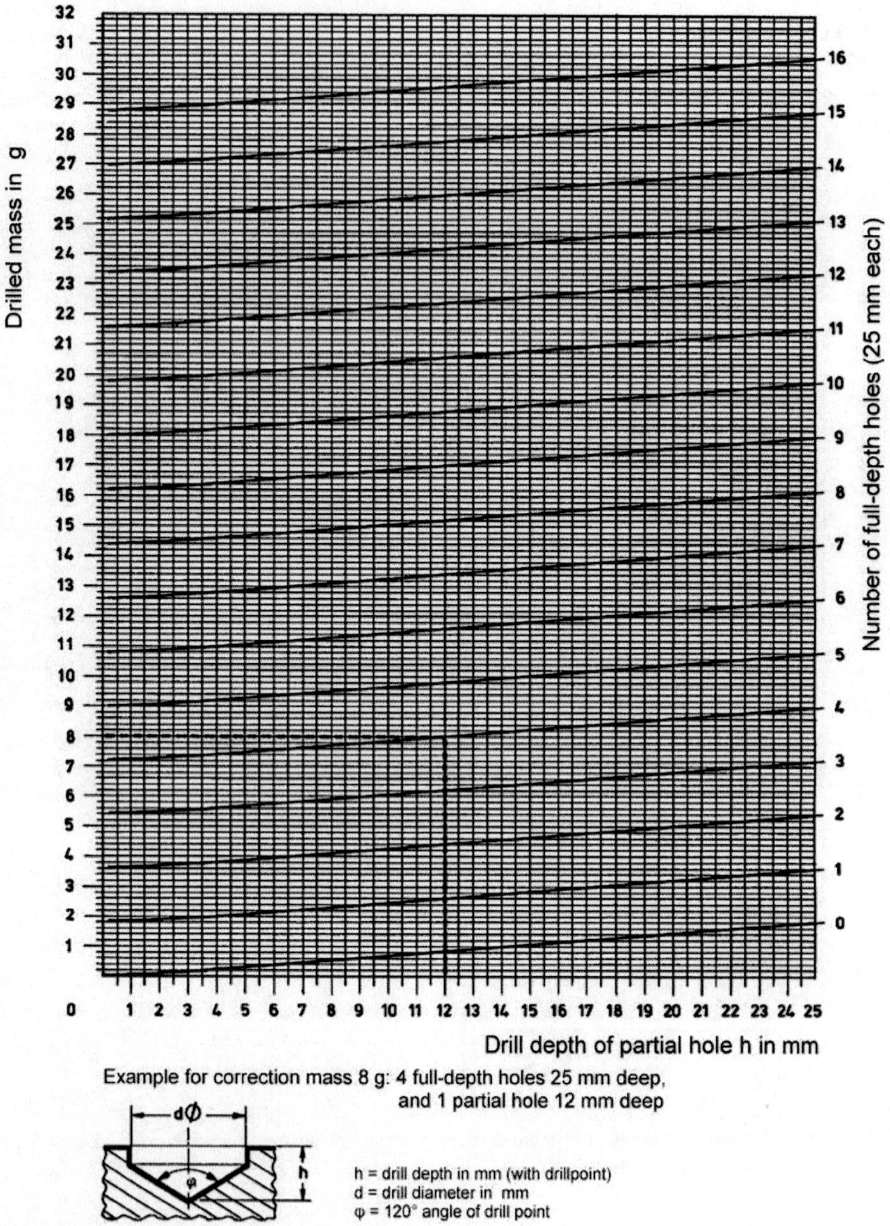

Fig. 20.22 Example of a drill panel for unbalance correction by several full-depth holes and one partial borehole. Drill diameter 6 mm in aluminium (density $\rho = 2{,}7$ g/cm^3)

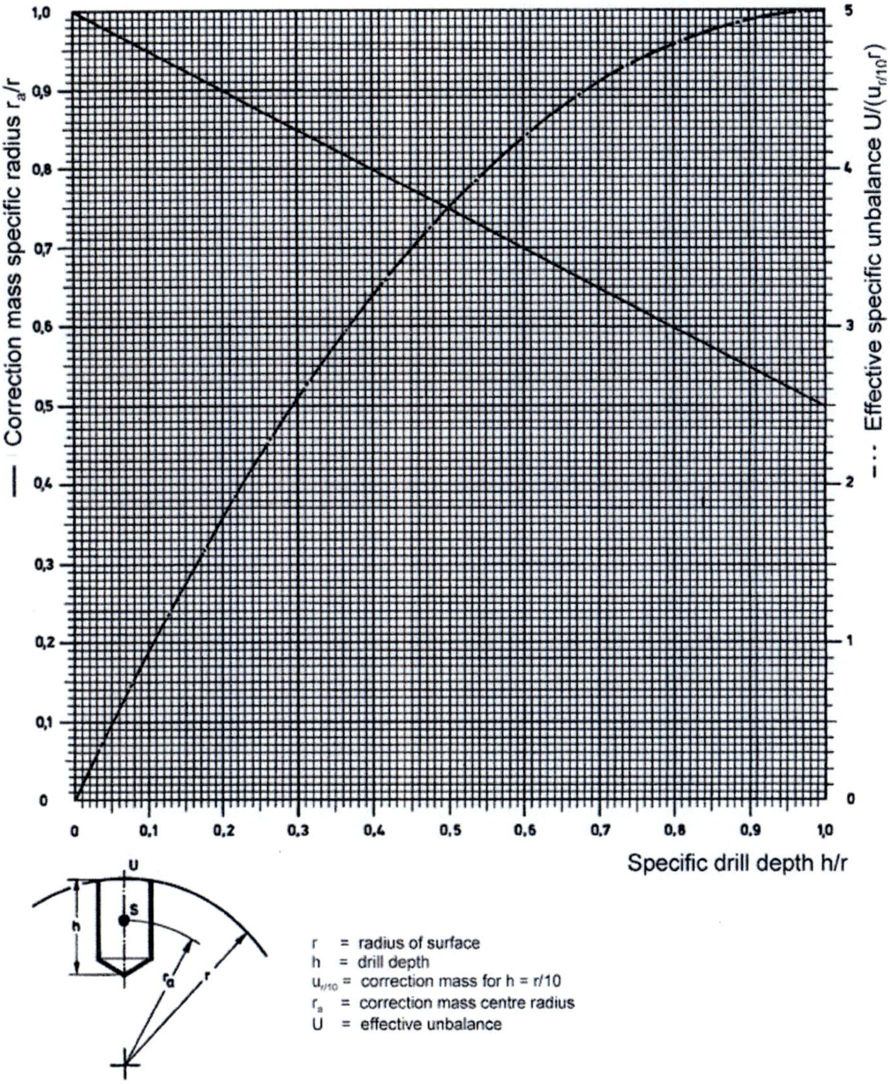

Fig. 20.23 Diagram to determine the effective radius and the effective unbalance for radial drilling

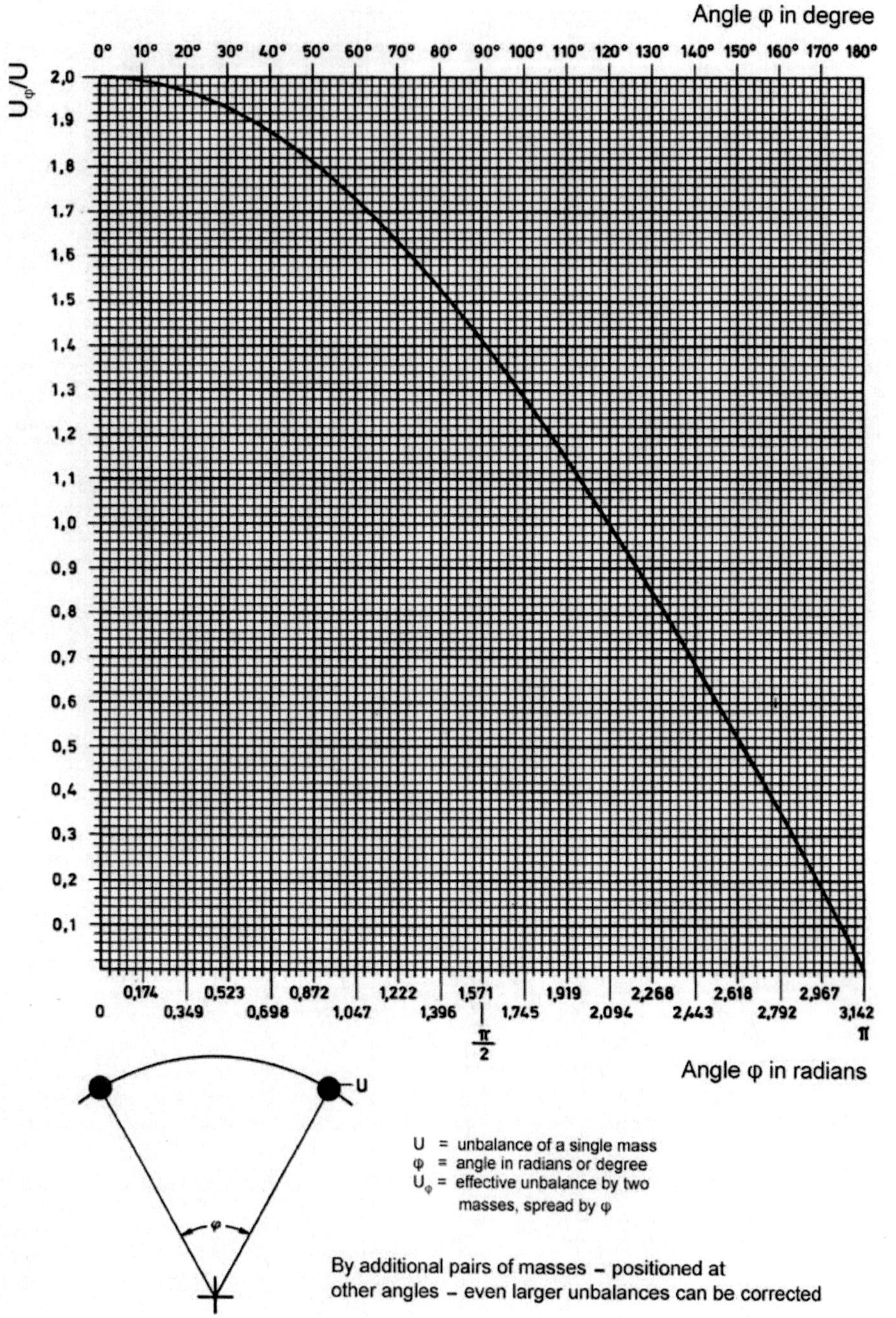

Fig. 20.24 Diagram to determine the effective unbalance with two equal unbalances spread by the angle φ

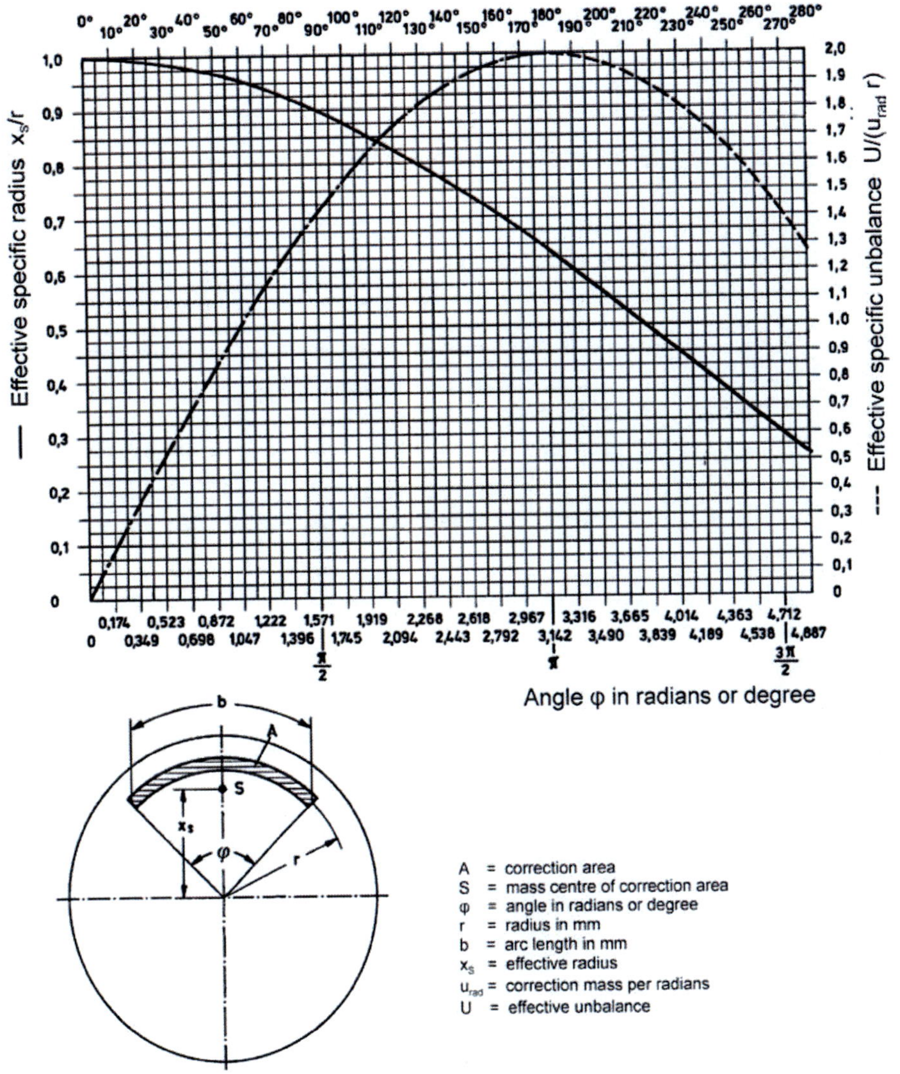

Fig. 20.25 Diagram to determine the effective radius and the effective unbalance for uniformly distributed correction over various circumferences

Image Sources

Photos

The photos in this book give a little impression of the wide range of balancing technology and its application in practice.

They were kindly provided by the following companies:

Fig	1.7	Schorch, Elektrische Antriebe und Maschinen, Mönchengladbach, D
	12.6	KKK, PGW Turbo, Leipzig, D
	12.9	Lütgert & Co. GmbH, Gütersloh, D
	12.10	Inometa, Maschinenbaugesellschaft mbH, Herford, D
	12.27	Voith Turbo GmbH & C. KG, Heidenheim, D
	12.29	ABB, Mannheim, D
	15.9	mtu, München, D
	17.7	Schorch, Elektrische Antriebe und Maschinen, Mönchengladbach, D
	17.8	KKK, PGW Turbo, Leipzig, D
The other photos:		
		Schenck RoTec GmbH, Darmstadt, D

Drawings, Tables

Some of the drawings and tables were taken from Schenck RoTec GmbH, D.

All others were created by the author or edited according to the particular situation in this book.

Drawings used from standards are named in the text.

H. Schneider, *Rotor Balancing*, https://doi.org/10.1007/978-3-662-66049-2

References

Monographs

1. Darlow, M.S. (1989): Balancing of High-Speed Machinery Springer Berlin Heidelberg New York
2. Federn, K.: Auswuchttechnik. Band 1 (1977, Nachdruck der 1. Auflage): Allgemeine Grundlagen, Messverfahren und Richtlinien. Springer, Berlin Heidelberg New York
3. Gasch, R.; Nordmann, R.; Pfützner, H. (2002): Rotordynamik. – Eine Einführung. Springer, Berlin Heidelberg New York
4. Kellenberger, W. (1987): Elastisches Wuchten Springer Berlin Heidelberg New York
5. Klotter: Technische Schwingungslehre, Band 1, 1978 heute als eBook): Einfache Schwinger Springer Berlin Heidelberg New York
6. Klotter: Technische Schwingungslehre, Band 2, 1960 heute als eBook): Schwinger von mehreren Freiheitsgraden Springer Berlin Heidelberg New York
7. Krämer, E. (1984): Maschinendynamik Springer Berlin Heidelberg New York
8. Lingener, A. (1992): Auswuchten – Theorie und Praxis Verlag Technik GmbH Berlin München
9. Stange, K. (1971, heute als eBook): Angewandte Statistik. Band 2: Mehrdimensionale Probleme. Springer, Berlin Heidelberg New York

Contributing works

10. Schneider, H. (2000): Balancing, in LIA Handbook of Laser Materials Processing Laser Institute of America and Magnolia Publishing Inc. Orlando, FL, USA

Journal articles

11. Schönfeld, H. (1973): Häufigkeitsverteilung der Unwucht in Großserien gefertigter Werkstücke Automobilindustrie Nr. 2 61 72
12. Schönfeld, H. (1987): Unwuchtkontrolle in der automatisierten Serienfertigung. Werkstatt und Betrieb, 120. Jahrgang Nr. 3, S. 223–228, Nr. 4, S. 309–313
13. Schönfeld, H. (1994): Prozessfähigkeit bei einseitig streuenden Messgrößen. wt-Produktion und Management Nr. 84, S. 203–205

H. Schneider, *Rotor Balancing*, https://doi.org/10.1007/978-3-662-66049-2

14. Schneider, H. (2016): Fortschritte bei Normen für die Auswuchttechnik: Normenreihe DIN ISO 21940 über „Mechanische Schwingungen – Auswuchten von Rotoren". DIN-Mitteilungen August 2016, S. 22–29, Beuth-Verlag, Berlin Wien Zürich

Special cases (theses, reports)

15. Federn, K. (1957): Grundlagen einer systematischen Schwingungsentstörung wellenelastischer Rotoren. VDI-Berichte Nr. 24: VDI, Düsseldorf S. 9–25
16. Menz, I. (2001): Drehzahlvariables Auswuchten starrer Rotoren in einer elastischen Wuchtmaschine. Diss. TU Berlin
17. Schneider, H. (1986): Balancing of integral gear-driven centrifugal compressors. 15th Turbomachinery Symposium. Texas A&M University, Corpus Christi, Tx, USA
18. Schneider, H. (2000): Systematische Bestimmung der zulässigen Unwuchten von Rotoren im wellenelastischen Zustand. VDI-Schwingungstagung 2000. VDI-Bericht Nr. 1550: Experimentelle und rechnerische Modalanalyse sowie Identifikation dynamischer Systeme, S. 729–742
19. Schneider, H. (2009): VDI-Richtlinie 3835 – Etwas Neues für das Auswuchten von Rotoren mit wellenelastischem Verhalten. SIRM 2009–8. Internationale Tagung Schwingungen in rotierenden Maschinen, Wien, AU
20. Schneider, H. (2011): Qualitätsmanagement von wellenelastischen Rotoren – Ist für das Auswuchten eine Schwingungsmessung als Gütekriterium verwendbar? SIRM 2011–9. Internationale Tagung Schwingungen in rotierenden Maschinen, Darmstadt, D
21. Wiese, D. (1994): Auswuchten von einfachen elastischen Rotoren ohne Testgewichtsetzungen. Diss. TU Berlin
22. Wiese, D. (1996): Evaluation of the state of unbalance of flexible rotors – vibration velocity versus modal unbalances. IMECHE, 6th International Conference on Vibrations in Rotating Machinery, Oxford, UK
23. Wutsdorff, P. (1975): Untersuchungen zum Auswuchten elastischer Rotoren des Turbomaschinenbaus. Diss. TU Berlin

Standards, guidelines and other regulations

DIN: German Standards (Deutsche Normen)

Vertrieb: Beuth Verlag, Berlin, D, www.beuth.de

Note: ISO has decided to issue all rotor balancing standards under No. 21940 as parts, arranged in groups, see Table 1.2 of this book. The latest editions of the DIN ISO balancing standards are therefore classified under this number.

DIN Standards (DIN Normen)

24. DIN 1301: Einheiten:
25. Teil 1 (2010): Einheitennamen, Einheitenzeichen
26. Teil 2 (1978): Allgemein angewendete Teile und Vielfache

27. DIN 1305 (1988): Masse, Wägewert, Kraft, Gewichtskraft, Gewicht, Last – Begriffe
28. DIN 1311: Schwingungen und schwingungsfähige Systeme:
29. Teil 1 (2000): Grundbegriffe, Einteilung
30. Teil 2 (2002): Lineare, zeitinvariante schwingungsfähige Systeme mit einem Freiheitsgrad
31. Teil 3 (2000): Lineare, zeitinvariante schwingungsfähige Systeme mit endlich vielen Freiheitsgraden
32. Teil 4 (1974): Schwingende Kontinua, Wellen
33. DIN 1315 (1982): Winkel – Begriffe, Einheiten
34. DIN 1319: Grundlagen der Messtechnik:
35. Teil 1 (1995): Grundbegriffe
36. Teil 2 (2005): Begriffe für Messmittel
37. Teil 3 (1996): Auswertung von Messungen einer einzelnen Messgröße, Messunsicherheit
38. Teil 4 (1999): Auswertung von Messungen, Messunsicherheit
39. DIN 13317 (1983): Mechanik starrer Körper – Begriffe, Größen, Formelzeichen
40. DIN 45661 (2016): Schwingungsmesseinrichtungen – Allgemeine Anforderungen und Prüfung
41. DIN EN 60034–14 (VDE 0530–14) (2019): Drehende elektrische Maschinen – Teil 14: Mechanische Schwingungen von bestimmten Maschinen mit einer Achshöhe von 56 mm und höher – Messung, Bewertung und Grenzwerte der Schwingstärke

DIN ISO Standards (DIN ISO Normen)

42. DIN EN ISO 6103 (2015): Schleifkörper aus gebundenem Schleifmittel – Zulässige Unwucht von Schleifscheiben im Lieferzustand – Statische Prüfung
43. DIN EN ISO 28927: Handgehaltene motorbetriebene Maschinen – Messverfahren zur Ermittlung der Schwingungsemission (mehrere Teile)
44. DIN ISO 1925 (2002): Mechanische Schwingungen – Auswuchttechnik – Begriffe
45. ersetzt durch DIN ISO 21940, Teil 2
46. DIN ISO 1940–1 (2004): Mechanische Schwingungen – Anforderungen an die Auswuchtgüte von Rotoren in konstantem (starrem) Zustand – Teil 1: Festlegung und Nachprüfung der Unwuchttoleranz
47. ersetzt durch DIN ISO 21940, Teil 11
48. DIN ISO 2954 (2012): Mechanische Schwingungen von Hubkolbenmaschinen und von Maschinen mit rotierenden Bauteilen – Anforderungen an Schwingstärkemessgeräte
49. DIN ISO 7919: Mechanische Schwingungen – Bewertung der Schwingungen von Maschinen durch Messungen an rotierenden Wellen (mehrere Teile)
50. DIN ISO 10816–1 (2017): Mechanische Schwingungen – Messung und Bewertung der Schwingungen von Maschinen – Teil 1: Allgemeine Anleitungen
51. Anm.: ist auch Ersatz für DIN ISO 7919–1
52. DIN ISO 10816: Mechanische Schwingungen – Messung und Bewertung der Schwingungen von Maschinen (mehrere Teile)
53. DIN ISO 10817–1 (1999): Messeinrichtung für die Schwingungen rotierender Wellen – Teil 1: Erfassung der relativen und der absoluten Radialschwingungen
54. DIN ISO 11342 (1999): Mechanische Schwingungen – Verfahren und Kriterien für das mechanische Auswuchten nachgiebiger Rotoren
55. ersetzt durch DIN ISO 21940, Teil 12

56. DIN ISO 19499 (2008): Mechanische Schwingungen – Auswuchttechnik – Einführung und Leitfaden für die Auswahl und Anwendung von Auswuchtnormen
57. wird: DIN ISO 21940: Mechanische Schwingungen – Auswuchten von Rotoren, Teil 1: Einführung
58. Normenreihe DIN ISO 21940: Mechanische Schwingungen – Auswuchten von Rotoren:
59. Teil 2 (2017): Begriffe. Ersetzt DIN ISO 1925
60. Teil 11 (2017): Verfahren und Toleranzen für Rotoren mit starrem Verhalten.
61. Ersetzt DIN ISO 1940–1
62. Teil 12 (2016): Verfahren und Toleranzen für Rotoren mit nachgiebigem Verhalten.
63. Ersetzt DIN ISO 11342
64. Teil 12: Verfahren und Toleranzen für Rotoren mit nachgiebigem Verhalten, Beiblatt 1 (2015): Verfahren zum Auswuchten bei mehreren Drehzahlen. Ersetzt E-VDI 3835
65. Teil 13 (2012): Kriterien und Sicherheitshinweise für das Auswuchten von mittleren und großen Rotoren am Aufstellungsort. Ersetzt DIN ISO 20806 (2010)
66. Teil 14 (2012): Verfahren zur Ermittlung von Abweichungen beim Auswuchten.
67. Ersetzt DIN ISO 1940–2 (1998)
68. Teil 21 (2012): Beschreibung und Bewertung von Auswuchtmaschinen. Ersetzt DIN ISO 2953
69. Teil 23 (2012): Verkleidungen und andere Schutzmaßnahmen für die Messstation von Auswuchtmaschinen. Ersetzt DIN ISO 7475 (2004)
70. Teil 31: keine deutsche Ausgabe vorgesehen, s. ISO 21940–31
71. Teil 32 (2012): Behandlung von Passfederverbindungen. Ersetzt DIN ISO 8821 (1991)

ISO: International Organization for Standardization

Vertrieb: ISO, Genf, CH, www.iso.org or Standards Organizations in several countries

Note: ISO has decided to publish all balancing standards as parts under No. 21940, arranged in groups, see Table 1.2 of this book. The latest editions of the ISO balancing standards are therefore classified under this number.

72. ISO 2041 (2018): Vibration, shock and condition monitoring – Vocabulary
73. ISO 10816–1 (2016): Mechanical vibration – Measurement and evaluation of machine vibration – Part 1: General guidelines
74. ISO 10816: Mechanical vibration – Measurement and evaluation of machine vibration – several parts
75. ISO 14694 (2003): Industrial fans – Specifications for balance quality and vibration levels
76. ISO 14695 (2003): Industrial fans – Method of measurement of fan vibration
77. ISO 14839: Mechanical vibration – Vibration of rotating machinery equipped with active magnetic bearings (several parts)
78. ISO 19499 (2007): Mechanical vibration – Balancing – Guidance on the use and application of balancing standards. Wird ersetzt durch ISO 21940, Part 1
79. ISO 21940 Series: Mechanical vibration – Rotor balancing –
80. Part 2 (2017): Vocabulary, replaces ISO 1925
81. Part 11 (2016): Procedures and tolerances for rotors with rigid behaviour, replaces ISO 1940–1
82. Part 12 (2016): Procedures and tolerances for rotors with flexible behaviour, replaces ISO 11342
83. Part 13 (2012): Criteria and safeguards for the in-situ balancing of medium and large rotors,

84. replaces ISO 20806
85. Part 14 (2012): Procedures for assessing balance errors, replaces ISO 1940–2
86. Part 21 (2012): Description and evaluation of balancing machines, replaces ISO 2953
87. Part 23 (2012): Enclosures and other safety measures for the measuring station of balancing machines, replaces ISO 7475
88. Part 31 (2013): Susceptibility and sensibility of machines to unbalances, replaces ISO 10814.
89. Part 32 (2012): Shaft and fitment key convention, replaces ISO 8821
90. ISO 80000–1 (2013): Quantities and units – Part 1: General

SAE International: Society of Automotive Engineers

Vertrieb: Warrendale, PA, USA, www.sae.org
91. ARP 587C (2012): Balancing Machines – Description and Evaluation, , Two-Plane, Soft Bearing Type for Gas Turbine Rotors
92. ARP 588C (2012): Balancing Machines – Description and Evaluation, Vertical, Single-Plane, Soft Bearing Type for Gas Turbine Rotors
93. ARP 4048 (2011): Balancing Machines – Description and Evaluation Horizontal, Two-Plane, Hard-Bearing Type for Gas Turbine Rotors
94. ARP 4050A (2017): Balancing Machines – Description and Evaluation Vertical, Two-Plane, Hard-Bearing Type for Gas Turbine Rotors
95. ARP 4162B (2017): Balancing Machine Proving Rotors
96. ARP 4163 (2016): Balancing Machines – Tooling Design Criteria
97. ARP 5323A (2017): Balancing Machines – Description and Evaluation Vertical, Single-Plane, Hard-Bearing Type for Gas Turbine Rotors
98. ARP 6217 (2015): Balancing Machines – Description and Evaluation Vertical, Single-Plane, Non-Rotating Type

VDI: Association of German Engineers (Verein Deutscher Ingenieure)

Vertrieb: Beuth Verlag, Berlin, D, www.beuth.deNormen
99. VDI 2057: Einwirkung mechanischer Schwingungen auf den Menschen (mehrere Blätter)
100. VDI 3831 (2012): Schutzmaßnahmen gegen die Einwirkung mechanischer Schwingungen auf den Menschen
101. VDI 3832 (2013): Körperschallmessungen zur Zustandsbeurteilung von Wälzlagern in Maschinen und Anlagen
102. VDI 3833 Schwingungsdämpfer und Schwingungstilger (mehrere Blätter)
103. VDI 3834 Messung und Beurteilung der mechanischen Schwingungen von Windenergieanlagen und deren Komponenten (mehrere Blätter)
104. VDI 3835 (Entwurf 2009): Auswuchten von Rotoren mit wellenelastischem Verhalten – 2 + N-Verfahren zum Auswuchten bei mehreren Drehzahlen,
105. ersetzt durch: Normenreihe DIN ISO 21940: Mechanische Schwingungen – Auswuchten von Rotoren:

106. Teil 12: Verfahren und Toleranzen für Rotoren mit nachgiebigem Verhalten, Beiblatt 1 (2015): Verfahren zum Auswuchten bei mehreren Drehzahlen
107. VDI 3836 (2012): Messung und Beurteilung mechanischer Schwingungen von Schraubenverdichtern und Rootsgebläsen (Ergänzung von DIN ISO 10816–3)
108. VDI 3838 (2004): Messung und Beurteilung mechanischer Schwingungen von Hubkolbenmotoren und -kompressoren mit Leistungen über 100 kW (Ergänzung von DIN ISO 10816–6)
109. VDI 3839: Hinweise zur Messung und Interpretation der Schwingungen von Maschinen (mehrere Blätter)
110. VDI 3840 (2004): Schwingungstechnische Berechnungen – Berechnungen für Maschinensätze

Index

H. Schneider, *Rotor Balancing*, https://doi.org/10.1007/978-3-662-66049-2